Archaeology and Anthropology of Salt: A Diachronic Approach

Proceedings of the International Colloquium,
1-5 October 2008
Al. I. Cuza University (Iaşi, Romania)

Edited by

Marius Alexianu
Olivier Weller
Roxana-Gabriela Curcă

BAR International Series 2198
2011

Published in 2016 by
BAR Publishing, Oxford

BAR International Series 2198

Archaeology and Anthropology of Salt: A Diachronic Approach

ISBN 978 1 4073 0754 1

COVER IMAGE *Salt spring of Zlatina in Vrâncioaia (Vrancea, Romania). Photograph Olivier Weller*

BAR Publishing is the trading name of British Archaeological Reports (Oxford) Ltd. British Archaeological Reports was first incorporated in 1974 to publish the BAR Series, International and British. In 1992 Hadrian Books Ltd became part of the BAR group. This volume was originally published by Archaeopress in conjunction with British Archaeological Reports (Oxford) Ltd / Hadrian Books Ltd, the Series principal publisher, in 2011. This present volume is published by BAR Publishing, 2016.

Printed in England

BAR
PUBLISHING

BAR titles are available from:

BAR Publishing
122 Banbury Rd, Oxford, OX2 7BP, UK
EMAIL info@barpublishing.com
PHONE +44 (0)1865 310431
FAX +44 (0)1865 316916
www.barpublishing.com

Table of Contents

Part III. Ancient Texts and Salt

Part IV. Historical Approaches

Part V. Linguistic and Philological Approaches

Foreword

In the era of industrialization salt has progressively become a trivial element. It ceased to be considered 'white gold' once the era of refrigeration dawned. The phenomenon of globalization helped in this process. Today's advanced societies forget that salt, the only edible mineral for Man, is absolutely necessary for animal life in general. They forget that many salt springs mark the location of Neolithic communities, and that such springs have always attracted animals and birds. (As a consequence, those places became hunting grounds, and the maintenance of control over salt unleashed bloody battles.) Mankind forgets that salt is divine (Homer), that it is the soul of dead flesh (Plutarch), that it ensures the preservation of food for non-productive seasons, and that it once determined the flourishing of Rome, or of Venice. We all forget that salt is a primordial reference.

A narrow group of specialists has taken on the task of giving back these essential truths to the present. From books espousing a synthetic approach which illustrated comprehensive visions, there was a gradual movement to increasingly analytical approaches, which were capable of explaining the most diverse aspects of the role played by the places which were rich in salt, or in brine all over the world. These places have left an imprint both on the life in human communities or on landscapes and also generated a specific (micro)toponymy and (micro)hydronymy. Salt has generated a complex symbolism, a whole spiritual world reflected in a remarkable lexical constellation in the world languages. Salt has influenced the environment and animal life; but its strongest effects were those on the life of Man. The anthropology of salt seems to be the precise syntagm which reunites, under a unifying name, the whole collection of studies within the humanities, preeminently, focused on the influence of salt on the development of human communities everywhere in the world.

If we limit ourselves to Europe, studies focused on salt from the perspective of studies within the humanities have been stimulated, over the last two decades, by the organization, in a sustained rhythm, of conferences whose focus was predominantly archaeological (Paris 1998, Liège 2001, Cardona 2003, Piatra Neamţ 2004, Arc-et-Senans 2006) or of those which had a predominantly historical focus (Halle/Saale 1992, Granada 1995, Cagliari 1998, Weimar 2001, Nantes 2004, Sigüenza 2006).

But this separation between the archaeological and historical spheres has mostly been a didactic one, as a result of the practical conclusion that the archaeological conferences included historical approaches, while the historical ones also included archaeological approaches. In other words, these conferences have progressively had an archaeohistorical character, and a more and more strongly anthropological one. Europe is also the host to some important research projects underway on the role of salt in the prehistory or the history, or both, of diverse zones of Europe. Extremely important research is going on in Asia and Mexico, and ethnological research in South America, Western Africa and Oceania. All this demonstrates more than convincingly that research centered on the theme of salt, irrespective of its nature, is gaining a more and more individualized place in the totality of humanities research in the whole world.

In Romania, extremely rich in salt resources, the first international colloquium on archaeological salt was organized in 2004 by the Piatra Neamţ History Museum, an institution well-known for its research into the Cucuteni culture. In 2008, a similar scientific event, more open to the anthropological approaches, took place at the "Al. I. Cuza" University of Iaşi, at which professors revealed pioneering work, which had begun in the 1950s, on the relationships between salt and human communities in Romanian prehistory.

Organization of this colloquium was made possible within the generous framework offered by a Romanian project[1], centered on traditional behaviors in an area with the oldest evidence for salt production in Europe, and probably worldwide.

For three days, between 2 and 4 October 2008, almost 30 researchers from Europe (Bulgaria, France, Italy, Portugal, Romania, Spain, United Kingdom) and America (Mexico) presented the results of their work in the field. A real balance between the traditional and interdisciplinary types of approach could be noticed, between the revaluations of old data and new records which were mostly archaeological. The organizers wished to capture as many perspectives on the complex phenomenon of salt, against the background of the elements which determined the prehistory and history of human communities. Clearly, the papers presented at this conference could not exhaust these intentions, as they tended to illustrate many possible approaches to a multidimensional phenomenon. We would like to emphasize that, beside the usual contributions from archaeologists, historians, and ethnologists, this colloquium also sought to stimulate a literary and linguistic approach, in order to underline fruitful ways of research into areas whose practitioners were otherwise unassertive, unsystematic and rather peripheral.

---

[1] *Salt springs in Moldavia: the ethnoarchaeology of a polyvalent natural resource*, (414/2007), financed by the *National Council for Scientific Research in Higher Education* (2007-2010).

This volume of proceedings starts with the talk held by Professor Nicolae Ursulescu from the Faculty of History of the University of Iaşi. This decision was made as a sign of respect to him as the author of the first Romanian analytical study devoted to a salt spring with archaeological vestiges from the first European Neolithic Age, and generally for his work in the field of salt archaeology in Romania. The 23 studies in this volume illustrate some distinct approaches from the point of view of topic and chronology.

Two case studies on salt springs in Romania and two on those in Mexico offer the possibility of contrastive analysis of an ethnoarchaeological nature; beyond the structural invariables of salt exploitation, there are notable differences at the level of the phenomenon under survey and all the types of approaches involved. Special mention must be made about the study on the primitive forms of salt mining still practiced today in the Carpathian bend zone.

The topic of archaeological salt exploitation starts with two studies which refer to the most recent discoveries in Bulgaria, at Provadia-Solnitsata. From the very beginning of research, this has become a highly important and interesting point of reference for the European archaeology of salt. Four studies illustrating different stages of research approach the problem of numerous salt springs or salt outcrops in Moldavia and Transylvania. These studies were either predominantly given in the form of a presentation, or they give models of interpretation. A study on the origins of salt exploitation in Spain offers the possibility of a contrastive analysis.

Ancient texts represent the starting point of some interdisciplinary analysis focussed on some salt uses in Ancient Egypt, on the quantification of salt needs in Greek and Roman Antiquity, as well on the complex reconstruction of the multiple dimensions connoted by a key term in Ancient Rome: *salina*. Clearly, by the nature of their strongly exploratory nature, these studies sometimes courageously answer the questions raised by ancient texts. These epistemic trials illustrate the difficulties created by the nonexistence of more detailed ancient texts and a distinct non-literary referent. We should not therefore be surprised that even if the researcher does have written references, paradoxically, after all, we are in the middle of prehistory more often than not.

Historical approaches include three studies which refer to the Iberian Peninsula and one about the Italian Peninsula. All these approaches aim at the fullest possible reconstruction of the past, and emphasize the need to refer to the results of other disciplines, from landscape archaeology, paleogeography and geomorphology to linguistics. We should note that all the studies in this section emphasize, in various degrees of intensity, the idea of continuity of the different forms of salt exploitation, either on the coast, or the exploitation of salt rivers or mines in continental areas. The need for dialogue between history and prehistory, between history and other disciplines of the Humanities comes out once again.

The final section, Linguistic and Philological Approaches, has been deliberately separated from the others precisely in order to emphasize the importance of reflexes in the language and mentality of salt in different ethno-cultural locations. One of the studies represents a critical synthesis at the level of common Indo-European and different Indo-European lexicons, while another study refers to Romanian halohydronomy and halotoponymy which reflect a zone of ethnic interferences. The degree of deep and diverse influence of salt on the world of Latin paremiology and that of different European peoples is another relevant topic for the symbolism of salt.

While all these approaches cover thematically, chronologically and methodologically different topics, so disparate a first sight, they have in common the complex relationship of humanity during its whole evolution to the *sine qua non* element of reference which is salt, a primordial element which left a strong imprint on everyday supply and consumption structures, and not just those alone. Technical behaviors related to this polyvalent mineral resource, the social, therapeutic, linguistic and spiritual implications are but some of the more important aspects of what we have come to know as the anthropology of salt.

M.A., O.W., R.-G. C.

Acknowledgements
The quality of this publication is due partially to the financial support of National Council for Scientific Research in Higher Education (Romania), the Commission for Foreign Excavations (French Ministry of Foreign Affairs) and the Daad Alumni Club of Iaşi (represented by Professor Lucreţiu Bîrliba).

The editors thank Mr. Christopher Lawson, M.A., for translating some of the articles and for the language improvement of the English versions of some contributions to the present volume.

We also thank Ana-Maria Buşilă, a Ph.D. candidate at the "Alexandru Ioan Cuza" University of Iaşi, for her editorial input to this work.

Welcoming Speech

Nicolae Ursulescu
Universitatea "Al. I. Cuza" Iaşi, Romania

My colleagues in the ETHNOSAL research project have asked me to deliver the opening speech at our conference. I consider this a great honour. It is also a privilege to have the opportunity to address such a distinguished audience, which assembles researchers from eight countries who are interested in the historical position and role of this apparently ordinary mineral.

The Romanian subsoil is extremely rich in salt reserves, capable, according to certain estimations (Stoica and Gherasie 1981), of providing for the needs of the planet for more than a millennium. But it was only relatively recently that researchers started to study its exploitation across prehistory.

This topic has attracted the attention of historians concerned with the Middle Ages (Ilieş 1956) and the modern era (Vitcu 1987), but not that of pre-historians.

As a direct participant in the opening of this 'Pandora's box', I would like to evoke a few memories and highlight some details about the early stages of the archaeological and ethnoarchaeological research of salt exploitation in Romania.

In 1952, Ion Şandru, a professor of geography at the "Alexandru Ioan Cuza" University in Iaşi, published a study of the 29 'salt springs' he had charted in southern Bukovina. At the same time he mentioned the presence of traces of old settlements around these springs. The traces included fragments of stemmed vessels, made of a paste containing organic materials, characteristic of the early Neolithic Starčevo-Criş culture (Şandru 1952, 414).

In October 2008, it was 40 years since my colleague Mircea Ignat, from the Museum in Suceava, with Professor Şandru's study as a starting point, initiated an investigation around the most famous salt spring, Slatina Mare, in the territory of the small town of Solca in Suceava County (Ursulescu 1977, 311, footnote 26). At my colleague's invitation, I participated in the final stage of the excavations, as, at that time, I was interested in the Starčevo-Criş culture.

Of the resulting findings, a category of ceramic fragments (pot bottoms) attracted our attention, since they had never been mentioned in the repertoire of the shapes of the prehistoric pots in Romania. After many years of bibliographical research, I managed to find an analogy with one of the forms described as 'briquetage vessels' (Déchelette 1913, 713-716; Filip 1969, 1194-1197). They were closely connected with the process of extracting salt by boiling salty spring water. At this time the best corresponding models came from Germany – the Halle region (Riehm 1960; Matthias 1961), Poland – the Cracow region, especially from Wieliczka (Jodlowski 1971; Bukowski 1985), and from France – particularly Normandy and Britanny (Gouletquer 1972; Nenquin 1961).

Unfortunately, no other categories of materials accompanied the pot fragments at Solca. Nor were any traces of dwellings discovered. This made us believe that the area was inhabited only seasonally, in close connection with salt exploitation for commercial reasons (Ursulescu 1977, 314-315).

In 1977, after reaching certain conclusions based on the analogies mentioned above, I published a first study about this discovery (Ursulescu 1977). As a consequence, I introduced a new category of materials to Romanian archaeological literature and I opened a new research direction for the reconstruction of the way of life of the prehistoric communities in our country.

The most difficult problems posed by this discovery were related to dating. The stratigraphic circumstances at Solca-Slatina Mare, Professor Şandru's observations, and the composition of the paste and the firing technique of these fragments, caused us to assign them to the Starčevo-Criş culture. The same belief was reinforced in the study published in the proceedings of the "Nature et culture" International Conference, held at Liège, 1993 (Ursulescu 1996). In spite of the appearance of some contradictory opinions over time (Dumitroaia 1994, 66; Monah and Dumitroaia 2007, 15; Nicola *et al.*, 2007, 35-56; Weller and Dumitroaia 2005; Monah 2008, 104-105), we continue to maintain this position today. The contradictory opinions mentioned above are based especially on the result of a new investigation carried out in 2003 at Solca-Slatina Mare by a team coordinated by Gh. Dumitroaia. This argued that no findings of the Starčevo-Criş culture were present in the new digging (Dumitroaia *et al.*, 2004). According to the analogies with the situation at Lunca, Neamţ County (Dumitroaia 1994; Weller and Dumitroaia 2005), the briquetage vessel bottoms should be exclusively assigned to the Cucuteni culture. Nevertheless, on different occasions I mentioned that the new investigation took place at a certain distance from the precedent of 1968, so that the situation of the discoveries and the stratigraphy could be different for that reason. This opinion takes into account the situation in 1968, when the first two sections, located above the salt spring well, recovered no findings (Ursulescu 1977, 311), while the third, dug right near the well, provided numerous and significant materials. That is why we consider that only extensive excavations at this site will solve the question of the correct assignment of the vestiges there.

Besides the chronological assignment, I also made an analogy between the archaeological situation at Solca and the traditional methods of salt water supply with the inhabitants in the neighbouring areas (Ursulescu 1977, 314 and footnote 34). My suppositions were confirmed by some pieces serving for salt water transport I saw in Toader Hrib's museum house in Arbore, near Solca. Hrib is a self-educated peasant. Thus, just like Molière's character, who involuntarily produced prose, I had started some ethnoarchaeological endeavours in the field of the traditional salt exploitation in Bukovina, even before it was established as an interdisciplinary domain.

Reserve first greeted the study of the findings about Solca-Slatina Mare, because of its novelty and because the statement (daring for that time) that the oldest European evidence of salt exploitation was to be found in the Romanian Oriental Carpathians. But, the publication of the discoveries at Lunca (Neamţ County) brought even more convincing evidence, owing to the more extensive excavations made there (Dumitroaia 1987; 1994). Likewise, Dan Monah's 1991 study rightly highlighted the importance of salt in prehistory, thus opening new directions of research. Marius Alexianu, Gh. Dumitroaia and Dan Monah's ethnoarchaeological research of salt exploitation continues this novel line (1992).

As a consequence of the increasing interest in this topic and in the central role salt has had for human communities across history, the last two decades have witnessed an expansion of the excavations concerning salt exploitation (Monah 2008). The research area has been extended towards other regions, such as the hills along the Carpathian Bend (Ciobanu 2002) or Transylvania (Cavruc and Chiricescu 2006). Congresses and volumes were dedicated to the archaeological and ethnoarchaeological research about salt (Cavruc and Chiricescu 2006; Monah *et al.*, 2007; Alexianu, Weller and Brigand 2007; Monah, Dumitroaia and Garvăn 2008).

Comprehensive national and international research projects were initiated, with the participation of specialists from France, Great Britain, Turkey and Romania. I note with satisfaction that the archaeological and ethnoarchaeological research about salt exploitation has become a favourite topic for prehistoric and ancient research in Romania, thus mirroring the research trend in other European countries.

Unfortunately, our interests, lately directed towards other areas, have made us drift apart from the issue of salt across prehistory. Yet careful observation of the problem and the accumulation of interesting materials suggest to us the perspective of two important future directions.

Thus, field and laboratory studies should result in the development of a complex typology of the salt springs, according to their concentration, composition and geographical situation, the existence or absence of a neighbouring fresh water source and the intensity of settlement traces during history. An authentic typology of this kind would allow for a more accurate understanding of the anthropical presence around certain springs.

Secondly, at the reconstitution the environment conditions we should consider the alternative sources of salt (salty ponds, salty grounds, halophyte plants, etc.), for the communities having no salt springs near their settlings. In fact, the idea that, for the prehistoric agricultural populations, a salt source was an essential condition for the establishment of a settlement has become very credible. Our own recent research at Isaiia (Iaşi County) has suggested this hypothesis (Ursulescu 2004).

Finally, a corollary of the interdisciplinary research concerning the place and methods of exploitation of salt during history would be the initiation of a salt museum, exhibiting scientific data (geological, geographical, chemical), as well as archaeological, historical, and ethnographical evidence, and evidence from folklore, about this domain.

Our colloquium provides a good opportunity to capitalize on the results in the field of salt research in various countries which possess rich resources and a certain tradition in the complex study of this mineral which is so vital for the history of humanity. With some nostalgia for the times when this issue captured my whole attention, let me wish for this meeting starting now in Iasi – the proceedings of the colloquium of salt anthropology – complete success!

References:

Alexianu, M., Dumitroaia, Gh. and Monah, D. 1992. Exploatarea surselor de apă sărată din Moldova: o abordare etnoarheologică. *Thraco-Dacica*, XIII, 159-167.

Alexianu, M., Weller, O. and Brigand R. 2007. *Izvoarele de apă sărată din Moldova subcarpatică. Cercetări etnoarheologice*. Iaşi, Casa Editorială Demiurg.

Bukowski, Z. 1985. Salt Production in Poland in Prehistoric Times. *Archaeologia Polona* 24, 27-71.

Cavruc, V. and Chiricescu, A. (eds.) 2006. *Sarea, timpul şi omul*. Sfîntu Gheorghe, Angustia.

Ciobanu, D. 2002. *Exploatarea sării în perioada marilor migraţii (sec. I-XIII e.n.) în spaţiul carpato-dunărean*. Buzău, Biblioteca Mousaios.

Déchelette, J. 1913. *Manuel d'archéologie préhistorique, celtique et gallo-romaine*, II, 2 (*Premier Âge de Fer ou époque de Hallstatt*). Paris, Picard.

Dumitroaia, Gh. 1987. La station archéologique de Lunca-Poiana Slatinii. In M. Petrescu-Dîmboviţa *et al.* (eds.), *La civilisation de Cucuteni en contexte européen*. Bibliotheca Archaeologica Iassiensis I, 253-258.

Dumitroaia, Gh. 1994. Depuneri neo-eneolitice de la Lunca şi Oglinzi, judeţul Neamţ. *Memoria Antiquitatis* XIX, 7-82.

Dumitroaia, Gh., Nicola, D., Munteanu, R., Preoteasa, C, Monah, D., Boghian, D. and Ignătescu, S. 2004. Solca, com. Solca, jud. Suceava. Punct: Slatina Mare. In *Cronica cercetărilor arheologice din România. Campania 2003*, Bucureşti, 314-315.

Filip, J. 1969. Salz und Salzgewinnung. In J. Filip (ed.), *Enzyklopädisches Handbuch zur Ur- und Frühgeschichte Europas*, II, 1194-1197. Prag, Tschechoslowakischen Akademie der Wissenschaften.
Gouletquer, P. L. 1972. Briquettagestättcn der französischen Atlantikküste. *Jahresschrift für mitteldeutsche Vorgeschichte* 58, 176-193.

Ilieş, A. 1956. Ştiri în legătură cu exploatarea sării în Ţara Românească pînă în veacul al XVIII-lea. *Studii şi materiale de istorie medie* I, 155-194.

Jodlowski, A. 1971. *Die Salzgewinnung in Kleinpolen in urgeschichtlichen Zeiten und im frühen Mittelalter*. Studia i materialy do dziejow zup solnych w Polsce IV. Muzeum Zup Krakowskich-Wieliczka.

Matthias, W. 1961. Das mitteldeutsche Briquetage-Formen, Verbreitung und Verwendung. *Jahresschrift für mitteldeutsche Vorgeschichte*, 45, 119-125.

Monah, D. 1991. L'exploitation du sel dans les Carpates Orientales et ses rapports avec la culture de Cucuteni-Tripolye. In V. Chirica and D. Monah (eds.), *Le Paléolithique et le Néolithique de la Roumanie en contexte eurpéen.* Bibliotheca Archaeologica Iassiensis IV, 387-400.

Monah, D. 2008. Sarea în preistoria României. *Arheologia Moldovei* XXX (2007), 87-116.

Monah, D. and Dumitroaia, Gh. 2007. Recherches sur l'exploitation préhistorique du sel en Roumanie. In D. Monah, Gh. Dumitroaia, O. Weller and J. Chapman (eds.), *L'exploitation du sel à travers le temps*, Bibliotheca Memoriae Antiquitatis XVIII, 13-34. Piatra Neamţ, Constantin Matasă.

Monah, D., Dumitroaia, Gh., Weller, O. and Chapman, J. (eds.). 2007. *L'exploitation du sel à travers le temps.* Bibliotheca Memoriae Antiquitatis XVIII, Piatra Neamţ, Constantin Matasă.

Monah, D., Dumitroaia, Gh. and Garvăn, D. (eds.) 2008. *Sarea de la prezent la trecut.* Bibliotheca Memoriae Antiquitatis XX, Piatra Neamţ, Constantin Matasă.

Nenquin, J. 1961. *Salt. A Study in Economic Prehistory.* Dissertationes Archaeologicae Gandenses VI. Brugge, De Tempel.

Nicola, D., Munteanu, R., Garvăn, D., Preotesa, C. and Dumitroaia, Gh. 2007. Solca-Slatina Mare (Roumanie). Preuves archéologiques de l'exploitation du sel en préhistoire. In D. Monah, Gh. Dumitroaia, O. Weller and J. Chapman (eds.), *L'exploitation du sel à travers le temps.* Bibliotheca Memoriae Antiquitatis XVIII, 35-56. Piatra Neamţ, Constantin Matasă.

Riehm, K. 1960. Die Formsalzproduktion der vorgeschichtlichen Salzsiedlstätten Europas. *Jahresschrift für mitteldeutsche Vorgeschichte* 44, 180-217.

Stoica, C. and Gherasie, I. 1981. *Sarea şi sărurile de potasiu şi magneziu din România.* Bucureşti, Editura Tehnică.

Şandru, I. 1952. Contribuţii geografico-economice asupra exploatării slatinelor în Bucovina de Sud. *Studii şi cercetări ştiinţifice* (Iaşi) III, 407-422.

Ursulescu, N. 1977. Exploatarea sării din saramură în neoliticul timpuriu, în lumina descoperirilor de la Solca (jud. Suceava). *Studii şi cercetări de istorie veche şi arheologie* 28, 3, 307-317.

Ursulescu, N. 1996. L'utilisation des sources salées dans le Néolithique de la Moldavie (Roumanie). In M. Otte (ed.), *Nature et Culture*, I, 489-497. Etudes et recherches archéologiques de l'Université de Liège 68. Liège.

Ursulescu, N. 2004. Archéologie et archéozoologie dans l'habitat de Isaiia (com. de Răducăneni, dép. de Iaşi). In L. Bejenaru (ed.), *Archaeozoology and Paleozoology Summercourses.* Socrates Intensive Programme, Academic Year 2002-2003, Iaşi, Editura Universităţii "Al. I. Cuza" Iaşi, 79-95.

Vitcu, D. 1987. *Istoria salinelor Moldovei în epoca modernă.* Iaşi, Editura Universităţii "Al. I. Cuza" Iaşi.

Weller, O. and Dumitroaia, Gh. 2005. The earliest salt production in the world: an early Neolithic exploitation in *Poiana Slatinei*-Lunca, Romania. *Antiquity* 79, 306, http://antiquity.ac.uk/projgall/weller/index.html.

Part I. Ethnographic Approaches of Salt

Salt Springs in Today's Rural World. An Etnoarchaeological Approach in Moldavia (Romania)

Marius Alexianu
Universitatea "Al. I. Cuza" Iaşi, Romania

Olivier Weller
CNRS, Laboratoire de Protohistoire européenne, UMR 7041-ArScAn
Maison de l'Archéologie et de l'Ethnologie, Paris, France

Robin Brigand
Université de Franche-Comté, UMR 6249
Laboratoire de Chrono-Environnement, Besançon, France

Roxana-Gabriela Curcă
Universitatea "Al. I. Cuza" Iaşi, Romania

Vasile Cotiugă
Universitatea "Al. I. Cuza" Iaşi, Romania

Iulian Moga
Universitatea "Al. I. Cuza" Iaşi, Romania

Abstract
The Subcarpathian area of Moldavia represents the ideal framework to perform extensive ethnoarchaeological research as there are here over 200 salt water springs near which are found archaeological deposits related to the exploitation of the salt water. Nowadays, these deposits are still exploited at an unexpected degree of intensity by the members of rural as well as of urban communities. The main research focuses on the identification of all salt springs in sub-Carpathian Moldavia and on the completion of complex ethnoarchaeological research (exploitation, use, distribution networks, commerce, hunting, halotherapy, social contexts, ethnoscience, symbolistics, etc.).

Keywords
salt springs, ethnoarchaeology, sub-Carpathian Moldavia

The sub-Carpathian zone of Moldavia (Eastern Romania), with its unusual wealth of salt springs (over 200), is of great interest for ethnoarchaeological research. This concentrates on the role of salt in the development of human communities for at least two reasons: 1) close to some salt springs, the oldest evidence for salt production in Europe, and probably worldwide have been discovered; 2) to a considerable extent, traditional practices for providing salt springs continue to be used up to the present time.

The radiocarbon dates obtained in *Poiana Slatinei*, Lunca-Vânători, in Neamţ county (Dumitroaia 1987, 1994), which indicate an exploitation of natural brine during Neolithic and Chalcolithic Times (6000-3500 BC), for almost 2,500 years, are very recent (Weller, Dumitroaia 2005), as moreover the reconstruction for prehistoric techniques from multidisciplinary studies (Weller *et al.,* 2008). But the mentions of traditional exploitation of salt springs are almost three centuries old. The oldest proof we know of has been left in manuscript (Peithner 1784; Ceauşu 1982), but a few decades after this, Western publications made known the fact that in Moldavia and Transylvania re-crystallized salts obtained from salt-water springs were produced through a uncommon process: 'It is worth remarking, that the present rude inhabitants of Moldavia and Transylvania, who live in the neighbourhood of salt spring, have the same method of procuring salt wich was common amongst the ancient Gauls and Germans; this was to pour gradually the salt water upon a wood fire.' (Townson 1797, 395; see also Brongniart 1807).

1. A short history of research

The research as such on the links between the prehistoric communities and the salt springs began as late as the 1960s and grew in intensity during the 1980s. This research has had, and still has, a pre-eminently archaeological character. It was a geographer from the „Alexandru Ioan Cuza" University of Iaşi who first pointed out, during the 1950s, the importance of salt water springs in two regions of Eastern Romania (the Cacica and Târgu Ocna-Slănic Moldova depressions) for the Chalcolithic communities of Precucuteni and Cucuteni (Şandru 1952 and 1961). That research direction was reopened 25 years later, when an archaeologist wrote the first study in the

Romanian literature in the field of archaeological discoveries near a salt spring, which was exploited from the Neolithic (Starčevo-Criş culture) to the Middle Ages (Ursulescu 1977).

The discovery of a Chalcolithic tell at Poduri (Bacău county) in an area rich in salt springs decided a group of archaeologists to initiate research on the possible relations between these natural springs and the complex dwellings of that tell (Monah *et al.,* 1980; 1987; 2003). The importance of the saliferous Moldavian sub-Carpathian area for the multiple development of the famous Cucuteni-Tripolye Chalcolithic complex was brought into prominence by the American researcher Linda Ellis (1984, 205) in a memorable statement: 'It is also no accident that the longest area of occupation for the Cucuteni-Tripolye culture (i.e. the Eastern Carpathians and sub-Carpathians) happens to be a region noted for one of the largest salt formations in Eastern Europe. Exploitation of, control over, and trading of this essential resource no doubt contributed to the stability of Cucuteni-Tripolye village life in the face of culture contact with Eastern steppe pastoralists, as well as enhancing the quality of food, storage, food consumption, and animal and human health'.

This hypothesis was reconfirmed by subsequent discoveries in an impressive site of exploitation at the *Poiana Slatinei*–Lunca salt spring (Neamţ county), where the prehistoric exploitation stratum, starting with the Starčevo-Criş culture, is up to 2.65m thick (Dumitroaia 1987; 1994). The rate of such discoveries was intensified by subsequent finds (Andronic 1989), which led to the first synthetic archaeological studies about the exploitation of salt springs (Monah 1991; 2002; Ursulescu 1995; Weller 2000). Taking into account the fact that the archaeological data looked promising, mainly in regard to their antiquity, researchers in international programs came to approach the problems of the Moldavian salt springs exploitation. Among these, we may mention the following: 1) two British-Romanian projects, *Research on trade and exchange in the Cucuteni-Tripolye Network* from 2001 until 2005, and, *Prehistoric salt exploitation in Romania and Anatolia* from 2002 until 2005; 2) two French-Romanian projects: *Aux origines de la production du sel en Europe: préhistoire et écologie des Carpates Orientales*, from 2003 until 2004, and, beginning in 2004 and always current, *Les eaux salées de la Moldavie roumaine: archéologie, histoire et écologie d'une ressource structurante du territoire*.

By resorting to actual dating methods, some archaeologists have shown that the oldest salt exploitation in Europe, and probably worldwide, so far known, dated between 6050-5500 BC, during the Early Neolithic, is the one at Lunca-Poiana Slatinei, Neamţ county, Romania (Weller, Dumitroaia 2005; Weller *et al.,* 2008). The production of ignigenous salt from salt springs continued even later, during the Cucuteni culture, at the end of phase A, but very intensely during Cucuteni B (Dumitroaia 1994; Nicola *et al.,* 2007). A series of more recent discoveries brought new evidence about the use of salt springs in Moldavia by the Chalcolithic population (Monah *et al.,* 2003; Chapman *et al.*, 2000; 2003; 2003-4; 2007; Andronic, Ursu 2003; Nicola *et al.*, 2007; Dumitroaia *et al.*, 2008; Munteanu 2006; Munteanu *et al.*, 2007; Weller *et al.*, 2007a; 2007b; 2010) and a synthetic GIS project is in progress (Weller *et al.*, 2007c; see Weller *et al.*, in this volume).

What differentiates this Moldavian sub-Carpathians area from other similar areas in Europe is the intense, unexpected continuity of traditional, non-industrial water supply coming from salt springs, with its multiple uses, in the villages, as well as in the cities. The intensity of salt exploitation in our days is an important benchmark for the understanding of the relationship between the rural communities and this natural resource. This phenomenon can be explained first of all by the obvious interactions in this area between the communities mentioned above and the environment. This interactions have been stimulated in a special way by the change in the economic system in 1989 which encouraged private initiatives related to animal breeding (animals need large quantities of salt, including liquid dispersal), and to production and conservation of food. Another element which facilitates the salt springs exploitation is the fact that these are generally situated on lands belonging to the state, or are owned as common property. Consumers can therefore decide to access them freely without the slightest financial, legal or administrative restrictions or regulations. But in some cases the local authorities tend to establish access taxes to salt springs. This meets with resolute refusal by the users to pay because they believe that salt water is 'a gift from God'.

Many traditional applications are specific to the area where we identified the oldest traces of salt exploitation in Europe. This constitutes a great opportunity for new ethnoarchaeological research. This statement was recently made: Ethnoarchaeology becomes 'a real science of reference for interpreting the past, if focused upon well-founded cross-cultural correlates, which link material culture with static and dynamic phenomena' (Roux 2007, 153; see also Morantz 1998; Gazin-Schwartz; Holtorf 1999; Damm 2005). As early as 1992, there was a focus (Alexianu *et al.*, 1992) on the importance of systematic ethnoarchaeological research in the area under discussion, where there are also elements of continuity in the chrono-topical system. These elements can only strengthen the credibility of the ethnographic analogy in our understanding of the complex archaeological situations related to salt exploitation.

This research has seen remarkable intensification under some Franco-Romanian programs begun in 2003 which are still in progress, and conducted by one of the authors (Weller, Dumitroaia 2005; Weller *et al.*, 2007a; 2007b; 2007c; 2008; 2010; Weller *et al.*, in this volume). Since 2007 research has been taking place at a faster rhythm with a Romanian project led by a small-sized team (dr. Marius Alexianu, manager, dr. Olivier Weller, expert researcher, dr. Roxana-Gabriela Curcă, dr. Vasile Cotiugă, dr. Iulian Moga). Outside the official team, PhD candidate Robin Brigand has made an important contribution to carrying out this project (Alexianu, Weller 2007; Alexianu *et al.*, 2007a; 2007b; 2008; Alexianu, Weller 2009).

2. Methodological aspects

Romanian ethnographers never systematically approached aspects linked to salt springs, although known since prehistory. This was why two of the authors (O.W. and M.A.) have designed three types of ethnographical questionnaires according to the location in the field (village or town, sheepfold, salt spring). The questionnaires were conducted according to standardized norms of ethnological research, and also included specifically archaeological issues. The three kinds of questionnaires were carried with the help of a spatial analyst (Laure Nuninger, UMR 6249, CNRS, Besançon), who treated the ethnological information in the light of a Geographical Information System (GIS). The creation of these working tools, which proved to be extremely efficient, ensured a systematic character to field research.

This was how exhaustive and intrinsic ethnographic surveys were carried out from archaeological standpoints. Among the items within the surveys mentioned above we might mention: 1) identification of salt springs; 2) chrono-topical dimension of the extraction, storage, manipulation and uses of the brine; 3) transportation and storage means in households; 4) sphere of use for salt water or for salt, as powder obtained through granulation or re-crystallization; 5) attraction exercised by salt springs on wild and domestic animals; 6) practice of hunting near the springs; 7) re-crystallization practices; 8) frequency of salt supplies for domestic use; 9) use in commerce and exchanges; 10) salt-related behaviours and ethnoscience; 11) symbolism and rituals centring on salt.

Since the beginning of our inquiries we have tried to ensure the credibility of accounts. Most frequently, the informants selected were not in the least influenced by their school education or by mass media. They were mostly elderly persons or participants of different ages who supplied salt water. We were surprised to find out that teen-agers or even children showed considerable knowledge about the most varied aspects of the relations between salt springs and present-day human communities in rural and urban areas. As far as situations which cannot be checked *de visu* are concerned, such as springs which disappeared or which are abandoned today, or salt-crystallization practices, the truthfulness of our recorded accounts was ensured by agreement between information deriving from at least three unconnected informants. As far as possible, we resorted to documentary sources which refer to the Moldavian space published by Ceauşu (1982), Curcă (2007b) or unedited sources (Peithner 1784). 143 ethnographical inquiries were conducted in 107 places by the date of drafting this paper (Figure 1), including an enormous database which requires a long processing time. Global analysis remains a future objective, involving the fact that the project begun in 2007 is under way. Nevertheless some modelling has been possible so far.

Ethnographic studies were paralleled by ethnohistorical research upon the texts referring to salt in works of Oriental (Moga 2008, 2009), Greek and Latin (Alexianu 2007, 2008; Curcă 2007a) authors, and also in Romanian medieval documents (Curcă 2007b) with the purpose to emphasize the trans-spatial and trans-temporal constants of human behaviour regarding salt applications.

The principal objective of this article has been to present the main patterns resulting from the ethnographical inquiries. These patterns can be useful in understanding the role played by the salt springs in the salt supply of human communities in archaeological time. On the other hand, we would like to draw attention to the fact that salt springs show more dimensions than those in foodstuffs and conservation to which archaeologists usually restrict their research. In fact, salt springs still exert a multi-dimensional influence, including in the spiritual plane. Ethnographical research shows three main modes of salt springs exploitation which can be classified according to the criterion of frequency: 1) sampling, transport and use of salt water as such; 2) sampling, transport and boiling of salt water to be used in the making of re-crystallized salt; 3) sampling, transport and use of naturally crystallized salt by the salt springs.

3. Exploitation of salt in liquid dispertion

Ethnographical research carried out so far show that three types of areas of distribution come from the salt water springs, according to the spatial expansion of settlements which used this salt water. This spatial extension was determined by the flow, salt concentration and purity of the salt spring, by the taste characteristics of the brine (Rmn. *slatină* or *saramură*), by its accessibility (by different transportation means or without transport). The three types have been defined in the following terms (Figure 2): 1) village area, with springs of strictly local importance, used by a maximum of three villages situated at a distance of at least five kilometres; 2) municipal or supra-village area, in the case of springs used by four villages situated at a distance up to 20km; 3) supra-municipal area, in the case of springs used by numerous rural and urban places which exert attraction at great distances.

The real dimensions of salt water supply in the Moldavian sub-Carpathians come out in moments when, for different historic reasons (wars, temporary crisis of supply in times of peace), the supply system with common salt is not operative. Ethnographic inquiries have shown that in these situations transport can reach distances of up to 100km.

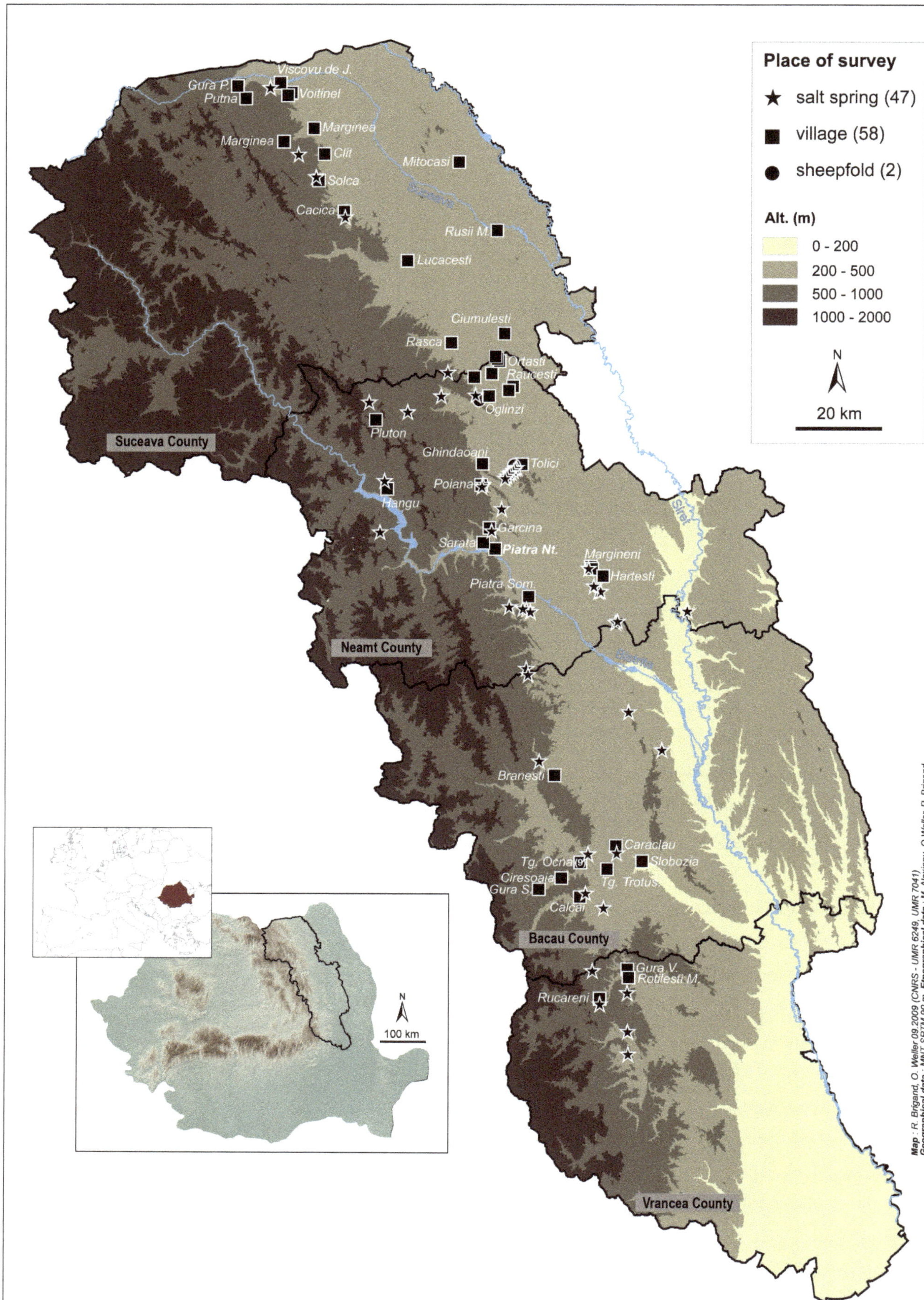

Figure 1. Study area and realized ethnographic inquiries (2004-2009).

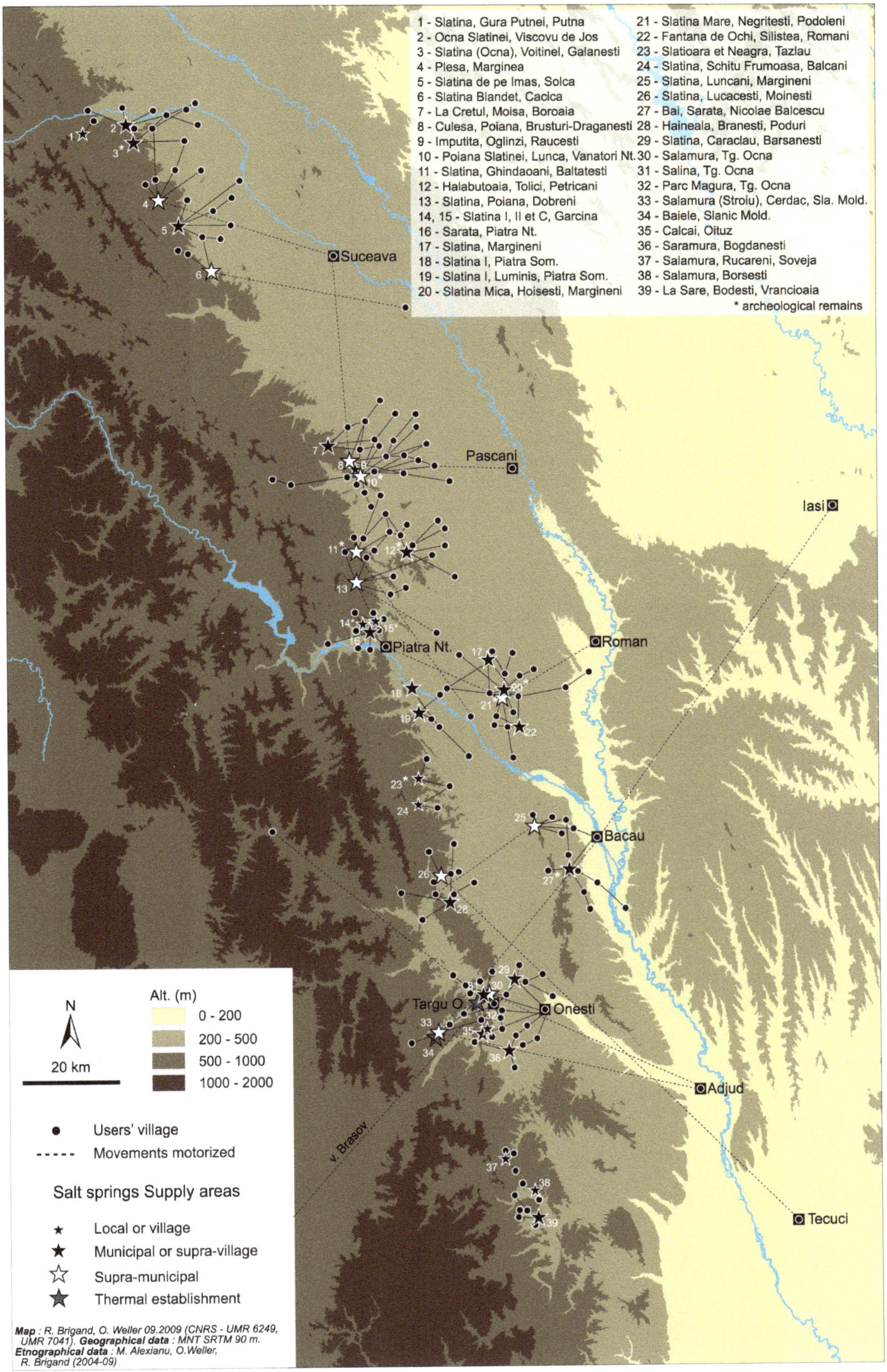

Figure 2. Use and supply areas of salt springs.

By applying the method of spatial analysis (Figure 2), we could conclude that the network of salt water in Moldavian sub-Carpathians, according to ethnographic inquiries carried out, actually cover the needs of all rural (and in some cases urban) localities in the area mentioned. In some cases of supply with brine some situations have been identified of the partial overlapping of distribution areas of two different springs. The distance between a spring and the locality which uses it is not decisive for the act of supply, so that some salt springs located too close to each other are only used by few village people; most people opt for a further spring whose flow, high salinity and taste qualities, retention capacity or access facilities are superior. In principle, we can admit the existence of a radial scheme of water distribution from a main salt spring in most of the encompassing habitat.

4. Salt water supply

Corroborating the ethnographic and archaeological information has shown that one has to operate the distinction, on one hand, between the simple supply and distribution of salt water, and, on the second hand, the production of re-crystallized salt (Rmn. *huscă*) from the salt springs and its distribution. As far as the first situation is concerned (salt water supply, Figure 2) on the basis of ethnographic questionnaires one could establish evidence of the following situations of relations between human communities and a salt spring.

1) Salt water supply point which practically corresponds with the area in direct proximity of salt spring: supply is short-term and depends on the total capacity of recipients used for transportation, on the flow of salt spring and on the number of persons involved in taking out salt water and pouring it into the recipients for transportation (Figure 3). It is a human activity which does not generate or leave traces, with the exception of some sherds from accidentally broken recipients (or now plastic bottles). In this category, there is all salt springs where sporadically fragments of archaeological ceramics have been found on the surface from one or several periods.

Figure 3. Salt water supply and rubbish (Coza, Tulnici, Vrancea; *Hălăbutoaia*, Ţolici, Petricani, Neamţ; *Poiana Slatinei*, Lunca, Vânători, Neamţ) (Photos O.W.).

Figure 4. Sheepfold and sweet cheese using salt water (*Hălăbutoaia*, Ţolici, Petricani, Neamţ; *Mătăhuia*, Căşăria, Dobreni, Neamţ) (Photos O.W.).

2) Dwellings which are supplied directly from a salt-water supply:

- seasonal dwellings of the sheepfold (Rmn. *Stână*) type; the salt spring is sometimes used to prepare sweet cheese, only as a foodstuff for shepherds; especially savoury are the pieces of sweet sheep cheese (Rmn. *caş*, from Lat. *caseus*) short time dipped in salt water before they are consumed (Figure 4). Identification of seasonal locations of this type when archaeological research takes place most certainly represents a very difficult task, but archaeologists must be warned about the possibility of the existence of such locations. Some agglomerations of ceramic fragments identical to those by a salt spring, situated at a distance of about 1 one or two kilometres from a spring, could indicate this kind of seasonal settlements for ovines and bovines.
- the settlements as such; ethnographic inquiries have noted that all villages around a spring use salt water as such. Two distinct situations could be identified. 1) With a single salt spring in a given area or with a salt spring of superior taste qualities, great flow and easy access in an area with several salt springs. Salt water supply today is carried out from settlements situated at a distance of about 10-15 kilometres from the respective spring (Figure 5); distances may reach 25-30 kilometres (Alexianu *et al.*, 2007, 144). 2) When there are several salt springs of similar taste, flow and access possibilities, situated at a distance of 5-6 kilometres between each other, each of them is used in groups of two to three villages situated at a distance of two to three kilometres from the spring (Monah *et al.*, 2003, 69-73). In this case the distance of supply from each spring decreases, but the supply area has parameters comparable with the previous situation.

3) Settlements supplied indirectly with salt water, located between 40 to 50 and approximately 100km from a salt spring. The distribution direction from direct users to settlements which use it is located in remote areas. According to inquiries carried out so far, salt water transport to settlements which are located so far away is more rarely attested; only in exceptional cases (the end of World War II and the following years, as a rule, salt water accompanied the transportation of re-crystallized salt obtained from a salt spring).

Figure 5. Salt water supply using cart (*Poiana Slatinei*, Lunca, Vânatori, Neamţ; *Hălăbutoaia*, Ţolici, Petricani, Neamţ) and ancient barrel for brine transport (Cucuieţi, Solonţ, Bacău) (Photos O.W.).

To conclude, use of salt water involves only the act of supply and generated a distribution network; more rarely a redistribution network. Today salt water supply is practised on a large scale and does not represent an indicator of poverty: it is used by different categories of economic and social status, mostly due to the quality of vegetable, cheese product and bacon conservation. With cheeses conservation even some (feta) cheese (Rmn. *telemea*) micro-production units use water from salt springs intensively.

5. Production and supply of ignigeneous salt

Before the modern production of crystallized salt (populary know as *huscă*), it is advisable to underline the very probable chronological continuity of the choices which led to salt exploitation of a spring rather than an other one. Indeed, salt springs with archaeological evidences for the Neolithic and the Chalcolithic exploitation are systematically an important water flow (or a well with a big capacity) and a very high salinity (Figure 6). It is also for these same reasons that this springs were, even recently, exploited for crystallized salt.

The practice used for re-crystallization of salt by boiling natural brine, which stopped in the 1995-1996, involved the following three main strategies:
1) production of re-crystallized salt (popularly known as *huscă*) in the proximity of the salt spring; 2) production of *huscă* in seasonal habitats such as isolated sheepfold in the mountains (Rmn. *stână*); 3) production of *huscă* in villages (in the courtyard or more rarely inside the houses). A cauldron on a support, sometimes suspended, was used for the brine evaporation (Figure 7).

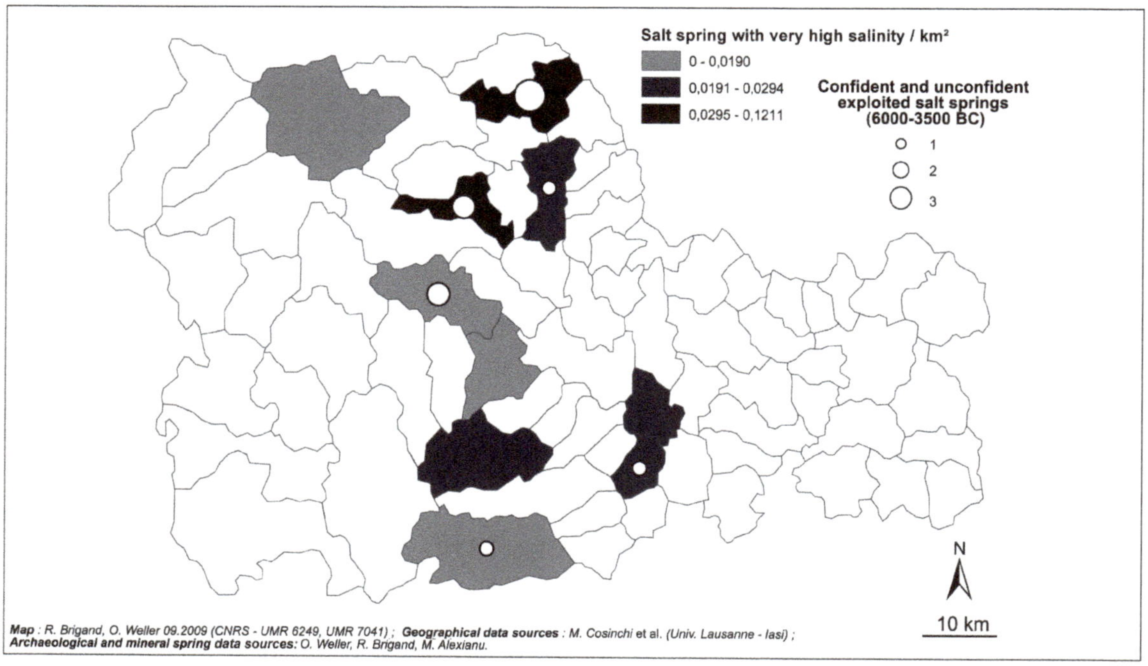

Figure 6. Modern and prehistoric salt exploitation in Neamt county. Confrontation between salt springs with high salinity, strong water flow, modern salt production (*huscă*) densities and prehistoric salt production.

Figure 7. Cauldron and fire tripod for salt production in the courtyard (Groşi, Brusturi, Neamţ; Cucuieţi, Solonţ, Bacău).

The following three stages in the use of re-crystallized salt from salt springs can be listed mostly in the northern half of the area investigated: 1) salt water supply, 2) production of re-crystallized salt and 3) its (re)distribution. The relationship between human communities and the salt spring becomes more complex. We can distinguish the following situations.
1) Salt spring – point of salt water supply; the salt water was transported across very short or longer distances. Ethnographic inquiries have shown that water taken from the spring was boiled either in the immediate proximity of the salt spring or at a short distance (30 to 50m), or in seasonal settlements such as *stână*, or in villages situated at a distance of 5-7km from the salt spring.

2) Point of production of re-crystallized salt by boiling natural brine, located near a salt spring, generally upstream, with a seasonal character. Most information has been obtained about these kind of production. This is a significant fact. The duration of the stay of a person or persons who produced *huscă* varied according to several factors (distance, accessibility and others). Most significantly it depended on the quantity of re-crystallized salt which was supposed to be obtained. The most common duration of stay took place in daylight (as in Sărata-Piatra Neamţ, Oglinzi, Boboieşti, Rucăreni-Soveja, Slătioara-Groşi, Slătioara-Tazlău). Otherwise, peasants might have stayed for 2-3 days, with a plan for each partner to obtain about 100 kilos of *huscă*, considering the fact that complete re-crystallization of brine (Rmn. *slatină*, *saramură*) in a cauldron needed six to seven hours of boiling (as with the people of Orţăşti after World War II or those in Râşca). As for the salt spring in Stroiu (Bacău county), the case of a family which stayed to obtain *huscă* for two to three weeks in an improvised seasonal dwelling was attested.

If the village was located relatively close to the spring, two to three people have been attested, as in Rucăreni-Soveja, as involved in the production of *huscă* at the spring. With localities farther from the spring, groups of producers of *huscă* have been found to be more numerous, as in the salt spring (Rmn. *slatină*) of Râşca, which shows concern over the economic profitability. But the existence of several *huscă*-producing groups at Neagra shows a constant micro-production carried out in the interwar period by the villagers of Tazlău located close to the respective spring, for trade reasons and selling in some villages and towns in Neamţ and Bacău counties. In the case of the same spring, inquiries have shown that *huscă* was produced both close to the spring and in the villages. *Huscă* producers are aware of the advantages and disadvantages of each production locality. With *huscă* production at the spring located at a more remote distance from the user settlement, the advantages involved the presence of fuel in that precise place and the transport of a larger quantity of re-crystallized salt; the disadvantage consisted of a longer stay, difficulties in supply and diseases generated by the low comfort of seasonal settlements. When *huscă* was produced in the user localities the advantage was that comfort specific to stable settlements could be organized, while the disadvantage lay in the transport of larger quantities of salt water and in the additional effort for fuel transportation.

The *huscă* was mainly destined to be bartered or sold in localities within a distance of 20-30km or at 70-200km.

3) Point of re-crystallized salt production by a seasonal settlement *stână* (sheepfold) type: the salt is used exclusively for the local needs, mostly for sheep.

4) Point of production of re-crystallized salt in a settlement as such: a. the salt is destined exclusively for the household needs; b. Salt is destined for partial household needs; c. Salt is destined partially for barter or sale in localities situated within up to 20-30km or localities at a distance of 70-200km.

Generally, we can conclude that the distribution territories of re-crystallized salt (Figure 8) are considerably larger that those of salt-water distribution, as a rule, up to 80-100km; but inquiries conducted in 2009 have shown longer routes of approximately 300km (e.g. Suceava – Galaţi).

As to the distribution of *huscă* for various exchange or sales, there are two situations depending on the permanent or sporadic character of *huscă* production (Figure 8):
- Under normal circumstances, when *huscă* is produced constantly, it is transported with a relatively regular frequency at short distances of 15-40km. So, inhabitants of Poiana-Negreşti who produced *huscă* would go to exchange 'salt-clods' (Rmn. *boţuri/bulgări de huscă*) with the Jewish merchants in the *Strada Mare* (i.e. the main street) in Piatra Neamţ. In exchange they would get olives, fish, carobs or peasant sandals (Rmn. *opinci*). *Huscă* was also sold directly in the market in Piatra Neamţ. Producers of *huscă* in Poiana-Negreşti would exchange cereals at the commission of settlers in neighbouring villages. As a rule, producers of *huscă* would transport it themselves and carry out the transactions. Specialized producers were both transporters and business people, a circumstance which contributed to increasing their profit. In one case *huscă*, this time as powder, obtained in the vicinity of *slatina* in Neagra and Slătioara (Tazlău) was usually sold; it was also bartered for oil with Jewish merchants in Piatra Neamţ, Buhuşi and Bacău towns and in the villages of Roznov and Rediu, which in their turn would trade them.

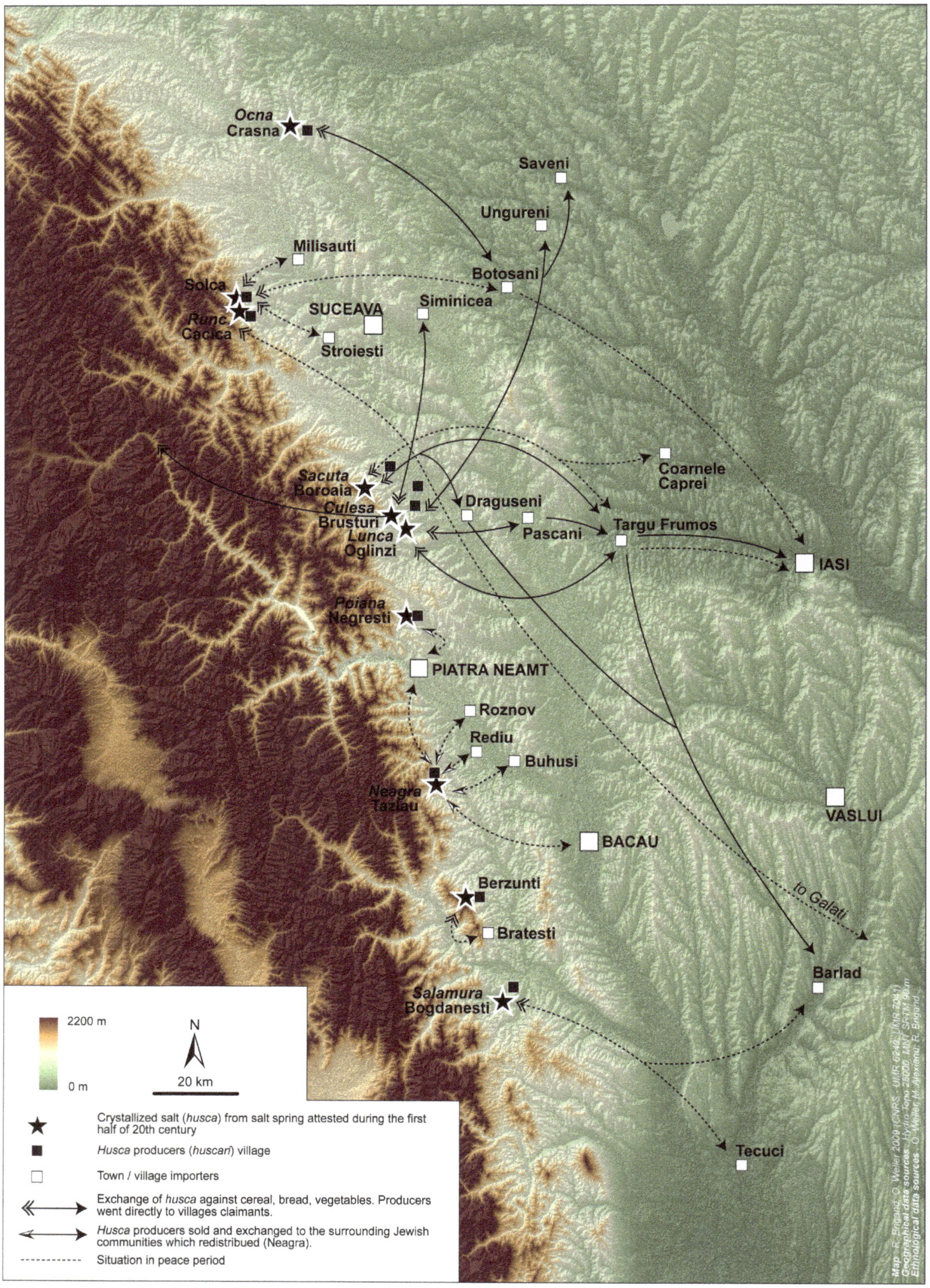

Figure 8. Salt crystallized production (*husca*) and barter during peace and war periods in the 20[th]century.

Under exceptional conditions, such as the disorganisation of supply with common salt at the end of World War II or during the drought in 1945-1946, we witness an intensification of re-crystallized salt production and a considerable extension of the distributional areal up to 100km, and even as far as 300km. For instance, *huscă* produced in the salt spring of Culeşa, village of Poiana, was transported at longer distances, between 70 and 100km, to Iaşi or Botoşani in carts hauled by horses or bulls (when horses had been requisitioned for the war). Fortunately, we were able to obtain information about the quantitative aspects of the barter. So, one kilo of *huscă* could be bartered for 2-3kg of wheat or 4 to 5kg of corn. But the circuit did not stop here at all, as sometimes a part of the quantity of wheat obtained in this way was sold in mountain areas in Valea Bistriței, where cultivation of cereals was almost impossible. In fact, the need for wheat and maize was satisfied by the exploitation of free natural resources, which also brought in the money necessary for other needs. We should underline the fact that the carts transported brine in barrels of a total capacity of 1000 litters.

Figura 9. Natural crystallizations of salt. Harvest and animal attraction (*Salamura*, Călcâi, Oituz, Bacău; *Zlatina*, Poiana, Vrâncioaia, Vrancea; *Hăineală*, Poduri, Bacău) (Photos O.W.).

Figure 10. Rock salt exploited for animal food (*Alghianu*, Poiana and Bodeşti, Vrâncioaia, Vrancea) (Photos O.W.).

The inquiries brought evidence of a similar model in several locations in the Moldavian sub-Carpathians. The barter with *huscă* had stopped in 1946 during the great famine, as on one hand, there were no cereals for barter, and on the other, the import from USSR of reddish common salt cake (5kg) started. Even the transport by train of small quantities of *huscă* (approximately 20kg), for very long distances during the drought, in order to solve by barter the needs of wheat and maize at the level of one family, have a certain relevance for archaeological time; an illustrative example in this respect can be the transport from Cucuieți-Solonț (via the railway station in Moineşti, Bacău, Eastern Romania) to the Banat (Southwestern Romania) or from Solca in the north down to Constanța in the south (Black Sea coast).

6. Use of naturally recrystallized salt

During the last period of ethnographic inquiries another salt exploitation technique was brought to light: the 'harvesting' of naturally re-crystallized salt around and downstream salt springs and its use as such in human and animal food, conservation, etc. (Figure 9). Even though the last century has not been in any way significant, this type of exploitation is extremely suggestive for the understanding of prehistoric situations. Our hypothesis is that prehistoric man first exploited salt water, and only later naturally re-crystallized salt, in the proximity of salt springs. Naturally re-crystallized salt offered a model for obtaining large quantities of salt by natural (solar) or artificial evaporation, by boiling. In other words, the ignigeneous process of obtaining salt results from the natural process of evaporation without any human intervention.

As a generalization, the data base regarding the *huscă*-cereal barter (Figure 8) indicates that the production of re-crystallized salt in Suceava, Neamţ and Bacău counties satisfied the needs in salt of the other counties in Moldavia, such as Botoşani, Iaşi, Vaslui and Galaţi. The model of *huscă* supply at such long distances presents some important interpretative implications for prehistoric times, and partially supports the hypothesis that a significant part of re-crystallized salt production was destined for long-distance barter.

Recent research carried out in Vrancea county where some salt springs and outcrops of rock salt still exploited today (Figure 10) have raised the question of relationships between the products of salt springs and rock salt. The conclusions of these results were surprising in the sense that production of re-crystallized salt by boiling was also usual in the territories with rock salt which is commonly used almost exclusively for animal food. By his superior taste qualities and his purer mineral composition, the salt re-crystallized by boiling is used first and foremost as a human foodstuff and only sporadically for animals.

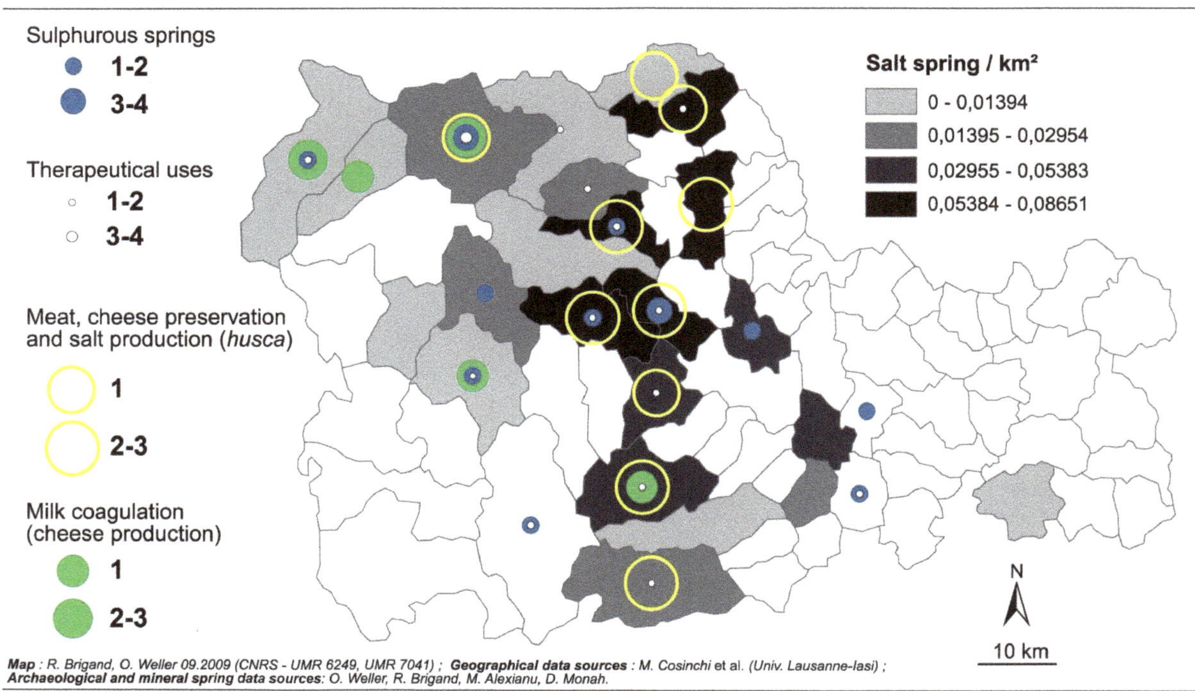

Figure 11. Salt springs densities and traditional uses in Neamţ county.

7. Varied uses

Ethnographic research has brought important precisions as far as the use of natural brine and salt in general (Figure 11). There are more uses than has been generally considered by archaeologists. Firstly, we would like to emphasize the fact that salt water is still used in large proportions by adding it directly into different dishes and foods. Salt water is used both for family and collective (in some restaurants, monasteries, etc.) consumption.

Figure 12. Salt vegetables, leaves, meat and mushrooms (Călcâi, Oituz, Bacău; Câmpeni and Băsăşti, Pârjol, Bacău) (Photos O.W.).

Figure 13. Traditional and modern therapeutic baths in salt water (Vizantea Mănăstirească, Vizantea Livezi, Vrancea; *Băile Sărata*, Nicolae Bălcescu, Bacău) (Photos O.W.).

Salt water has a generalized use in different mixtures of food, particularly for pig food. Forage given to cattle is sprinkled with salt water. Salt water is generally used mostly for the conservation of bacon (Rmn. *slănina*) and pork, of various types of cheeses, of various vegetables or greens (Figure 12).

Important quantities (500-3000l) are used by microenterprises, mostly by cheese factories which produce feta-type cheese (Rmn. *telemea*).

Some surprising results have been the use of salt water and salt as a remedy in various diseases. Peasants in the Moldavian sub-Carpathians use a wide variety of procedures, some of which are rather singular in the European space (such as heating stones for salt water in the case of treatment of rheumatism in large wooden baths) (Figure 13). Recent research has shown that a considerable part of halotherapeutic practices identified by ethnographic inquiries carried out in Moldavia are to be found in the Greek and Latin world (Jouanna 1994; Gil 2004). The common clinical spectre of ancient and modern (but traditional) therapies includes gum and dental diseases, skin burns, headaches, angina, tonsillitis, boils, inflammations of the skin and dermatosis, kidney and stomach pains, lumbar and leg pains, joint pains, dog or cat bites, frostbite, mouth and ear diseases, bleeding (Curcă 2007; Sandu *et al.*, 2009; 2010). The existence of some common therapies in such different chronological and cultural spaces implies that salt has had a strong therapeutic dimension in prehistory, a reality commonly neglected by archaeologists.

Game and hunting

Another interesting aspect is the attraction of dry or humid salt micro-zones exerted by the salt springs to some wild animals and birds. But some wild animals do not come here to satisfy their need of salt: so, the wild boars come to take baths in mud micro-zones, and the birds (doves, cranes, storks) are attracted by the micro-fauna. Stags and hinds lick dry and damp salted micro-zones. In some cases, when a damp salt zone is located in the proximity of a village not only wild animals lick it, but also domestic animals, first of all the bovines. Frequent visits by some animals and birds to the micro-zones situated nearby brought them, in numerous cases, to the attention of hunters who had come to improvise shelters to facilitate stalking.

The symbolism of salt

Our inquiries confirmed previous research on the customs which involve a certain symbolism of salt; sometimes some unexpected aspects and nuances appeared. The use of salt in the rituals of house foundation is of particular interest. For example, at Solca (Suceava county), when a house is being consecrated, through a ceremony performed by the orthodox priest, salt and a coin are placed in each of all four orifices from the corners of the house. Salt is placed there to assure prosperity for the family living in that house (Figure 14). By way of a general remark, we note that in the whole area of research salt presents only positive symbolism and connotations (Alexianu *et al.*, 2007b, 123-127).

8. Salt springs and their linguistic reflexes

Salt springs, brooks and puddles have generated numerous halotoponyms or halohydronyms which in fact represent the linguistic reflex of some zones characterized by the presence of salt in liquid dispersion. In their turn, the latter represent the linguistic indicator of saliferous resources. In this zone, which has known across history processes of ethnic interference, hydronyms and toponyms related to salt are mostly Latin or Slavic. Lat. Ac. *salem* > Rmn. *sare*, gave the adjective *sărat* (salted) generated simple (*Sărata*, *Săratu*) or compound halohydronyms (*Balta Sărătura*, *Pârâul Sărat*, *Pârâul Săratului*); also, Rmn. *saramură-salamură* (brine), which also names the salt water and the salt spring is of Latin origin. *Slatină* is of Slavic origin, with the meaning of salt spring or natural brine (Ciorănescu 2001 *s.v.*; Iordan 1963, 125-126), and also *Slătior* (salted brook), *Slătioară* (salted pool), *Soloneț* or *Solca*. Sometimes *Slatină* can be the name of a brook (Grămadă, 1996, 97). Salt springs can sometimes be qualified in a way which emphasizes their characteristics, such as *Slatina Mică* (small), *Slatina Mare* (Big), *Slatina Veche* (Old) or as *Slatina Rea* (Bad) when the spring is not suited for use. Some examples of compound halohydronyms of Slavic origin: *Dealul Slatinei* (Hill), *Drumul Slatinei* (Road), *Fântâna de Slatină* (Well), *Izvorul cu Slatină* (Spring), *Pârâul Slatinei* (Brook), *Poiana Slatinei* (Grove). For the linguistic consciousness of common inhabitants, the halohydronyms and the halotoponyms of Slavic origin, unlike those of Latin origin, are not etymologically transparent (Curcă 2008).

There is one more, indirect, linguistic indicator of the presence of salt water which refers to ignegenous salt. As we have already mentioned, re-crystallized salt obtained by boiling the brine is called *huscă* (Bulgăr, Constantinescu-Dobridor 2002, 213), a regionalism which has a quick development to the status of an archaism, replaced today, mostly in the south of Moldavia, by the descriptive syntagm *sare uscată*, dry salt. A Ukrainian etymology Puşcariu (1976, 291) or Old Germanic (Poruciuc 2007) have been suggested for this word, found by our research as the variant '*uscă*'. This also gave some derivatives, *huscari* which names those who produced the salt from the water of salt springs, including the anthroponym *Huscariu* (Iordan 1983), or toponymic indices such as *Huscani* or *Huscărie*. A rather rare second (family) name *Huscă* can be also found.

Figure 14. Ritual of house foundation using salt and coins (Solca, Suceava) (Photo O.W.).

Preliminary conclusions

The information obtained on the basis of ethnographic questionnaires offers a general image, which is far from exhaustive, of a whole world linked to the presence of salt springs in the life of communities in Moldavian sub-Carpathians. The information related to short- and long-term supply strategies, to various uses of natural brine, to barter exchange with salt water and/or re-crystallized salt against agricultural products, to specialized producers (and sometimes of sellers) in *huscă*, to the *huscă*-obtaining practices, to attraction in the closest zones of salt springs for wild animals and birds, to symbolism of salt, represent seldom recorded evidence for the areas with numerous archaeological artefacts of natural brine exploitation.

The unity of place with archaeological remains, historical evidence and recent practices (in the last century) represents a serious argument in support of ethnographic analogies for a better understanding of archaeological situations, which is difficult to reject. In this respect, we would like to emphasize the persistent character of supply behaviours with brine coming from salt springs. Thus, the Austrian report already quoted (Peithner 1784) of over two centuries ago illustrates this to be the case despite political, economical and social changes in Romania (including joining the EU in 2007) everyday activities meant to satisfy the needs of salt have not changed radically; they represent real behavioural and economic patterns. Sometimes, mostly in halotherapy or animal foodstuffs, but also with techniques (as, for instance, the egg test for the brine salinity assessment used in the conservation of some cheeses), the practices have hardly evolved since Antiquity, as attested in Greek and Latin texts. The questionnaires applied in the Moldavian sub-Carpathians (Suceava, Neamţ, Bacău, Vrancea counties) supplied an impressive quantity of ethnographical informations even before the end of our project. Corroborated with other kinds of data (such as the chemistry of salt water, distribution and density of population, the road network, etc.), these results, modelled and interpreted in their whole dimensions, will furnish a solid referential and unique means of approach for the understanding of the history of exploitation of traditional salt springs, which today is actually extinct in Europe. Indeed, the study of this still vital economy of an ingenious aspect, which parallels the regulated economy, can answer many problems raised by archaeological research and the reading of ancient texts (Alexianu 2007), often too short, which omit or overlook many precious details.

Acknowledgements

We would like to acknowledge the Romanian government for their support (project CNCSIS IDEI 414/2007), the department of archaeology and social sciences of the French Ministry of Foreign Affairs (Franco-Romanian archaeological mission Lunca-Ţolici from 2004) and the CNRS Human and Social Sciences Department (interdisciplinary program about water 2003-2005).

We would like to thank the team led by dr. Vasile Cotiugă, from Arheoinvest Platform ("Al. I. Cuza" University of Iaşi): Andrei Asăndulesei, Bogdan Venedict, Radu Balaur, Cristi Nicu.

We are, also, greatly indebted to all subjects of our ethnographical inquiries, for providing us all the precious informations, with such great solicitude.

References:

Alexianu M. 2007. L'exploitation de l'eau salée chez les Autariates. In A. Fíguls, O. Weller (eds.), *1a Trobada internacional d'arqueologia envers l'explotació de la sal a la prehistòria i protohistòria. 1st International archaeology meeting about prehistoric and protohistoric salt exploitation, Cardona, desembre 2003*, 281-290. Barcelona, IREC.

Alexianu, M., Dumitroaia, Gh. and Monah, D. 1992. Exploatarea surselor de apă sărată din Moldova: o abordare etnoarheologică. *Thraco-Dacica* 13/1-2, 159-167.

Alexianu, M. and Weller, O. 2007. Recherches ethnoarchéologiques sur le sel: les enquêtes de 2004 et les premiers résultats obtenus dans la zone de Poiana Slatinei à Lunca (dép. Neamţ, Roumanie). In D. Monah, Gh. Dumitroaia, O. Weller and J. Chapman (eds.), *L'exploitation du sel à travers le temps*. Bibliotheca Memoriae Antiquitatis XVIII, 299-318. Piatra Neamţ, Constantin Matasă.

Alexianu, M., Weller, O. and Brigand, R. 2007a. Approche ethnoarchéologique de l'exploitation des sources salées de Moldavie: les enquêtes récentes (2005). In N. Morère Molinero (ed.), *Las salinas y la sal de interior en la historia: economía, medio ambiente y sociedad. Inland saltworks and salt history: economy, environment and society*, 407-434. Madrid, Dykinson.

Alexianu, M., Weller, O. and Brigand, R. 2007b. *Izvoarele de apă sărată din Moldova subcarpatică. Cercetări etnoarheologice*. Iaşi, Casa Editorială Demiurg.

Alexianu, M. 2008. Une catégorie d'esclaves thraces: les halônetoi. In A. Gonzales (ed.), *La fin du statut servile? (affranchissement, libération, abolition)*, Actes du 300 colloque du Groupe International de Recherches sur l'Esclavage dans l'Antiquité (GIREA) - Besançon, 15, 16, 17 décembre 2005, vol. II, 487-492. Besançon, Presses universitaires de Franche-Comté.

Alexianu, M., Weller, O. and Brigand, R. 2008. Usages et enjeux autour des sources salées de Moldavie précarpatique, Roumanie. In O. Weller, A. Dufraisse and P. Pétrequin (eds.) *Sel, eau et forêt: hier et aujourd'hui*, 49-72. Cahiers de la MSH Ledoux 12 (coll. Homme et environnement, 1). Besançon, Presses Universitaires de Franche-Comté.

Alexianu M., Weller O. 2009. The Ethnosal project. Ethnoarchaeological investigation at the Moldavian salt springs. *Antiquity*, vol. 83, issue 321, September 2009. *http://antiquity.ac.uk/projgall/weller321/*.

Andronic, M. 1989. Cacica – un nou punct neolitic de exploatare a sării. *Studii şi Cercetări de Istorie Veche şi Arheologie* 40/2, 171-177.

Andronic, M. and Ursu, E. 2003. Voitinel, com. Gălăneşti, Marginea, com. Marginea, jud. Suceava (punct 218). *Cronica cercetărilor arheologice din România. Campania 2002*, 31.

Brongniart, A. 1807. *Traité élémentaire de minéralogie*, I. Paris, L'imprimerie de Crapelet.

Bulgăr Gh. and Constantinescu-Dobridor Gh. 2002. *Dicţionar de arhaisme şi regionalisme*, I. Bucureşti, Editura Saeculum Visual.

Ceauşu, Şt.M. 1982. Contribuţii la istoricul exploatării sării la sfârşitul secolului al XVIII-lea şi în prima jumătate a secolului al XIX-lea în Bucovina, *Suceava, Anuarul Muzeului Judeţean*, Suceava, IX, 377-392.

Chapman, J., Monah, D., Dumitroaia, Gh., Armstrong, H., Millard, A. and Francis, M. 2000. The exploitation of salt in the prehistory of Moldavia, Romania. *Archaeological Reports for 1999/2000* 23, 10-20.

Chapman, J., Monah, D., Dumitroaia, Gh. and Francis, M. 2003. The exploitation of salt in the prehistory of Moldavia, Romania. *Archaeological Reports for 2001/2002* 24, 7-11.

Chapman, J., Dumitroaia, Gh. and Monah, D. 2003-2004. The earliest salt exploitation in the world: excavations at Lunca – Poiana Slatinei, Neamţ County, Moldavia, Romania. *Archaeological Reports for 2002* 25, 34-48.

Chapman, J. and Monah, D. 2007. A seasonal Cucuteni occupation at Silişte-Prohozeşti, Romania. In D. Monah, Gh. Dumitroaia, O. Weller and J. Chapman (eds.), *L'exploitation du sel à travers le temps*. Bibliotheca Memoriae Antiquitatis XVIII, 71-88. Piatra Neamţ, Constantin Matasă.

Cioranescu, G. 1958-1966. *Diccionario etimologico rumano*. Tenerife Madrid, La Laguna (7 fascicules).

Curcă, R.-G. 2007a. La halothérapie – textes antiques et pratiques actuelles dans la Roumanie orientale. In D. Monah, Gh. Dumitroaia, O. Weller and J. Chapman (eds.), *L'exploitation du sel à travers le temps*. Bibliotheca Memoriae Antiquitatis XVIII, 259-269. Piatra Neamţ, Constantin Matasă.

Curcă, R.-G. 2007b. Halotoponymes et halohydronymes dans les documents médiévaux de la Moldavie (XIVe-XVIIe s.). In N. Morère Molinero (ed.), *Las salinas y la sal de interior en la historia: economía, medio ambiente y sociedad. Inland saltworks and salt history: economy, environment and society*, 777-784. Madrid, Dykinson.

Damm, Ch. 2005. Archaeology, ethno-history and oral traditions: Approaches to the indigenous, *Norwegian Archaeological Review*, 38, 2, 73-87.

Dumitroaia, Gh. 1987. La station archéologique de Lunca-Poiana Slatinei. In M. Petrescu-Dîmboviţa, N. Ursulescu, D. Monah, V. Chirica (eds.), *La Civilisation de Cucuteni en contexte européen*, Iaşi, 253-258. Bibliotheca Archaeologica Iassiensis, I.

Dumitroaia, Gh. 1994. Depunerile neo-eneolitice de la Lunca şi Oglinzi, judeţul Neamţ. *Memoria Antiquitatis* XIX, 7-82.

Dumitroaia, Gh., Munteanu, R., Weller, O., Garvăn, D., Diaconu, V. and Brigand, R. 2008. Un nou punct de exploatare a apei sărate în preistorie: Ţolici-Hălăbutoaia, jud. Neamţ. In D. Monah, Gh. Dumitroaia and D. Garvăn (eds.), *Sarea de la prezent la trecut*. Bibliotheca Memoriae Antiquitatis XX, 203-224. Piatra Neamţ, Constantin Matasă.

Ellis, L. 1984. *The Cucuteni-Tripolye Culture. A Study in Technology and the Origins of Complex Society*. Oxford, British Archaeological Reports 217.

Gazin-Schwartz, A. and Holtorf, C.J. (eds.) 1999. *Archaeology and Folklore*. London, Routledge.

Gil, L. 2004. *Therapeia. La medicina popular en el mundo clasico*. Madrid, Triacastela.

Grămadă N. 1996. *Toponimia minoră a Bucovinei*, vol. I, Bucureşti, Anima.

Iordan I. 1963. *Toponimia romînească*, Bucureşti, Editura Academiei RPR.

Iordan I. 1983. *Dicţionar al numelor de familie româneşti*, Bucureşti, Editura Ştiinţifică şi Enciclopedică.

Jouanna, J. 1994. L'eau, la santé et la maladie dans le traité hippocratique des Airs, eaux, lieux. In R. Ginouvès, A.-M. Guimier-Sorbets, J. Jouanna and L. Villard, *L'eau, la santé et la maladie dans le monde grec*. Athènes-Paris, De Boccard.

Moga, I. 2008. Symbolic Functions of Salt and Related Similes in Oriental Curses and Blessings. *Arheologia Moldovei*, 29, 95-130.

Moga, I. 2009. Salt Extraction and Imagery in the Ancient Near East. *Journal for Interdisciplinary Research on Religion and Science*, No. 4, January, 175-213.

Monah, D. 1991. L'exploitation du sel dans les Carpates orientales et ses rapports avec la culture de Cucuteni-Tripolye. In V. Chirica and D. Monah (eds.), *Le Paléolithique et le Néolithique de la Roumanie en contexte européen*, 387-400. Iaşi, Bibliotheca Archaeologica Iassiensis.

Monah, D. 2002. L'exploitation préhistorique du sel dans les Carpates orientales. In O. Weller (ed.) *Archéologie du sel. Techniques et sociétés dans la Pré- et Protohistoire européenne*, 135-146. Actes du colloque international, XIVe congrès UISPP, Liège (Belgique), 2001. Rahden Westfalen, Verlag Marie Leidorf GmbH. Internationale Archäologie, ASTK, 3.

Monah, D., Antonescu, S. and Bujor, A. 1980. Raport preliminar asupra cercetărilor arheologice din comuna Poduri, jud. Bacău. *Materiale şi Cercetări Arheologice*, 86-99.

Monah, D., Popovici, D., Dumitroaia, Gh., Cucoş, Ş. and Bujor, A. 1987. Raport preliminar asupra săpăturilor arheologice de la Poduri-Dealul Ghindaru (1984-1985). *Memoria Antiquitatis* XV-XVII (1983-1985), 9-20.

Monah, D., Dumitroaia, Gh., Monah, F., Preoteasa, C., Munteanu, R. and Nicola, D. 2003. *Poduri-Dealul Ghindaru. O Troie în Subcarpaţii Moldovei*. Piatra Neamţ, Constantin Matasă.

Morantz, T. 1998. The past and the future of ethno-history. *Acta Borealia*, 59-77.

Munteanu, R. 2006. Cucuieţi, com. Solonţ, jud. Bacău, Punct: Slatina Veche. In V. Cavruc and A. Chiricescu (eds.) *Sarea, timpul şi omul*, 62-64. Sfîntu Gheorghe, Angustia.

Munteanu, R., Garvăn, D., Nicola, D., Preoteasa, C. and Dumitroaia, Gh. 2007. Cucuieţi-Slatina Veche (Romania). Prehistoric Exploitation of a Salt Resource. In D. Monah, Gh. Dumitroaia, O. Weller and J. Chapman (eds.), *L'exploitation du sel à travers le temps*. Bibliotheca Memoriae Antiquitatis XVIII, 57-70. Piatra Neamţ, Constantin Matasă.

Nicola, D., Munteanu, R., Garvăn, D., Preoteasa, C. and Dumitroaia, Gh. 2007. Solca – Slatina Mare (Roumanie). Preuves archéologiques de l'exploitation du sel en préhistoire. In D. Monah, Gh. Dumitroaia, O. Weller and J. Chapman (eds.), *L'exploitation du sel à travers le temps*. Bibliotheca Memoriae Antiquitatis XVIII, 35-56. Piatra Neamţ, Constantin Matasă.

Poruciuc, A. 2007. Romanian "huscă" ("Salt obtained by evaporation of Salt Water") explained as an Old Germanism. In D. Monah, Gh. Dumitroaia, O. Weller and J. Chapman (eds.), *L'exploitation du sel à travers le temps*. Bibliotheca Memoriae Antiquitatis XVIII, 319-324. Piatra Neamţ, Constantin Matasă.

Poruciuc, A. 2008. Denumirile sării şi implicaţiile lor în plan (indo)european. In D. Monah, Gh. Dumitroaia and D. Garvăn (eds.), *Sarea, de la prezent la trecut*, 133-144. Piatra Neamţ, Constantin Matasă.

Puşcariu S. 1976. *Limba română, I, Privire generală²*. Bucureşti, Minerva.

Roux, V. 2007. Ethnoarchaeology: A Non Historical Science of Reference Necessary for Interpreting the Past. *Journal of Archaeological Method and Theory*, Vol. 14, 2, 153-178.

Sandu, I., Alexianu, M., Curcă, R.-G., Weller, O. and Pascu, C. 2009. Halotherapy: from ethnoscience to scientific explanations. *Environmental Engineering and Management Journal*, 8, 6, 1331-1338.

Sandu, I, Poruciuc, A., Alexianu, M., Curca, R.-G. and Weller, O. 2010. Salt and Human Health: Science, Archaeology, Ancient Texts and Traditional Practices of Eastern Romania. *The Mankind Quarterly* L, 3, 225-256.

Şandru, I. 1952. Contribuţii geografico-economice asupra exploatării slatinelor în Bucovina de Sud. *Studii şi Cercetări Ştiinţifice* 1-4, III, 407-428.

Şandru, I. 1961. Contribuţii geografice economice asupra evoluţiei aşezărilor omeneşti în depresiunea subcarpatică Oneşti. *Analele Ştiinţifice ale Universităţii "Al. I. Cuza" din Iaşi*, VII, fasc. 1, 215-230.

Townson, R. 1797. *Travels in Hungary with a short account of Vienna in the year 1793*. London, Paternoster-Row.

Ursulescu, N. 1977. Exploatarea sării din saramură în neoliticul timpuriu, în lumina descoperirilor de la Solca (jud. Suceava). *Studii şi Cercetări de Istorie Veche* 28/3, 307-317.

Ursulescu, N. 1995. L'utilisation des sources salées dans le néolithique de la Moldavie (Roumanie). In M. Otte (ed.) *Nature et Culture*, 487-495. ERAUL, 68.

Weller, O. 2000. Produire du sel par le feu: techniques et enjeux socio-économiques dans le Néolithique européen. In P. Pétrequin *et al.* (eds.), *Arts du feu et productions artisanales*, 565-584, XXe Rencontres Internationales d'Archéologie et d'Histoire d'Antibes, 1999. Antibes, APDCA.

Weller, O. and Dumitroaia, Gh. 2005. The earliest salt production in the World. An Early Neolithic exploitation in Poiana Slatinei-Lunca, Romania. *Antiquity* 79, 306. *http://antiquity.ac.uk/ProjGall/weller/index.html.*

Weller, O., Dumitroaia, Gh., Monah, D. and Nuninger, L. 2007a. L'exploitation des sources salées en Moldavie: un exemple de ressource structurante du territoire depuis le Néolithique. In D. Monah, Gh. Dumitroaia, O. Weller and J. Chapman (eds.), *L'exploitation du sel à travers le temps*. Bibliotheca Memoriae Antiquitatis XVIII, 99-113. Piatra Neamţ, Constantin Matasă.

Weller, O., Brigand, R. and Alexianu, M. 2007b. Cercetări sistematice asupra izvoarelor de apă sărată din Moldova. Bilanţul explorărilor din anii 2004-2007 efectuate în special în judeţul Neamţ. *Memoria Antiquitatis* XXIV, 121-190.

Weller, O., Brigand, R., Nuninger, L., Dumitroaia, Gh. and Monah, D. 2007c. Analyses et modélisation spatiale autour des sources salées de Moldavie précarpatique durant la Préhistoire. In N. Morère Molinero (ed.), *Las salinas y la sal de interior en la historia: economía, medio ambiente y sociedad. Inland saltworks and salt history: economy, environment and society*, 165-184. Madrid, Dykinson.

Weller, O., Dumitroaia, Gh., Sordoillet, D., Dufraisse, A., Gauthier, E. and Munteanu, R. 2008. Première exploitation de sel en Europe: Techniques et gestion de l'exploitation de la source salée de Poiana Slatinei à Lunca (Neamţ, Roumanie). In O. Weller, A. Dufraisse and P. Pétrequin (eds.) *Sel, eau et forêt: hier et aujourd'hui*, 205-230. Cahiers de la MSH Ledoux 12 (coll. Homme et environnement, 1). Besançon, Presses Universitaires de Franche-Comté.

Weller, O., Brigand, R. and Alexianu, M., 2010. Recherches systématiques autour des sources salées de Moldavie. Bilan des prospections 2008-2010. *Memoria Antiquitatis*, XXV, 437-504.

New Ethnoarchaeological Investigations upon the Salt Springs in Valea Muntelui, Romania

Dan Monah
Institutul de Arheologie, Iaşi, Romania

Gheorghe Dumitroaia
Complexul Muzeal Judeţean Neamţ, Romania

Dorin Nicola
Complexul Muzeal Judeţean Neamţ, Romania

Abstract
Ethnoarchaeological researches regarding the salt springs in Moldavia already have a certain tradition in Romania. The investigations began at the end of the 19he80s, and provided important and interesting information upon the current use of brine from the salt springs in the Moldavian Subcarpathians. One of the recent areas investigated by the authors is Valea Muntelui, in the Neamţ Mountains, near the Bicaz Reservoir. Here we found salt springs whose brine can coagulate milk. The piece of information was confirmed through a simple experiment which produced a light cheese, excellent for children.
The authors also formulate a hypothesis which could explain the existence of an isolated Chalcolithic settlement, that of Hangu, in a mountain depression, which is totally unusual for a settlement belonging to the Cucuteni culture.

Keywords
salt springs, coagulation, milk, cheese, Valea Muntelui, the department of Neamţ

The ethnoarchaeological investigations in Moldavia were determined by major discoveries at Solca (Ursulescu 1977; 1995), Lunca-*Poiana Slatinei* (Dumitroaia 1987; 1994) and in the hinterland of the Poduri-*Dealul Ghindaru* tell (Monah, Antonescu and Bujor 1980). These discoveries led to the idea of an intense salt exploitation in the Neolithic and Chalcolithic and of the great importance of this substance in intercommunity exchanges. At the beginning, the investigators only wanted to get information in order to explain the purpose and characteristics of the prehistoric settlements near salt springs. Nevertheless, the program was diversified during the researches, and we also recovered some important ethnographic information.

Like other countries in the Balkans, Romania has the reputation – as John Nandriş said– of a living laboratory of the perishable material (Nandriş 1985, 260). Of course, this precious heritage should not be lost, mostly given the fact that the rapid modernization of the Romanian society is leading to the disappearance of ancestral habits and traditions (Nandriş 1987, 202). The discovery in Moldavia of major and consistent archaeological deposits near salt springs stimulated archaeologists to start a series of investigations. Their subject was the current populations and their purpose to characterise the everyday practices of exploiting salt springs, their regime and the use of brine. As ethnographic' information had only partial answers concerning the archaeologists' preoccupations and questions, they had to direct their studies according to the problems suggested by the archaeologists. The result was, paraphrasing the first Romanian ethnoarchaeologist, the creation in Moldavia of a new ethnoarchaeology subdiscipline – the ethnoarchaeology of brine (Alexianu, Dumitroaia and Monah 1992, 159). We have to admit and to emphasize that the ethnoarchaeological research in Romania was initiated and developed thanks to John Nandriş, one of the creators in Europe of this subdiscipline. In Moldavia, the main disciple of John Nandriş is Marius Alexianu, who realized, in collaboration with archaeologists, ethnographers and specialists in other fields numerous investigations and ethnoarchaeological surveys, showing a good theoretical background and a remarkable skill in field inquiries (Alexianu, Weller and Brigand 2007).

The first ethnoarchaeological investigations regarding the exploitation of salt springs in Moldavia were carried out toward the end of the 80s in the following regions: the Berzunţi Mountains, the Moineşti Depression, the city of Piatra-Neamţ and the surroundings, as well as the surroundings of Tg. Neamţ. The surveys in the area, achieved by a team which included Marius Alexianu, Dan Monah, Gheorghe Dumitroaia and Elena Florescu, were completed in time by a series of surveys and ethnoarchaeological micro-inquiries. We have to mention that there were difficulties specific to the Communist period, such as the difficulties of finding transport and the problematic approval and equipments necessary to organize field inquiries. Thus, the *ad hoc* ethnoarchaeological nucleus found the solution of doing the ethnoarchaeological surveys mostly in the areas with archaeological excavations. Thus, D. Monah, during the excavations at Poduri, discovered several springs in the hinterland of the Chacolitihic tell Dealul Ghindaru. Nearby there were seasonal Cucuteni settlements (Monah *et al.*, 1987, 14; Monah *et al.*, 2003, 69-73). Gh. Dumitroaia did the same thing: while directing the excavations at Lunca-*Poiana Slatinei* and

Oglinzi-*Băi*, he investigated the Lunca-Oglinzi-Răuceşti area, and also the surroundings of Tg. Neamţ. There, he found numerous salt springs, still used by the local population, and also by prehistoric communities (Dumitroaia 1994, 7-8, Figure 1).

In 1986, the first ethnoarchaeological expedition took place, meant to investigate the Berzunţi area, in the department of Bacău. The team had as members M. Alexianu, D. Monah and E. Florescu. Even though there has not much time for the research, we obtained important information regarding the salt springs and the way of using the salt water in the area. Unfortunately, as we did not have much time and local informants, we did not identify archaeological depositions relating to exploiting the salt springs. The action was more successful in the Poduri area, where we identified six Cucuteni seasonal settlements, at less than 500-600m from a salt spring (Monah *et al.*, 2003, 69-73).

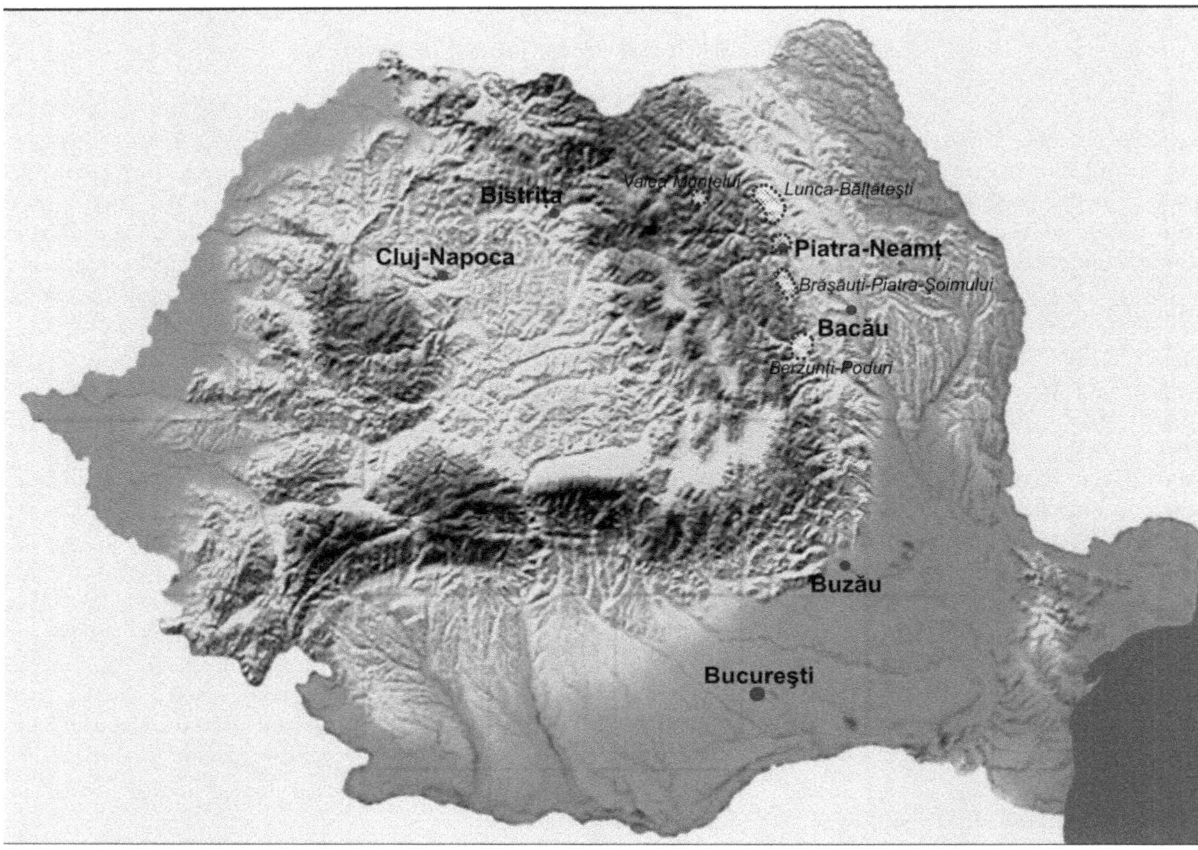

Figure 1. The areas within the ethnoarchaeological investigations (*apud* Alexianu *et al.*, 2007).

Later, Gh. Dumitroaia and M. Alexianu also investigated the Piatra-Neamţ area, extending the research to the whole saliferous region in the department of Neamţ. The field inquiries and the surveys led to the discovery of new salt springs and archaeological sites related to them. We obtained empirical information regarding the NaCl concentration of certain springs (Alexianu, Dumitroaia and Monah 1992, 160-164; Dumitroaia 1994). During the researches at the Poduri Chalcolithic tell we also did archaeological micro-surveys through which we eluded numerous aspects related to the use of brine in salt springs, the supplying frequency, the means of transporting, etc. We also obtained information regarding the crystallization of salt and its use within exchanges.

In the area between the towns of Tg. Neamţ and Piatra-Neamţ the ethnoarchaeological documentation was completed by Gh. Dumitroaia, who did numerous micro-surveys at Vânători-Lunca-Răuceşti, and then at Bălţăteşti and Ţolici. We also investigated the Piatra-Neamţ area, where we identified several salt springs, some of them related to well-known Chalcolithic settlements and sometimes archeologically investigated. The researchers Gh. Dumitroaia and M. Alexianu also investigated the Roznov - Piatra Şoimului region (former village of Calu) and obtained information from many persons (Alexianu, Dumitroaia and Monah 1992, 160-161; Alexianu, Dumitroaia and Monah 2007, 300-301). As we do not intend here to make a history of ethnoarchaeological investigations in Moldavia, we shall not develop this subject any further. Our intention is to inform the reader as regards the documentation and difficulties during the first ethnoarchaeological study upon the exploitation of salt springs in Moldavia (Alexianu, Dumitroaia, Monah, 1992).

*

In the first study, dedicated to the ethnoarchaeological research of salt springs in Moldavia (Alexianu, Dumitroaia and Monah 1992, 161), there was an important piece of information, provided by Maria Apetrei, biologist and museum assistant at the Museum of Natural Science in Piatra-Neamţ. Coming from a village in the mountain area Valea Bistriţei, also known as Valea Muntelui, Maria Apetrei told us that in the villages of Borca, Sabasa, Pârâul Pântei, Farcaşa and Stejaru there are salt springs whose water can coagulate milk, obtaining cheese without using curds (Alexianu, Dumitroaia and Monah 1992, 161). In the same publication we mention that the brine in these springs was used for the short-time preservation of cottage cheese and bacon (Alexianu, Dumitroaia and Monah 1992). The moment we obtained this piece of information, we did not have many possibilities to do ethnoarchaeological surveys in such an isolated region as Valea Muntelui. This is why we included the information in the study without any field inquiry (Figure 2).

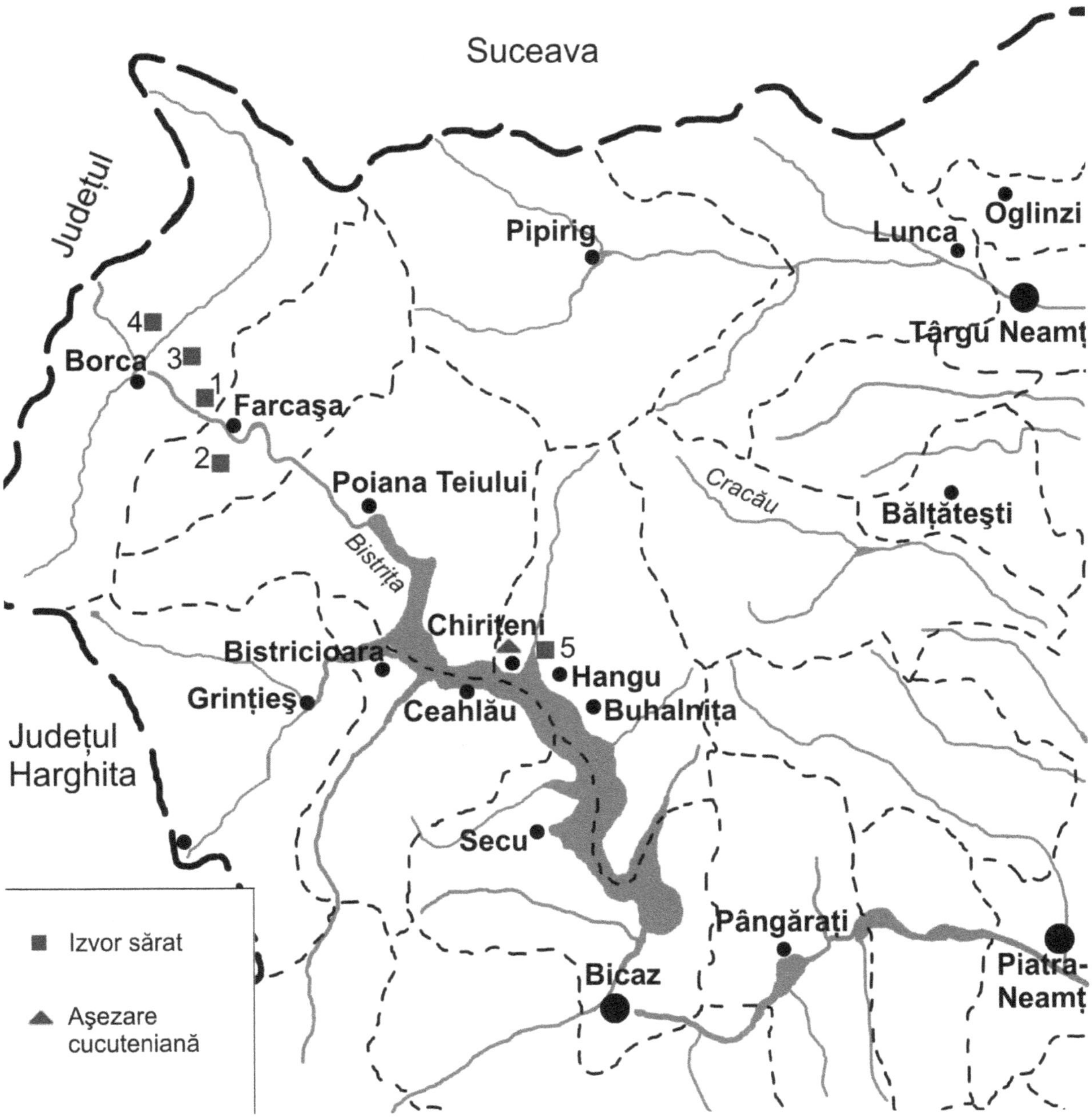

Figure 2. Salt springs with brine which coagulates milk (1. Farcaşa, 2. Poiana Pântei, 3. Stejaru, 4. Sabasa and 5. Hangu), the Cucuteni A settlement at Hangu-Chiriţeni (■ salt spring; ▲ Cucuteni settlement).

Starting with 2003, thanks to a Franco-Romanian – Brancusi IAP (Integrated action program), the researches regarding salt exploitation in Romanian prehistory multiplied and diversified. On this occasion, we took over the ethnoarchaeological investigations on salt springs as well, which had not been totally abandoned. In the years that followed we did numerous archaeological surveys and ethnoarchaeological inquiries, especially in the region near the Lunca site and in several perimeters in the departments of Neamţ and Bacău (Alexianu, Dumitroaia and Monah 2007; Weller *et al.*, 2007; Alexianu and Weller 2007). Currently, for Central and Northern Moldavia, we have almost complete ethnoarchaeological documentation.

Nevertheless, we still have to explore the South of the department of Bacău and the whole department of Vrancea, where there are big halite deposits, with numerous outcrops and thus, with salt springs. We do not believe that the ethnoarchaeological investigation of these areas would much change the big picture so far, but, nonetheless, it is important to fill out ethnoarchaeological picture.

A micro-region needing a complete ethnoarchaeological investigation was of that of Bistrița Mountains. For this micro-region, we had information regarding the existence in the villages of Hangu, Farcaşa, Stejaru, Pârâul Pântei, Sabasa and Borca of salt springs whose brine has the capacity to coagulate milk, thus obtaining a special cow cheese (Alexianu, Dumitroaia and Monah 1992, 161). Naturally, we had as purpose assessing the information that we already had. We thus organized two expeditions, on August 22 2007 and November 7 2008. We shall present the results of these investigations.

The village of Farcaşa, com. of Farcaşa, dep. of Neamț

In these villages, there are two brines with similar properties. Because of the bad weather, we only studied the brine in the village, often used by the inhabitants. The other brine is not generally used as it is far from the village and not taken care of.

- *The 'La Ionel' salt spring* (Figure 3)
GPS coordinates (Garmin): N 47°09'79"; E 025°50'75"; altitude 612 m. Date: 22.08.2007.

Figure 3. The 'La Ionel' salt spring at Farcaşa (Photo D. Nicola).

The salt spring is in a small orchard on the mountain, belonging to a local called Gheorghe Ionel. The access is both through the backyard and through the back, on a minor mountain track. The fence in the back of the house, of the orchard with a few apple trees, has a stile which makes it easier for the people to get to the brine. The spring is captured through a wood trunk with an oval section and 49 x 31cm diameter. The brine has an average salinity. The wood trunk and the brine on it are both bricky-reddish, probably because of the iron oxides. The front side of the trunk was cut and carved to facilitate access to the spring when the debit is low. The back side is also carved and even has two circular perforations with no practical use. It seems that it is a decorative idea of the owner. Even though it is on a private property, access is free and the owner has made efforts (installing the chopped trunk and the stile) for people to be able to take brine from the spring. There were no archaeological depositions, a fact also confirmed by our informant Dumitru Gavril (aged 78 – Figure 4).

Figure 4. Dan Monah talking to Dumitru Gavril of Farcaşa (Photo D. Nicola).

The village of Stejaru, com. of Farcaşa, dep. of. Neamț

- *The salt spring in the village* (Figure 5)
GPS coordinates: N 47°09'20"; E 025°49'01"; altitude 606 m. Date: 22.08.2007.

Figure 5. 'The salt spring in the village' at Stejaru (Photo D. Nicola)

The spring is on a street in the centre of the village. Initially, it was in a villager's yard, but the fence was modified for everybody to have access to the brine trunk. The chopped trunk, slightly oval, is made of an oak and has 53 x 54cm diameter. The water has an average salinity and probably contains iron oxides, as the trunk and the stones with brine deposits are reddish.

The brine is used to make cow cheese, and in the past it was also used to prepare polenta and other dishes.

The village of Pârâul Pântei, com. of Borca, dep. of Neamț (Figure 6)
- *The salt spring in the forest*
GPS coordinates: N 47°12'13"; E 025°49'17"; we could not determine the altitude as there was bad weather and we could not get proper signals because of the clouded sky. Date: 7.11.2008.

The spring is near a forest track, used for the exploitation of conifers. It was often used by the inhabitants because of the high salt concentration. Because of the large amount of rain that year, when we did the inquiry, the spring was covered with bed materials and grit. As at Sabasa, at the level of the spring, there are many red traces, probably from iron oxides. It brine was mostly used to make cow cheese, obtaining a light, sweet cheese. We did not find any archaeological material.

Figure 6. The destroyed salt spring at Pârâul Pântei (Photo D. Nicola)

The village of Sabasa, com. of Borca, dep. of Neamț (Figure 7)
- *The salt spring in the glade*
GPS coordinates: N 47°12'13"; E 025°49'22"; altitude 626 m. Date: 7.11.2008.

To get to the spring, situated on the mountain near the village, we hadto pass through the property of Mihai Florescu, at the road. There are no improvements of the spring. In fact, it is just a hole with a few big stones around and a stick with a mug to take the brine. There are also cans used to transport the brine. The earth is red on the portion of the brine fall, probably from the iron oxides in the water. The water reaches the surface, and thus there is easy access to the brine. The spring is in a glade within a coniferous forest, on an abrupt bend. The spring is 25-30cm in depth. The brine is concentrated and rather clear. As at Farcaşa, it is used to curd the milk. We did not find any archaeological material.

*
* *

For the brine springs in the Farcaşa-Borca region, we gained information from two inhabitants: Dumitru Gavril (aged 78) and Vasilica Simion (aged 38), both from Farcaşa. They told us that the brine is exclusively used to make cow cheese. The cow cheese obtained is used especially for children. The brine is not used for pickles or food/bacon conservation. It is not used for polenta or other dishes, except for the cases when the inhabitants have run out of salt. The cow milk with brine is not sold, as the city market is far. The distance to Bicaz is over 20km, and the distance to Tg. Neamț is even bigger, thus it is not profitable to sell cheese, even though there are many milk cows in the village. We were told that many cattle are sent, during the summer, to the sheepfold; nevertheless, there are also cattle fed around the household or on the grazing land (Vlăduțiu 1973, 195). At the sheepfold, the cow milk is combined with sheep milk to make more cottage cheese. This proceeding started during the Communist period because of the food crisis, but lately there is a tendency of not combining them anymore and using only sheep milk.

Figure 7. 'The salt spring in the glade' at Sabasa (Photo D. Nicola)

*

Vasilica Simion was kind enough to show us how to prepare cheese with brine. She had a litre of milk cheese boiled and then she added 100 ml brine and took it off. In less than 2-3 minutes, the milk began to coagulate and, after several more minutes, the cheese was under a colander. Afterwards, the cheese may be consumed as such or put into a spherical mould (Figure 8). Decades ago, the cow or sheep cheese or the mixed variety was put in a 'pouch' made of hemp fibres (Kogălniceanu 1973, 373, Figure 203). Usually, for a family, cheese is made from 4-5 liters of milk, and even more, for a few days' consumption. Because of the salt, it is well preserved for a few days, but it is usually put in a cold place to keep it fresh.

Old houses did not have a cellar, and, in order to conserve potatoes people dug a hovel, but the perishable aliments were usually kept in the house (Vuia 1973, 112, 151). Nowadays people generally use refrigerators. The cheese prepared at Farcaşa is sweet and slightly salty, very tasty, light and healthy.

Even though we did not make chemical analyses, we assume that the brine in the springs at Valea Muntelui has, besides iron oxides, calcium, which leads to the coagulation of the milk. During the Communist period, the city women prepared, from the milk bought in markets, a light cheese for small children, putting calcium pills in the boiled milk. The taste is identical, but the cheese made with brine is slightly salty.

Figure 8. Milk coagulation with brine from 'La Ionel' salt spring (Photo D. Nicola).

According to the information in the literature, such a brine as seen at Valea Muntelui is also found in the village of Neguleşti, com. of Piatra Şoimului, dep. of Neamţ. In this village, too, the brine is used to coagulate milk and to make cheese. The interesting fact is that people in this village use the same quantity of brine to make cheese (0.5l for 8-9l of milk) as those at Valea Muntelui (Alexianu, Dumitroaia and Monah 2007, 104). Nevertheless, it seems that, at Neguleşti, there are two recipes, the second consisting of five litres of milk (we were not told what kind of milk) and only 2-3 spoons of brine (Alexianu, Dumitroaia and Monah 2007, 104). The quantities of cheese made with brine at Valea Muntelui and at Neguleşti do not seem to have wider economic purposes, and the cheese is made for domestic use. Nevertheless, the observations and information obtained at Valea Muntelui, and also at Neguleşti, make us see that the villagers know very well the environment and the natural resources and also the fact that they are very inventive.

*

We shall continue the chapter with the region that really interests us - Valea Muntelui - with the communes of Poiana Teiului, Farcaşa and Borca, and then with the commune of Hangu. The Hangu-Bicaz region was studied in detail by historians and archaeologists when researching the Reservoir during 1954-1958. Thanks to these complex investigations, we have much archaeological, historical and ethnographic information (Petrescu-Dîmboviţa and Spinei 2003b, 11-26; Cihodaru 2003, 145-203). Valea Muntelui was populated relatively late, probably in the 18th and 19th centuries. Before the constitution of villages, there were only sheepfolds and hunting groups. It seems that in the 18th century there were small settlements that did not have a seasonal character anymore. The exploitation of salt springs probably began with the foundation of the first stable settlements. The fact that there are no remains of glass or ceramic may also be explained by the fact that most of the recipients used until modern times were made of wood. Up to the middle of the 20th century, the economy in the region was mostly made of raising livestock, forest exploitation and some kind of agriculture, especially corn. In the modern period, people began to develop forest exploitation and cribbing, but also continued practicing traditional activities such as carpentry, coopery and clapboarding. In the Communist period, there were also the building works, especially at the Bicaz hydroelectric plant, and even mining, for a short period. As the region is rather isolated, the traditional way of living continued up to the last decades.

*

The commune of Hangu and mainly the village of Hangu benefit from a more generous documentation. Besides documentary evidence, we also have archaeological information obtained during the rescue excavations of 1955-1957 at Chiriţeni and Piciorul Bocăncenilor (Petrescu-Dîmboviţa and Spinei 2003b, 18-19). At Chiriţeni, we excavated a Cucuteni A settlement, identifying 20 habitation complexes and two pottery kilns. The settlement, with its dispersed habitations, seems to have the characteristics specific to Cucuteni (Petrescu-Dîmboviţa and Teodor 2003a, 128-133). Nearby, at Piciorul Bocăncenilor, we identified a settlement, probably seasonal, belonging to the Cucuteni A phase. At Chiriţeni, on the place of the Chalcolithic settlement, we also discovered a settlement from Bronze Age, and we analyzed two habitation complexes (Petrescu-Dîmboviţa and Teodor 2003b, 137-138), and at Hangu-Cetăţuia a settlement with sporadic traces since the Second Iron Age (Niţu and Teodor 2003, 139-143). We also quote the discovery of some remains dating from the pre-feudal period (probably the AD 10th-11th centuries). We shall only discuss here the significance of the Cucuteni habitations (Petrescu-Dîmboviţa and Teodor 2003a, 133). Unfortunately, since the osteological remains have not yet been analyzed by archaeo-zoologists, we do not know the role of livestock raising and hunting for the Chiriţeni settlement.

Even though the Cucuteni A phase takes place in a warmer climate, we do not believe that the cultivation of cereals (wheat and barley) was possible and that it could not have sufficed for the entire community at Chiriţeni. If we apply the same calculation formula as for the other Cucuteni settlements (Monah and Cucoş 1985, 48-49), the Chiriţeni communities must have had around 200 individuals. Even with a more conservative estimate, there couldn't have been less than 150 individuals. Such a community needed important resources which could indeed be provided by raising livestock, hunting, fishery and gathering. But, if we take into account an economy based upon raising livestock and the large consumption of secondary products, especially milk and cheese products, the livestock breeders needed large quantities of salt.

If we accept the fact that such a community, with around 150 inhabitants, had 45 cattle and 225 sheep (Chapman and Gaydarska 2003, 207, tab. 5), then we estimate that people needed a minimum of 0.75kg of salt a day for them and 1.125kg a day for the livestock (for the cattle: 0.45kg/day and for the sheep: 0.67kg/day), and the annual consumption reached 273.75kg/year for people and 410,62kg/year for livestock, which gives us a minimum annual consumption of 684,43kg. The maximal consumption, according to the same source, reached to 810kg/year for people and 1215kg/year for sheep and cattle. The total of maximal consumption for the Cucuteni communities of Hangu-Chiriţeni should be of 2025kg/year. We did not include pigs in the calculation, although they are frequent in Cucuteni settlements. No matter how relative the calculations or the estimation of the population in the settlement brought by our English/Bulgarian colleagues would be (Chapman and Gaydarska 2003, 207-208), we must admit the fact that the salt needed for the Hangu-Chiriţeni Cucuteni community was truly a large quantity. Without salt, we cannot talk about the wealth of people and livestock (Monah 1991, 387-389). In conclusion, the community at Hangu could not have survived without an everyday salt supply.

The settlement at Hangu has a rather strange position, being situated on the upper Bistriţa river, deep in the mountains (Monah and Cucoş 1985, Figure 1, no. 488). The site is at over 40km away from the closest contemporary Cucuteni settlement. It is obviously the most isolated Cucuteni community we know. The community that founded the Chriţeni settlement and generated the seasonal one at Piciorul Bocăncenilor had its reasons to found an isolated site and live in a region less favourable for the Cucuteni lifestyle. The fact that the settlement has 20 households, more or less contemporary, makes us reject the idea of a group that came here from an 'overpopulated' region. Its very isolation excludes the idea of settlement hiving-off from a lowland village. The communities that migrate found their settlements within the tribal territory, or, if they come to a non-inhabited place, they want to be close to their 'mother-community', mostly for protection. If the community came from the South-East, from further down the valley of Bistriţa (the most logical theory), there were many other places (the Pângăraţi-Tarcău region, the Bicaz or Potoci regions) with more favourable conditions for the Cucuteni lifestyle. Being closer to communities with the same culture, people who left the communities had the possibility to keep in touch with the people they knew. The reasons for the migration and the isolation in case of the Hangu community were probably 'political'. The community could go to the mountains in case of a military conflict, even though we may not exclude a religious cause, harder to identify.

As we already said, downstream in the Bistriţa valley, from Tarcău to Hangu, there are rather large fields proper for agriculture and from where connections were still possible with relatives. Still, the 'exiled' group preferred to go deep into the mountains, founding a settlement at Hangu. Obviously, the region had certain opportunities: rich and diversified hunting and excellent fish (trout, salmon and other species). Plant gathering in the beech and coniferous forests offered great resources of mushrooms and other vegetables. Moreover, the Bistriţa meadow, over whose resources the Chiriţeni inhabitants did not have competition, offered a wide range of ampleresources. Nevertheless, we believe that Hangu attracted the Cucuteni-culture people because of the salt springs, which gave them the much-needed water. This resource compensated for the relative isolation of the community at Chiriţeni. We should also note that, after the Cucuteni A phase, we have no documentation as regards the prolongation of settlement into the Cucuteni A-B and B phases. This could be an extra argument sustaining our hypothesis. Only during the Bronze Age and the Second Iron Age, the region was sporadically inhabited (Petrescu-Dîmboviţa and Teodor 2003b, 137; Niţu and Teodor 2003, 139).

*

A Franco-Romanian team undertook a series of investigations in the Hangu, identifying certain mineral and salt springs. The Valea Glodului spring (E 26°01'38.5''; N 47°04'39.1''; altitude 758m – Figure 9) is near the Chiriţeni resort (Weller *et al.*, 2007, 139-140). The springs have been known for a long time, being mentioned in several works (see Ghiorghiu 1904, 19-20). According to the information received from the inhabitants, we mention the existence of four chopped trunks for salt springs in a rather small territory. We also identified two catches with oak trunks, but only one is still being used. The brine is used for 'food conservation and making cheese' and also to salt the forage for cattle and sheep. In the springs' area we did not find archaeological depositions (Weller *et al.*, 2007, 140). In the commune of Hangu, we mention two other mineral springs mildly salted and a lost salt spring in a forest near Buhalniţa (Ghiorghiu 1885, 20, 369; Weller *et al.*, 2007, 140).

The information suffices for our matter – in the hinterland of the Cucuteni A settlement at Hangu, there were several salt springs to provide the Chiriţeni community with enough salt for people and livestock. Without any doubt, salt springs were also attractive for wild animals, making it easier to the Cucuteni people to capture them. We also mention that five pebble arrowheads were discovered at Chiriţeni (Petrescu-Dîmboviţa and Teodor 2003a, 135). For us, it is obvious that the reason that the Cucuteni community decided to settle at Hangu were the existence of salt springs and the possibilities of easy exploitation.

Figure 9. The salt springs at Hangu (Photo R. Brigand).

Our hypothesis, be it less conformist, may explain the exceptional position of the Cucuteni community at Hangu-Chiriţeni. We do recommend, though, further inquiries for Hangu, regarding the inhabitants of Chiriţeni and Piciorul Bocăncenilor, and also the areas with salt springs.

Acknowledgements
We thank our colleague Robin Brigand for allowing us to use the very expressive photo of the salt spring at Hangu. We also thank our colleague Rodica Popovici for the essential information regarding the population in the Hangu – Borca region.

References:

Alexianu, M., Dumitroaia, Gh., Monah, D. 1992. Exploatarea surselor de apă sărată din Moldova: o abordare etnoarheologică. *Thraco-Dacica* XIII/1-2, 159-167.

Alexianu, M., Weller, O. and Brigand, R. 2007. *Izvoarele de apă sărată din Moldova subcarpatică. Cercetări etnoarheologice.* Iaşi, Casa Editorială Demiurg.

Alexianu, M., Dumitroaia, Gh. and Monah, D. 2007. The Exploitation of the Salt-Water Sources in Moldavia: an Ethno-Archaeological Approach. In D. Monah, Gh. Dumitroaia, O. Weller and J. Chapman (eds.), *L'exploitation du sel à travers le temps.* Bibliotheca Memoriae Antiquitatis XVIII, 279-299. Piatra-Neamţ, Constantin Matasă.

Alexianu, M. and Weller, O. 2007. Recherches ethnoarchéologiques sur le sel: les enquêtes de 2004 et les premiers résultats obtenus dans la zone de Poiana Slatinei à Lunca (dép. Neamţ, Roumanie). In D. Monah, Gh. Dumitroaia, O. Weller and J. Chapman (eds.), *L'exploitation du sel à travers le temps.* Bibliotheca Memoriae Antiquitatis XVIII, 299-318. Piatra-Neamţ, Constantin Matasă.

Chapman, J. and Gaydarska, B. 2003. The Provision of Salt to Tripolye Mega-sites In *Tripolian Settlements-Giants. The International Symposium Materials*. Kiïv, Korvin Press.

Cihodaru, C. 2003. Regiunea Hangu-Bicaz în secolele XIV-XIX. In M. Petrescu-Dâmboviţa, V. Spinei (eds.), *Cercetări arheologice şi istorice din zona Lacului de Acumulare Bicaz*, Bibliotheca Memoriae Antiquitatis XII, 145-203. Piatra-Neamţ, Constantin Matasă.

Dumitroaia, Gh. 1987. La station archéologique de Lunca-Poiana Slatinii. In M. Petrescu-Dîmboviţa, N. Ursulescu, D. Monah, V. Chirica (eds.), *La civilisation de Cucuteni en contexte européen* BAI, I, 253-258. Iaşi, Editura Universităţii "Al. I. Cuza".

Dumitroaia, Gh. 1994. Depunerile neo-eneolitice de la Lunca şi Oglinzi, judeţul Neamţ. *Memoria Antiquitatis*, XIX, 7-82.

Ghiorghiu, C. 1885. *Dicţionar geografic al judeţului Neamţ*. Bucureşti.

Ghiorghiu C., 1904. – *Ape minerale. Locuri istorice ale judeţului Neamţ*, Bucureşti.

Kogălniceanu, V. 1973. Hrana. In R. Vuia *et al.* (eds.), 1973. *Etnografia Văii Bistriţei*, Piatra-Neamţ.

Monah, D. 1991. L'exploitation du sel dans les Carpates Orientales et ses rapports avec la culture Cucuteni-Tripolye. In V. Chirica, D. Monah (eds.), *Le Paléolithique et le Néolithique de la Roumanie en contexte européen*, BAI IV, 387-400. Iaşi, Editura Universităţii "Al. I. Cuza".

Monah, D., Antonescu, S. and Bujor, A. 1980. Raport preliminar asupra cercetărilor arheologice din comuna Poduri, jud. Bacău. *Materiale şi Cercetări Arheologice*, 86-99.

Monah, D. and Cucoş, Şt. 1985. *Aşezările culturii Cucuteni din România*. Iaşi, Junimea.

Monah, D., Popovici, D., Dumitroaia, Gh., Cucoş, Şt. and Bujor, A. 1987. Raport preliminar asupra săpăturilor arheologice de la Poduri-Dealul Ghindaru (1984-1985). *Memoria Antiquitatis*, XV-XVII (1983-1985), 9-19.

Monah, D., Dumitroaia, Gh., Monah, F., Preoteasa, C., Munteanu, R. and Nicola, D. 2003. *Poduri-Dealul Ghindaru. O Troie în Subcarpaţii Moldovei*, Bibliotheca Memoriae Antiquitatis XIII. Piatra-Neamţ, Constantin Matasă.

Nandriş, J. 1985. The Stina and the Katun: foundation of a research design in European Highland Zone Ethnoarchaeology. *World Archaeology*, 17/2, 256-268.

Nandriş, J. 1987. Romanian Ethnoarchaeology and the emergence and development of Cucuteni in the European Context. In M. Petrescu-Dîmboviţa, N. Ursulescu, D. Monah, V. Chirica (eds.), *La civilisation de Cucuteni en contexte européen*, BAI I, 201-222. Iaşi, Editura Universităţii "Al. I. Cuza".

Niţu, A. and Teodor, D. Gh. 2003. A doua epocă a fierului. Hangu-'Cetăţuia'. In M. Petrescu-Dîmboviţa, V. Spinei V. (eds.), *Cercetări arheologice şi istorice din zona Lacului de Acumulare Bicaz*, Bibliotheca Memoriae Antiquitatis XII, 139. Piatra-Neamţ, Constantin Matasă.

Petrescu-Dîmboviţa, M. and Teodor, D. Gh. 2003a. Eneoliticul. In Petrescu-Dîmboviţa M., Spinei V. (eds.), *Cercetări arheologice şi istorice din zona Lacului de Acumulare Bicaz*, Bibliotheca Memoriae Antiquitatis XII, 128-136. Piatra-Neamţ, Constantin Matasă.

Petrescu-Dîmboviţa, M. and Teodor, D. Gh. 2003b. Epoca bronzului. Hangu. 'Chiriţeni'. In M. Petrescu-Dîmboviţa, V. Spinei (eds.), *Cercetări arheologice şi istorice din zona Lacului de Acumulare Bicaz*, Bibliotheca Memoriae Antiquitatis XII, 137-138. Piatra-Neamţ, Constantin Matasă.

Petrescu-Dîmboviţa, M. and Spinei, V. (eds.), 2003a. *Cercetări arheologice şi istorice din zona Lacului de Acumulare Bicaz*, Bibliotheca Memoriae Antiquitatis XII. Piatra-Neamţ, Constantin Matasă.

Petrescu-Dîmboviţa, M., and Spinei, V. (eds.) 2003b. Cuvânt înainte. In M. Petrescu-Dîmboviţa, V. Spinei (eds.), *Cercetări arheologice şi istorice din zona Lacului de Acumulare Bicaz*, Bibliotheca Memoriae Antiquitatis XII. Piatra-Neamţ, Constantin Matasă.

Ursulescu, N. 1977. Exploatarea sării din saramură în neoliticul timpuriu, în lumina descoperirilor de la Solca (jud. Suceava). *Studii şi Cercetări de Istorie Veche şi Arheologie*, 28/3, 307-317.

Ursulescu, N. 1995. L'utilisation des sources salées dans le Néolithique de la Moldavie (Roumanie). In M. Otte (éd.), *Nature et Culture*, Colloque de Liège (13-17 décembre 1993), 68, 487-495. Liège, E.R.A.U.L.

Vlăduţiu, I. 1973. Creşterea animalelor în zona Bicaz. In R. Vuia *et al.* (eds.), 1973. *Etnografia Văii Bistriţei*, Piatra-Neamţ.

Vuia, R. 1973. Curtea şi gospodăria. In R. Vuia *et al.* (eds.), 1973. *Etnografia Văii Bistriţei*, Piatra-Neamţ.

Vuia, R. *et al.* (eds.), 1973. *Etnografia Văii Bistriţei*, Piatra-Neamţ.

Weller, O., Brigand, R. and Alexianu, M. 2007. Cercetări sistematice asupra izvoarelor de apă sărată din Moldova. Bilanţul explorărilor din anii 2004-2007, efectuate în special în judeţul Neamţ. *Memoria Antiquitatis* XXIV, 121-183.

Traditional Methods of Salt Mining in Buzău County, Romania in the 21st Century

Doina Ciobanu
Muzeul Județean Buzău, Romania

Abstract
At the Carpathian Bend, '**popular**' salt mining has been and is still practised today. This mining is the oldest form of salt mining. It precedes the mining ordered by the authorities and was carried before the feudal state was formed, when salt production was controlled by village communities. It had the form of a privilege, and later became the remnant of a former order.

The information about this sort of mining dates from the second half of the 17th century, but it applies to an older state of affairs. This right was granted to the population of the counties of Buzău, Râmnicul Sărat and Saac (north of the counties Buzău and Vrancea), entitled to extract salt for their own needs only, in exchange for a tax called '**sărărit**' – '**salting**'. Often control of these matters would be lost and trials and safety measures would be required. This primitive form of salt extraction is still practised today in Bisoca – Sări, where a network of galleries has been dug. Apart from electric lighting, the tools used in the process are no different from the ones in the Dacian era.

The present work is based on specific data on primitive salt extraction in the 21st century.

Keywords
salt mining, traditional methods, Romania

Salt is a vital element. It can be found in soil, water and air, spread unequally throughout the world. Romania is one of the privileged areas for salt resources; there are over 300 salt massifs (Atudorei, Bocante and Miclea 1971, 29).

Romania ranks 7th among European countries and 12th worldwide for salt resources. Here, salt can be found very near the surface, which is why a 'popular' form of salt extraction has always been witnessed. It was made very easily, with simple household tools.

In times of need, some of the inhabitants would focus on salt extraction only, thus serving the locals and people from neighbouring villages. In the area of the Carpathian Mountains Bend (today the lands of Vrancea and Buzău), where salt layers are found in the open and inhabitants have been raising and still raise animals, salt has always been extracted by primitive methods.

Apart from the organized mining of salt layers (an undertaking which was supervised by the ruling authority – that had a monopoly on salt production), a popular form of salt extraction was carried. This began in the 17th century, under the guardianship of officials appointed by the ruling authority.

The popular mining of salt was made in the mountain areas of Buzău and Vrancea, within the counties of Buzău, Râmnicul Sărat, Saac and the land of Vrancea, where salt can be found very near the surface, forming so called '*salt banks*'.

The land of Vrancea, situated at the crossing point of the three Romanian provinces (Wallachia, Moldavia and Transylvania), had a particular political status: its inhabitants had been granted an ancient privilege of extracting salt, but only 'for their own needs', in exchange for which they had to prevent the transportation of salt over the border (from Transylvania or Wallachia) into Moldavia. They were also forbidden to sell local salt inside or outside the borders. If this agreement were broken, they would risk paying twice the value of the salt they sold as damages to the entrepreneur and even losing their former privilege (Vitcu 1987, 67).

The rights of the inhabitants of Vrancea to extract their own salt unofficially from the local resources had been known since ancient times (as tradition says, since Stephen the Great) as 'sărăritul munților' –'salting of the mountains'. In exchange for this, the ruling authority levied a tax from the beneficiaries, paid equivalently in kind. In time, this was replaced by the duty of 'guarding' salt (Vitcu 1987, 67).

In the mountain area of Buzau, Saac and Râmnicul Sărat, the inhabitants had the right to extract salt for their own needs only, in exchange for a small contribution called 'sărărit' – 'salting' or 'sărăritul munților'- 'salting of the mountains', levied by 'salters' (Ilieş 1956, 198).

The information about this sort of salt extraction is recent, dating from the second half of the 17thcentury, but it applies to an older state of affairs as well. This right was granted to the village population, even to rroma, who paid 44 'bani' – 'pennies' a year --- in exchange. This tax amounts to 66 bani (Ilieş 1956, 155-194; Giurescu 1973, 20; Romanian Academy Library, Doc. 5389, f 8) within the following century. Here is a quote from a register in 1700 belonging to ruler Constantin Brâncoveanu, well-known for his sophisticated book-keeping: 'the salting shall be written along the villages south of Râmnic, south of Buzău and south of Saac, where there is the custom of salting, for they are free to take salt from those mountains of salt. So they shall write for every man, be he servant, tax collector, footman, shield man, knight or pedestrian, or priest, deacon, monk, or gipsy or any other sort of man, the sum of 44 bani' (Ilieş 1956, 155-194; Romanian Academy Library, Doc 5389, f 8).

From an official decree in 1784 we learn that the salt banks were guarded by 'the men of the salt keepers' (officials charged with keeping the mine) in order 'not to make waste and carry salt by cart or by horse, apart from those who pay for their salting, and whom we know, but even those must only be allowed to carry salt in sacks, for their domestic use...to take from them one lamb's skin, cheese and 6 'oca' (capacity measure of 1288-1520ml) of fur or 5 thalers each from those who shall have none of the former. And as for those who shall be caught stealing salt with a cart or with a horse, let all their cattle be taken as fine, and they shall be sent according to custom to receive the punishment of the mine.' (Ilieş 1956).

These forms of salt mining carried on without interruption until the 20th century.

Nemuţ Nicolae, age 90, former hammer man at the 'Salt bank', still lived in Valea Sării (Valea Sării village, Vrancea county) in 2005. He had been a worker in the mine since the age of 16. He dug for salt when required and drilled new wells when needed. His most prolific activity was carried out during the Second World War, when salt was being sent to Brăila. The mine functioned with the approval of the Village Hall. Two diggers opened the bank of Poiana Boului. The chamber was large and had 'two pillars in front and two in the back'.

The hammer men extracted salt lumps which buyers would load at the mine entrance and carry away. The buyers, who were sent by neighbouring villages, would pay a tax to the Village Hall of Valea Sării, and 2 *lei* (Romanian monetary unit) for each block to the hammer men.

The tools they used were very simple, made by the local smith: pick-axes, 5-6kg sledgehammers, chisels, hammers and even long axes, owned by each digger.

In the 1990s, after the control of the salt layer area of Buzău was lessened, the inhabitants of villages such as: Vipereşti, Lopătari, Mânzăleşti, Bisoca, Sări, moved on to seasonal extraction of the surface salt, 'according to their needs'.

In the summer of 2008, in the village of Băltăgari, Bisoca, there is still one salt mine where salt is extracted in an organized manner. The mine is located at 585m altitude, N - 45°52'454'' latitude, E - 26°74'869'', on the slope of the Băltăgari hill. The main coordinator of this activity is a digger named Dragomirel Dobroiu (aged 35). They are digging for 'foam' salt (white salt), which can be reached by descending on wooden bars through a vertical channel of 10-12m depth and a diameter of approximately 1m, its walls consolidated with wood pieces and twings. In the lower part there is an apparently disorganized gallery, which follows the white salt seam ('foam', as the digger calls it). The diameter of the gallery varies between 2 – 4m, having additional holes as well. The mine has one 'mouth' (entrance). The lighting is electrical, carried through a roughly made wire drawn from Dragomirel's house, which is situated 50 – 70m from the mouth of the mine.

Some tools are rough – hammers, sledges, pick – axes, shovels and chisels made at the smithy from Chiojdeni, Vintileasa, Vrancea county. Salt is lifted using a pulley installed at the entrance, from where the carriers take them to the carts. Those who require salt are usually rroma from Sări – Bisoca, neighbouring villages, who sell it at the fairs in Râmnicul Sărat and Focşani to the small farmers. Salt is mostly required in autumn for making pickles and in winter for preparing meat for Christmas and as food for stable animals. The price of salt is 20 *lei* a bag at the mine for the digger, and 30 *lei* a bag at fairs.

There are no security measures, and the digging goes on 'until soil is reached'.Then the path is changed according to the seam. The mine has been functioning for two years.

In the 90s there was another mine here, but it collapsed, 's-a ocnit' – 'it mined'.

The current mine is causing very visible external changes of the soil, changes which were remarked between 26th June and 1st September 2008, when a hole appeared in front of the plateau at the mine entrance. This did not create any impression on the workers.

The description of the two forms of primitive salt extraction in the 20th – 21st centuries is a faithful model of 'salting' and can undoubtedly be viewed as an experiment of salt extraction from historical periods that are less widely known and less rich in records.

References:

*** Romanian Academy Library, Doc. 5389, f 8.

Atudorei, Costică, Bocante, Emil and Miclea, Pavel 1971. *Cercetarea, exploatarea şi valorificarea sării.* Bucureşti, Editura Tehnică.

Giurescu, Dinu C. 1973. *Ţara Romănescă în secolele XIV-XV.* Bucureşti, Editura Ştiinţifică.

Ilieş, Aurora 1956. Exploatarea sării în Ţara Românească până în secolul XVIII. *Studii şi materiale de istorie medie,* vol. I, 155-194.

Vitcu, Dumitru 1987. *Istoria salinelor Moldovei şi epoca modernă,* Iaşi, Editura Universităţii "Al. I. Cuza".

El Salado-Ixtahuehue and Benito Juárez-Soconusco: an Ethno-Archaeological Study of Salt Pre-Industries of Southeast Veracruz, Mexico

Jorge A. Ceja Acosta

Universidad Autónoma de México, Mexico

Abstract

Various authors have carried out investigations to understand how different groups in the past took advantage of salt (Andrews 1983; McKillop 1995 y 2002; Parsons 2001 and Williams 2003). Many of these archaeological investigations have highlighted only the technological aspect of the production of salt. For this reason it is important to pursue new theoretical perspectives, in order to understand the relationships between social agents, technology and social practice.

Studies of oral history and ethnoarchaeological observations at two salt works of Southeast Veracruz have allowed me to consider evidence of pre-Hispanic salt production, going beyond the technological process (Ceja Acosta 2007a and 2008). The social practice in these two examples calls for our attention, because the technology is tied to social relations such as gender, constructions of the use of space, and perceptions of the world.

Ethnographic data obtained in these two communities allow me to reflect on the importance of salt production in the municipalities of the Tuxtlas and Acayucan in the current state of Veracruz, Mexico. The salt-production process has a close relationship with activity and social identity. Therefore, a technological description of how the saltmakers engaged in the process of salt production is still essential, as are how to define and know the world which has been made for them. Therefore it is essential to understand the relations of the ancient saltmakers, in the context where they conduct their daily activities in the saltworks.

Keywords

salt, everyday life, social practices, domestic rites and technology

Introduction

The salt pre-industries of El Salado-Ixtahuehue and Soconusco-Benito Juárez are located in the regions of San Andres Tuxtla and Acayucan Veracruz (Figure 1), respectively. These are sites with different topographies. The first of them, located on the side of a slope of the Coxole hill, has a rugged relief. The second saltworks operation was located on a plain with low slopes subject to seasonal flooding.

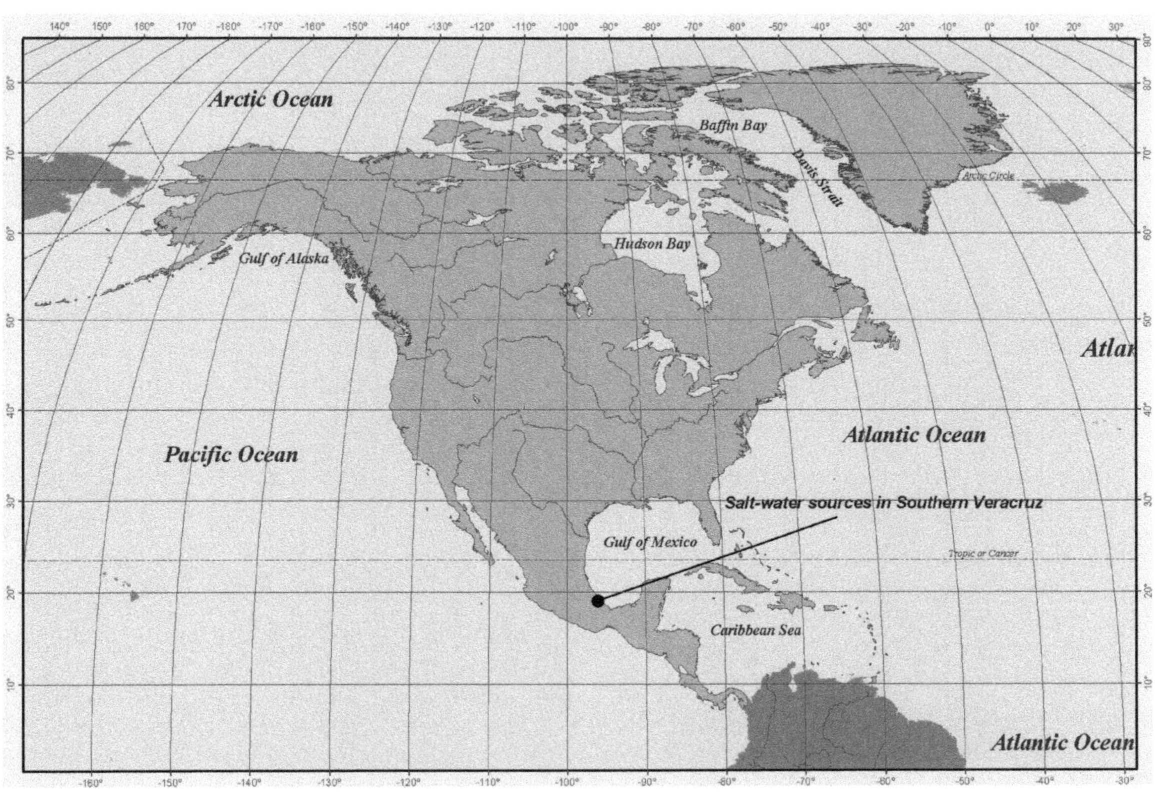

Figure 1. Location of the Salineras of the Veracruzano Southeastern (Google Earth in 2007).

Both sites have continued the tradition of boiling salt. Only Salado-Ixtahuehue, near the town of San Andres Tuxtla, has abandoned this practice. However there is still a very small number of people who occasionally go to the site to collect salt crusts.

In recent years these two zones have been objects of ethnoarchaeological investigations (Ceja Acosta 2007a and 2008). The idea of studying these two areas is based on the enormous ethnographic potential of these communities, as the continuity of historical and ideological practice allows the documentation of potential technical and social variables in the past.

These investigations began in 2004. The main goal was to determine whether the current methods of salt production were similar to those in ancient times. Now, however, I would say that the theoretical and methodological model obtained from this first foray is a functional interpretation, and that the development of salt production is a social practice that could change over time and space. The data obtained in the etnoarchaeological Project of Strategies of Salt Procurement (EOS in Spanish; Ceja Acosta 2007a) have allowed me to reflect on several things: how people give meaning to the activities they perform, how socially defined spaces govern the way in which personal relationships develop; and moverover, how socially defined spaces construct the notions of the saltmakers.

I have also been able to approach subjects treating social identity with greater depth. My goals have been to differentiate the technological decisions and the social propositions from the saltmakers; to define to what extent they identify themselves with this label, and to understand when social meaning gives sense to the cohesion of the group.

Salt works of one part of the Veracruzan Southeast

In order to begin to discuss saltmakers, it is necessary to briefly describe the actions that they carry out in the saltworks during the dry season.

At the beginning of the season, the people move from the settlements up to the location of the salineras. In the case of Soconusco community, people go to the salt spring, located near the present town of Benito Juárez, while Ohuilapan community goes to the salt spring which is near the railroad tracks.

The men who take charge of cleanliness of the area come first to the salt spring because throughout the year the place has remained abandoned since the dry season. Later, family groups, loaded with domestic utensils, begin to come to the cleaned area, which are used both in salt procurement and in food preparation. The meeting of the different families in the place begins with indispensable conversation on the achievements and births of the past season, as well as the memorialization of those community members, an essential part of the community who have passed away[1].

An important moment in salt procurement, in the case of Soconusco's community, is the rites carried out before the cross (Ceja Acosta 2007a, 88). These actions consist of praying, burning copal and depositing offerings of food next to the cross (Figure 2). Nevertheless, local people have commented to me that the offerings are also buried near community limits.

Figure 2. Offerings to the Cross in the Saline Water Well.

The salt production process in both communities begins with the transportation of brine to the place where the pots are put by the fire. While pots and pans are being placed, other community members carry out other activities. For example, they collect firewood, construct hearths, maintain tools, and prepare the space where they are going to spend the night.

Brine is poured into large metal cazuelas, then boiled until the salt begins to solidify. While boiling, the salt forms a crust on the surface. The crust is then broken and falls to the bottom of the metal cazuelas.

When this happens, people make sure the fire is not extinguished. Children perform this task. When it is necessary to put more wood on the fire, it is adults who perform the action.

These tasks are divided into groups according to the gender and age of each person. That is to say, while women are occupied with the salt work, men go in search of firewood. Meanwhile children are divided into two smaller groups to perform other tasks (they may possibly be separated due to the concept of social maturity).

Gender and age are important categories in the negotiation of space in the saltworks, because the older age groups direct the actions of younger people. For women, the elderly are closer to the salt spring while younger women are located in the periphery. This data has been observed only with women. I have no knowledge as to whether men have the same kind of hierarchy within the salt work. It is important to note that the archaeologist Maria Luisa Martell

[1] The arrival of the family groups to the salt is carried out by communities, in this way. When a community of boiling salt ends, another one will replace it. This continues until the arrival of the rainy season.

will further develop this information. She develops this topic from the perspective of gender and geography of space (in this volume).

Apparently the distribution of family groups is also a working strategy, as it takes up more space. The surveillance of boiling salt is therefore lower (Figure 3). Although the above seems obvious it is not, as in the case of individualization of practice in communities of Soconusco (Ceja Acosta 2007a) and Ohuilpan (Ceja Acosta 2007b) where social cohesion is broken under the eyes of the observer.

At home in the community of Soconusco, each family boils salt. They combine these actions with domestic activities, with the result that extended- family ties tend to weaken. However, at the salt water spring they recognize blood ties. At home they are accepted as a family although they no longer share the same social setting that united them. In the second community, social distancing by blood begins with the creation of new wells. Each family has its own well and no longer sees the need to share it. However, in this last example it is difficult to determine if the wells are covered once the season ends. This action may be a way of defining the ownership of the well by each family. In Soconusco, because there is only a salt spring, when a family finishes boiling salt, it is replaced by another.

An important part of the process of salt procurement is deciding what is to be achieved as a final product. The production of coarse salt only requires maintaining the fire until all the water evaporates. In contrast, for solid salt is necessary to keep the fire at a low, constant temperature.

This serves as contrast to what seems to be a historical difference in the way of producing salt in the Ohuilapan community (Figure 4) over time. In this case, from an analysis of the distribution of archaeological material, a change in the type of vessels used was detected on the surface at the ancient saltwork operation of El Salado-Ixtahuehue.

Taking this into account, it is arguable that there is a moment in the history of saltworking, in which the saltmakers of that site shift from a predilection for the barrel pots to using pans for cooking salt (Ceja Acosta 2007c). These data were observed in the archaeological case of Mesoamerica and particularly in the southeastern Veracruz area among post-classic-period communities to the recent historical period. It is also necessary to emphasize that in the Soconusco community this difference in the preference in the use of vessels has not appeared.

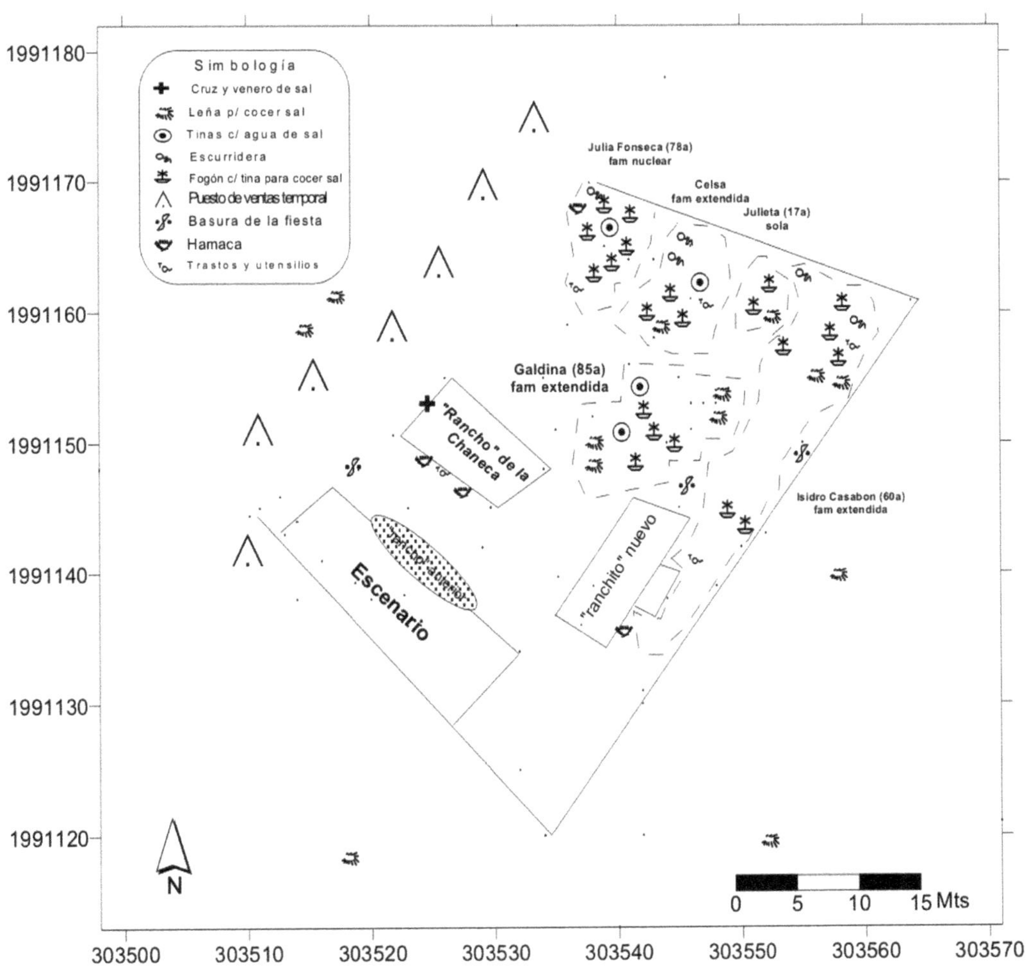

Figure 3. Distribution of Families in the Saline Water Well.

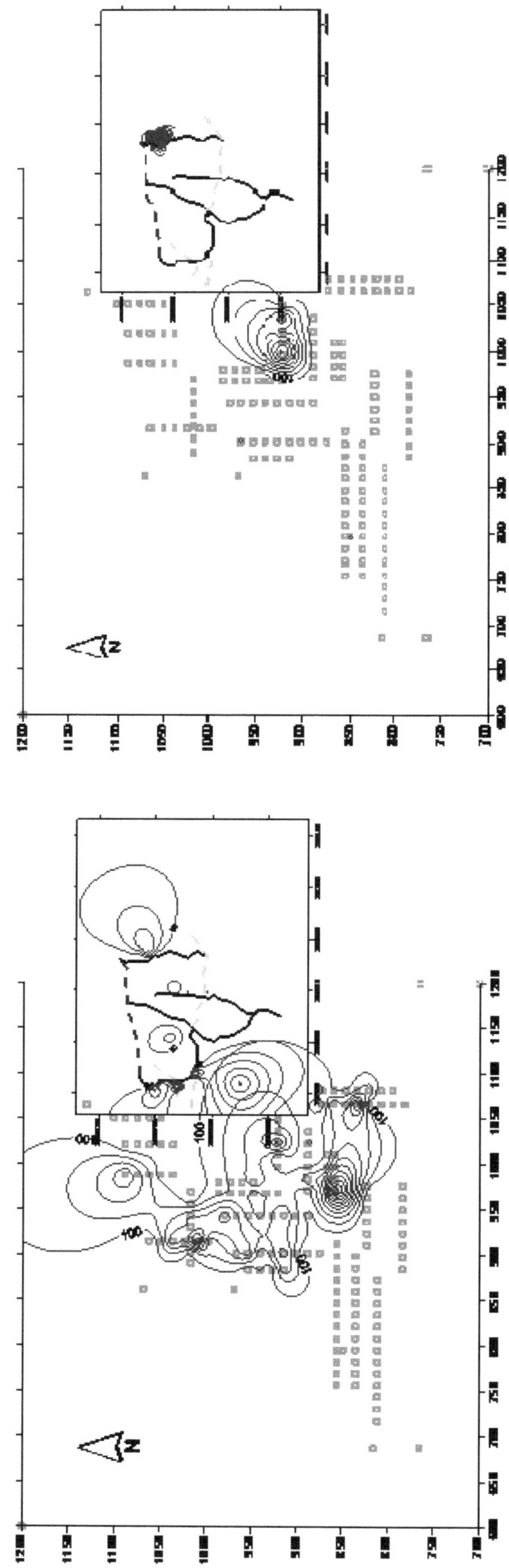

Figure 4. Distribution of Two Types of Pots in the Salty Salinera in the Ohuilapan Region, San Andrés Tuxtla Veracruz.

At this point I think that the preference between different forms of vessels may be related more to a technological decision (Lemonnier 2002) than a ceramic mode. But I do not think this technical decision directly relates to the functionality of each vessel in the process of salt procurement. These distinctions are essential for the understanding of everyday life. Perhaps they even allow more clarity in understanding the saltmakers' perception of the world in this area of southeastern Veracruz. Better yet, they may helps to determine whether in fact they share the concept of being a saltmaker, or whether each community has its own vision.

Later, when the salt has already been solidified, people begin to scrape the metal tubs with a ladle, then put salt in a plastic colander to remove remaining moisture. However, as the water content has not been completely removed, the salt is placed on a larger draining board where it remains for several hours until completely dry.

Saltmakers of the Soconusco and Ohuilapan communities have commented that they must remove residual moisture. It causes the salt to taste bitter. For this reason people of these communities hang small bags of salt near the hearth.

As the salt kept in sacks begins to drain, when the boiling salt tub is finished they begin with another. In this way they add kilos upon kilos of salt bags. It is important to mention that the bags are always put on bases that have no contact with the ground. In this way the people avoid any contamination of the salt and eliminate as much residual moisture as possible from the sacks.

New practices in the process of salt production in Soconusco community are to be found in the festivities of the salt fair. The festival is both a dance and huapango[1]. The first takes place in the main settlement, the second at the well. During this holiday a *samo* king and salt queen are designated[2].

Throughout the season of salt production, as the days go by, the people prepare food, and fill physiological needs, when they are transported to nearby villages to visit their relatives, wash their clothes in the nearby stream, bathe and then return to the well to discuss and review the progress of the boiled salt.

The speed of boiled-salt production depends on the quality of fuel, the distance at which the fuel can be gathered and the ability to charge for transportation. The level of heat and air control on the hearth is maintained during boiled-salt production. Therefore, these data become important to define whether the technique in the salt boiled in Soconusco community is part of a technological decision.

Formerly, salt boiling was done in clay pots. Obtaining ceramic temper was thus an important endeavor. Some women have told me that in former times, to go to boil salt, surplus pots were necessary. The development of this activity required a large number of vessels put to the fire (Figure 5). The average life of each of the pots was low (as a result of structural fatigue), as the pots needed to be constantly exposed to fire. Then they started to develop cracks until they finally broke in half[3].

Figure 5. Utensils for the Obtaining of Salt in Soconusco-Benito Juaréz.

It is important to add a discussion of the production of solid salt, *samo* (Figure 6), to the description above. This type of salt requires the intentional breaking of the container within which is the 'loaf'. The *samo* requires three to four days over low constant heat to form. This causes the salt loaves to stick to the walls of the vessel, so that destroying the container is the only way of obtaining the loaf of salt (Figure 7).

The two issues raised in the previous paragraph relate to a strategy in advance, plus a choice about the type and amount of salt that is desired. For once the vessels used to make salt loaves are used, they cannot be repaired. They can only be replaced by new vessels.

From the observations made in the community of Soconusco (Ceja Acosta 2007a) and the recovery of oral history in the community of Ohuilapan (Ceja Acosta 2008): It is important to note that just as in the past, families boiled salt using a large number of vessels. As in the archaeological record, I identified vessels intended for cooking and those that were used to boil salt, while others might even be used for both activities.

[1] Mexican popular celebration that gives name to a dance.
[2] The king of the "salt festivity' is a male entity who is related to the solid salt (samo Nahuatl name for salt solidified), while the queen has been associated with loose salt (personal communication with archaeologist Martell 2007).

[3] The average life span for the pots in Soconusco was 3-4 days, depending on the place where production took place, while the average number of vessels per season for each family varied from twelve to twenty.

Figure 6. Solid salt of the Soconusco- Salt place Benito Juárez.

Figure 7. Metal Container Used for the Obtaining of Salt, constructed in such way as to allow the removal of the block.

However, the pottery used for cooking activities are not the only tools found in the salt work. Some household items can also be used as tools for the production of salt. This is the case for the strainers and plastic cups, as well as for the pokers and ladles.

Having presented some data on salt production in Ohuilapan and Soconusco communities, it is necessary to reveal the archaeological implications of this research, while showing the progress that has been obtained in the Ethno-archaeological research project, Strategies for Sal Procurement (EOS).

Exploring the life of the containers used to boil salt is a valid approach. The information in the salt work of Soconusco reveals that prolonged exposure of the pots on the fire causes a short service life. This runs contrary to the idea of Shott and Williams (2001) who propose that large vessels have a longer lifespan than other tools.

From the data obtained from the experiment conducted on structural fatigue in pots (Ceja Acosta 2008), I would like to point out that the life of the vessels is related to the size of the vessel, but also to the activity for which the vessel will be used. The above is not a causal relationship, since the same type of vessel can have different functions. It is however necessary to extend these observations and to recognize their potential in the archaeological record.

Another aspect which must be considered is the manner in which production takes place within the pots and the final surface treatment (Fournier 1996). In the case of the Soconusco community pot production is carried out in the traditional way. This involves people constructing vessels by hand. These are put into the fire along with others, and later placed in a hole in a ground which serves as a combustion chamber. This technique is juxtaposed with closed kilns which control the entry of oxygen. There may be more production errors with the latter technique. For the Soconusco community, the production of pots is done in the traditional way. There may be failures in the process with this latest production technique.

Based on the data obtained so far on the life of the pots in the Soconusco community, vessels spend five days in salt work and eight or more days in the household on the same activity. The difference lies in the time spent on each activity. Salt work involves constantly filling the vessels with brine. In order to solidify, salt production must be constant and regular. For households surveillance of the pans with brine is sporadic. The lifespan of the vessels is usually therefore extended (Ceja Acosta 2007a).

Through observations of the salt work of EL Salado-Ixtahuehue, some areas have been defined, such as where the density of fragments is high (compared to the total size of fragments in the collection). For this reason, two possibilities present themselves: 1) The potsherd densities represent de facto garbage, abandoned there (or hidden, as some reported in the Ohuilapan community, until people return the next season). 2) Fractured potsherds are directly discarded (as a result of structural fatigue in the vessels).

Either of these perspectives has validity. Both positions have to do with waste practices, as might be expected from the ethnoarchaeological model (Ceja Acosta 2007a). To identify the difference between the two propositions, it is necessary to define the type of fracture. At this point I believe that structural-fatigue fractures have a regular pattern: they are larger fragments, pieces with sharp angles and the possibility of reconstructing complete pieces. Hence they could have been scattered over a larger area. Whereas the pieces discarded by deliberate fracture could be smaller, they would have been exposed to trampling and therefore to fragmentation. Hence they could have been scattered over a smaller area.

Currently, materials collected are being compared to address these two models. The typological analysis has been shelved, as the data obtained by Santley and Philip (1996) has been well defined. Moreover, the focus of this research project has been shifted to understand the actions that took place in the salt work.

It is not therefore necessary to correct the ceramic chronology.

From the data presented above, the tradition of boiling salt in the Ohuilapan community is still in force. Although it is no longer practical to procure salt in the salt work, the oral history of the people teaches us about salt production.

It is worth noting that many of the data provided by the informants on salt production were obtained in the community of Soconusco. However, it should also be noted that there are small differences at the technical level, for example, in the way they carry salt for distribution.

The information obtained in Ohuilapan communities was designed to understand the local process of salt production. We also wanted to assess whether the responses expressed by people were generally consistent, or if it was only possible to obtain particular perceptions about the process of salt procurement.

As in the Ohuilapan community, in Soconusco salt is boiled in salt pans and dried (to extract the moisture). As yet, I do not know how they remove the excess water from the salt, whether they use some kind of draining board or simply put small bags near the hearths.

In Soconusco the salt is placed on the hearth once it has completed the process of runoff. The salt is put on 'thenequens'. According to some informants, it is also put in intentionally fractured or perforated pans just for this purpose.

Wrapped-type salt-working also differs between the two communities. In the community of Soconusco, samo (disc-shaped solid salt) or salt is wrapped in a *benequen*. In contrast, in the community of Ohuilapan, I observed that salt is wrapped in the form of *tamale*.

Many of these data help to clarify the use of space in salt work in El Salado-Ixtahuehue. Subsequent research on salt will be carried out to further refine research goals. Some of these goals may clarify my doubts about other technical aspects of the production of salt.

I do not know much about ritual activity related to the production of salt. I am uncertain whether people in the past carried out a ritual in the salt-Ixtahuehue El Salado, or if people made offerings that may be observed in the archaeological record. This idea follows the observation of one informant, who 'had many major archaeological pieces.' Unlike the Soconusco community, in Ohuilapan I do not know with certainty how activities are distributed in the salt work. There could be a link between how space is conceived. For this reason it is necessary to identify the categories of gender and rituals related to this particular point.

There is evidence that figurines were recovered in the course of previous research in salt work (Santley 2004). The location is known but the physical appearance of the figurines is not. With this information it would be possible to know how they define themselves as social actors, or how they were perceived by agents outside the practice.

If the main objective of this research is to define the agency of the saltworks, this is a fundamental point. As agency is related to identity, this has implications for the way in which people observed the ordinariness of their actions and how they defined themselves through ethical and moral values.

Archeological implications

So far, the ethnographic data obtained in the saltworks of Ohuilapan and Soconusco have allowed me to observe social constructions in the archaeological record. Although actions taken by people at the wells are mostly domestic in nature, this should not be confused with the results of everyday activities.

Data from the Soconusco community allowed me to point out that the procurement of salt is seasonal, and that this influences many social practices variably, in this as in other examples of saltworks. Thus in the Soconusco community, as in the community of Ohuilapan, the saltmakers carry out other tasks outside the dry season.

While it is true that there is a surplus in production of salt, this does not exceed 140 kilos per family. Many of the people who store this quantity of salt in their homes do not leave to sell it. On the contrary, other inhabitants who come to look for salt in the very homes of the saltworkers.

Although this investigation began recently, some items have already been analyzed. Some comparisons can be made between the two communities. One example is the spatial distribution of archaeological materials related to the procurement of salt (Ceja Acosta 2008).

From the collections made by Santley (2004) in El Salado-Ixtahuehue, it is possible to define a pattern of behavior in the waste materials in the western part of the salt work, in a north-south axis. This can be interpreted as a way of using space. While this may be very subjective, it is necessary to mention that in the first part, Santley (2004) noted that the procurement of salt in the Salado-Ixtahuehue was performed by full-time specialists. Procurement of salt exceeded 115 tons per annum (Santley 2004, 214-215; 218-219). However, based on the distribution of materials on the surface there is no evidence of households, or full-time specialists.

First, following an analysis of salt collection, the materials are fairly homogeneous. All can be related to the practice of boiling salt. Although it appears that, based on my previous remarks, I supported the idea of Santley (2004) on specialized production, we must not lose sight of the fact that his remarks were based on the presence of only one ceramic type, with a high density, setting aside the interpretation of other archaeological materials. But incorporating these archaeological materials allows us to have a broader picture of the activities carried out in the El Salado-Ixtahuehue salt work (for a review of the distribution of materials see Ceja 2007d).

As I mentioned, the production of salt in the two salt-water sources of southeastern Veracruz is temporary and seasonal. The method of making salt in both is by the positioning of several pots on the fire. For this reason, the presence of a ceramic-type diagnosis may be related to the intensity of use of the vessel (which could indicate the activity of storing brine). My previous assumption aims at understanding how the structural fatigue of the vessels is related to daily practice and, at the same time, to technological decision-making.

The variable of temporality in salt production required salt-makers to develop ways of maximizing the storage of the brine or improving the resistance of the vessels (if you want to boil a large quantity of salt).

When the ethnographic data is taken into account, if the saltmakers of Soconusco were performing domestic activities in the salt spring, there is no evidence of permanent residence. For this reason it is expected that the distribution of elements of domestic character would be minimal.

Santley (2004) mentions that the salt work El Salado-Ixtahuehue was a community of people permanently dedicated to the production of salt. However, the sampling is only macroscopic. That is, he interprets only the distribution of archaeological materials, but he puts more emphasis on the estimation of salt production in the area.

Ethnoarchaeological research and studies of activity areas have shown that while the identification of areas of waste is important for the interpretation of the archaeological record, chemical analysis must also be carried out to understand the wider picture of the processes carried out there.

This is precisely the next objective of my research, since it is important to understand how the saltmakers used the inner space of the salt work. Because, in the ethnographic example, several families go to boil salt at once, part of household practices is mixed with the practice of salt production. But many domestic activities are performed there, so a more detailed record of the tasks they perform in the salt spring is necessary.

This observation has several implications. One of them is that these data could be interpreted as the economic strategy of each group, but could also be understood as how they relate the task of boiling salt with the world. If Dobres (2000) is correct, and the development of any process is related to social constructions, then the salt ritual practices must be identified.

One example of these ritual practices is the presence of an entity called 'the *chaneca'*, a character which governs and punishes the behavior of people in the salt work (Figure 8). At least a part of the ritual aspect of salt production can be detected archeologically. At this point, there is a difference between salt production as a mixed subsistence strategy (shared with other tasks), and specialized production (a singular mode of subsistence).

Currently institutionalized salt works are mentioned rarely. One of them, perhaps the best-known in Mexico, was written by Friar Bernardino Sahagún. However, this work is misused when used as a pure ethnographic analogy for archaeological comparison. The group of people who are involved in salt production of salt cannot be understood just by reading about them. In the case of the Veracruz region, the salt workers practiced a cult of the deity called *Huixtocihuatl*.

For this reason, Ohuilpan and Soconusco communities have no direct ethnic identity relationship with the saltmakers mentioned by Sahagun. This commentary, a part- observation of the data provided by the same friar in his *Historia General de las Cosas de la Nueva España* (Garybay 1975), as compared with field data, is not compatible in terms of technology (for further discussion see Ceja Acosta 2007b). While the difference above could be related to the way in which the saltmakers carry out the activities, I must respond that the Soconusco community does not recognize the image of the goddess of the salt of the Central Highlands of Mexico, although they identified an entity closer to its historical reality.

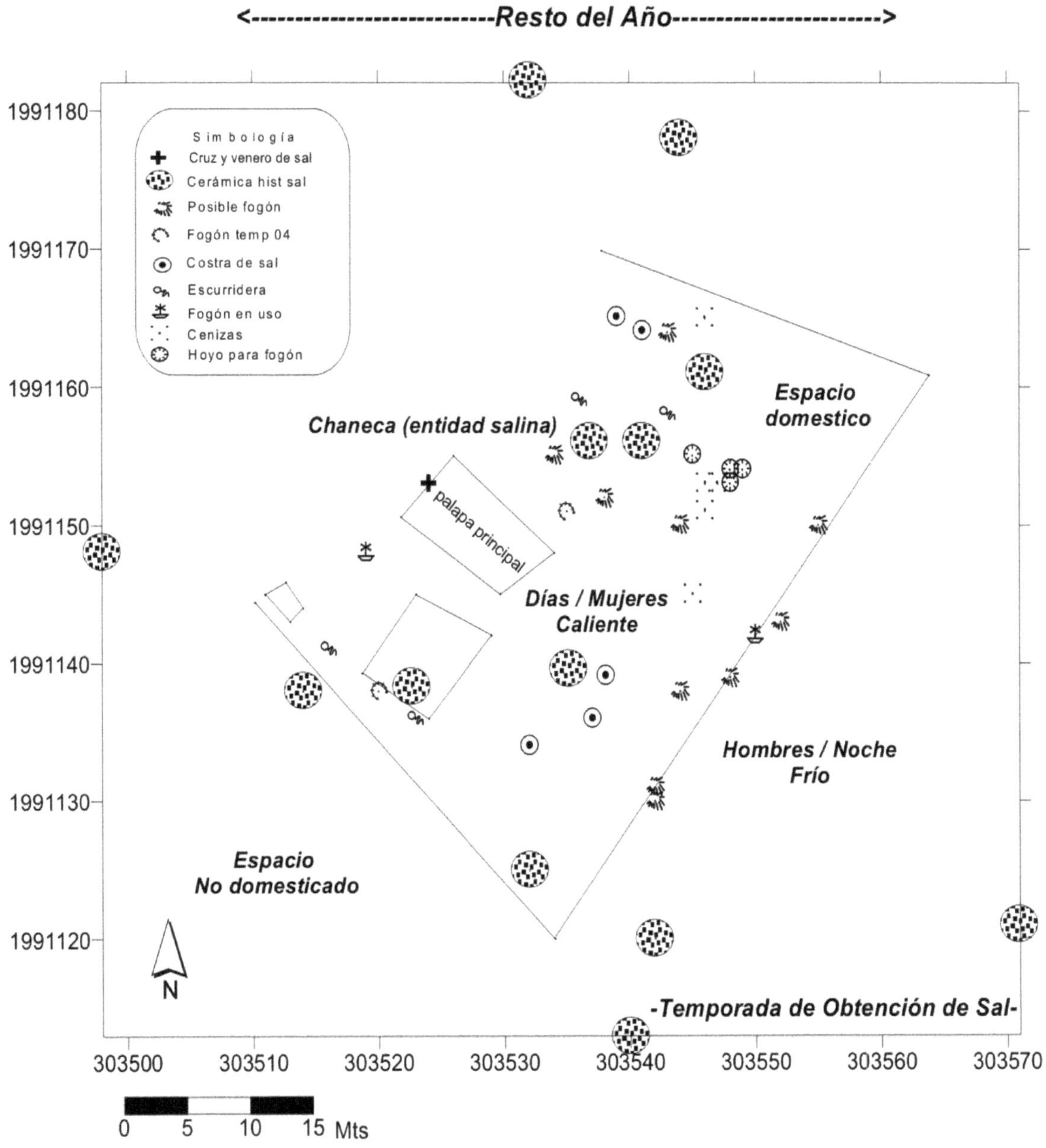

Figure 8. Gender Distribution of the Spaces.

In another paper, I mentioned that there are at least three groups in the contemporary scene in the production of salt (Ceja Acosta 2007a). One of them is trying to hook the other into a proposal to 'preserve our cultural practices' (Figure 9).

In the salt sources of Ohuilpan and Soconusco there are two types of archeological sites. In the first community, small archaeological structures were built in direct relation to the salt spring. However, in the case of Soconusco, the settlement is located 300m away and has at least 10 archaeological structures. At Ohuilapan, this is a group of four archaeological structures arranged around the salt spring.

The architectural assemblage in the Soconusco community contains a few archaeological pieces, including a sculpture that relates to land animals, and stone in the shape of a skull. These elements could be related to a ritual setting designed specifically for the saltworks. Although there were so few of them, there are still enough pieces to relate the modern saltworks with the old process of salt production.

If small social groups carried out the production of salt in Ohuilapan and Soconusco, how could we identify the presence of various groups from other communities? A tentative answer lies in the identification of ceramic micro-styles. At this time this possibility exceeds the expectations of this research. There is no archaeological sample covering the ancient population of both communities.

Other examples of pre-industrial salt

The main objective of this research is to define the agency of salt working and its implications for the definition of space and gender. If agency is related to identity, this means that the way in which people view salt in their daily lives partially defines them in accordance with their moral and ethical values. As a result, I would like to illustrate how other saltworks outside the study area show similarity in the social, symbolic and technological aspects.

Figure 9. Concept on the Obtaining of Salt on the part of the Municipal Presidency of the Place.

In the example of salt source of Soconusco, the *chaneca looked after the salt.* She dictates the rules of conduct in the place. Pomeroy (1988, 133 and 137) in the highlands of Ecuador, Reina and Monaghan (1981, 17) in Guatemala and Pétrequin, Pétrequin and Weller (2000) in New Guinea have data on inland salt sources which are related to gendered entities.

In the case of Ecuador, the gender of the salt worker depends on the relationship between salt and fresh water. If there is a higher concentration of salt to fresh water she is female, male if there is a higher concentration of salt to salt water. It is worth noting that most of the work was done by women, while men and children do not participate much in this activity (Pomeroy 1988, 138).

In Guatemala, the salt source of a town had two *nagual*es[1], one masculine and one feminine, but due to a flood many of them went elsewhere. The feminine nagual left. Formerly the men were the only group which developed the activities in the salt spring. Leaving her anima, women could integrate themselves into the production of salt (Reina and Monaghan 1981, 17 and 18).

In New Guinea, the various tribes have also linked the saline springs with a female entity (Pétrequin, Pétrequin and Weller 2000). However, the activity is sometimes carried out by men, sometimes by women. Pétrequin, Pétrequin and Weller (2000) concluded that the procurement of salt in New Guinea incorporates various purposes, including social cohesion, relations with other groups, as well as ritual activities.

As for the technological aspect, salt production takes the form of boiling salt brine in clay pots. (This is one of the most effective ways to accelerate the production process.) However, a technological variant lies in temperature control from the kilns[2]. In San Mateo Ixtatan, Guatemala, there are two ways of boiling salt. The first is done at home, with the making of large pots on a hearth. The second is placing the pots on a group of stones (Navarrete 2008, picture on page 79). Navarrete said that to produce the salt, it must be carried out in broken pots (Navarrete 2008, the next picture on the same page).

Ethnoarchaeological research carried out in the salt mines of southeastern Veracruz point toward the idea that pots will be cracked when the final product desired is solid salt (Figure 8). To collect loose grains of salt, there is no such need. On the other hand, solid salt is crystallized at low constant temperatures, and execution time is longer.

Reina and Monaghan (1981, 31, Figure 39) in Sacapulas, Guatemala, have data that indicates that vessels are broken to produce solid salt. In both cases in Guatemala, the pots and jars were destroyed in relation to the process of the final product. From the data mentioned above and noted in the salt work of Soconusco, there is a close relationship between the saltmakers and the people who craft ceramic vessels[3].

However, in some parts of New Guinea, salt production is done without the use of pots or bowls (Pétrequin, Pétrequin and Weller 2000, 555-560), although fire is needed for the cooking and the crystallization of the salt[4].

The production of salt loaves is important. Of the two basic forms of salt production (solid or loose grain), I noted the following information in the communities of Ohuilapan and Soconusco. Usually, loose salt was used locally, while solid salt was sold for its hardness and ease of transportation. Pomeroy (1988, 132) has noted that in the highlands of Ecuador, balls were made of salt called *pucus* or *hormas*. Ewald (1997, 386, footnote 55) reports a similar case. She mentions that in La Concordia Chiapas, people receive salt in the form of crosses or Stars of David.

Solidification of salt may be related to transport. If left unsolidified, it may be an indicator of exchange. Ceja

[1] The *naguales* in the Mesoamerican tradition are spirits, in some cases are "owners' of the places that live.

[2] Is there a link on technological choices between the quality of fuel to generate heat, and kilns as a means to control the temperature.

[3] In Mesoamerican domestic situations many tasks are carried out by women. The same applies in the case of obtaining salt in the southeastern Veracruz.

[4] The use of fire as an accelerator for the production of salt may be related to the rainy season in the region. In the rain, the percentage of salt in saline springs is reduced.

Acosta (2007a and 2008), Navarrete (2008, 148) and Weller (2006, 58) observed that, in many cases, small amounts of salt become a gift and a matter of prestige. In some cases, social relationships with external groups, rather than trade, is the principal goal.

Strict control is exercised over the exploitation of saline springs. The salt sources of southeastern Veracruz (Ceja Acosta 2007, 152-161), San Mateo Ixtatán (Navarrete 2008, 146-148) and New Guinea (Weller 2006, 55, Figure 2) show that there is no single owner of the salt source. The communities surrounding the salt springs can access the brine without problems. The only restriction surrounds the order in which they access the salt spring. In this regard there are two levels. The first is that older people tend to be closer to the source of saline water. The second is that, in the most traditional villages they have the right to come first and stock up on brine. The more remote communities have no apparent order of access.

There are two types of strategies in social organization. The first has to do with what is done in the salt spring. The second has to do with what takes place in the home. The two types of organizations have social and technological implications. Up to the present time, Fabrica San José-Oaxaca (as reported by Drennan (1976) and San Mateo Ixtatan- Guatemala (as reported by Navarrete (2008) show similarities of organization to those of the salt sources of southeastern Veracruz.

However, much more information needs to be uncovered in order to understand the dialectic between the different strategies for salt production. At the technological level, there may be more ways of producing salt in both communities. On the other hand, ritual practices having to do with salt have declined and we must understand the implications.

In light of this situation a rupture can be observed in the social cohesion of a group. First, salt is obtained in the saline spring, while later activity takes place in the home. All cases, of course, are not similar. Nor do all the examples have to follow this order. For this reason, I believe studies on this topic must be conducted, in order to understand how each family carries out its saltmaking activities.

Conclusion

It was the goal of this discussion to determine whether the ethnoarchaeological model studied in the salt sources of Ohuilapan and Soconusco could be applied to pre-Hispanic practices of salt production in Mesoamerica, and elsewhere in the world. In this regard Weller (2006, 52) has mentioned that, in the example of New Guinea, social organization for salt production defines the identity of the people who perform the task. At the same time, in carrying out this research I learned the value of ethnographic studies, in the course of my work with communities at Ohuilapan and Soconusco. The value of the data and its implications for understanding the archaeological record are enormous. However, I must end my comments here, first, because this project is still in development, and secondly because the variables discussed still require a broad spectrum of interpretation, in further reviewing the data and detailing information more clearly. Much that I have demonstrated in ethnoarchaeological work has not yet been excavated in the archaeological contexts of saltworks. However, the initial work here will, I believe, serve as a precedent for other research on the topic of identity. It will also reassess the conventional idea that saltmakers were a monolithic social group. Finally, data presented here offer ways of addressing the social territory and the complexity of salt production, even when this is considered only as a domestic task (Figure 10).

Figure 10. Family of Saltmakers in the Company of Archaeologist Maria Luisa Martell.

Acknowledgments

Dr. Marius Alexianu, Dr. Vasile Cotiugă and Dr. Gabriela Curcă for our invitation to this colloquium. We thank you for your hospitality and miss the beautiful landscapes of Romania and the warmth of your people.

For all his comments on the obtaining of salt in other countries to Dr. Olivier Weller, as well as his helpful information about the interchange and the social value of salt.

To Dr. Dan Monah and Mr. Bernard Moinier for their assistance and their comments on our subject of study.

To Blas Castellon, Yoko Siugura and Annick Dannels for their valuable comments about the obtaining of salt in Mesoamerica. Our thanks to all of them for the patience with which they listened to my comments and for their contribution of new ideas to this investigation. Special thanks to Shanti Morel-Hart for her assistance with the translation of the text, also to Iulian Moga for his initial comments on the text. Without his comments it would not have been possible to produce the final version.

References:

Andrews, A. P. 1983. *Maya Salt Production and Trade*. Tucson, University of Arizona.

Ceja Acosta, Jorge Alejandro 2007a. *Estrategias de Obtención de Sal en la Región de Soconusco, Acayucan, Veracruz: un Modelo Etnoarqueológico de Obtención de Sal en Grupos Contemporáneos no Industrializados*. Tesis de Maestría no publicada. México, Escuela Nacional de Antropología e Historia.

Ceja Acosta, Jorge Alejandro 2007b. El Deterioro de la Cohesión Social en la Obtención de Sal en la Región de San Andrés Tuxtla, el Individualismo de la Práctica Social. In Cristina Corona Jamaica (coord.), *6to Coloquio del Posgrado de Arqueología de la Escuela Nacional de Antropología e Historia*. México, Escuela Nacional de Antropología e Historia.

Ceja Acosta, Jorge Alejandro 2007c. Análisis de la Distribución de Materiales Cerámicos de El Salado-Ixtahuehue: Algunas Reflexiones Metodológicas, sobre los Procesos de Abandono del Clásico Tardío al Postclásico e Histórico Reciente. In Yamile Lira (coord.), *Cincuenta Años de Antropología en la Universidad Veracruzana*, 81-100. Xalapa, Universidad Veracruzana.

Ceja Acosta, Jorge Alejandro 2008. *Proyecto Arqueológico: Agencia Salinera de una Parte del Sureste Veracruzano (1era Fase de Investigación San Andrés Tuxtla; Veracruz), Informe Técnico Temporada 08 Enero-Febrero*. Informe Dirigido al Instituto Nacional de Antropología e Historia. México, Universidad Veracruzana-Instituto Nacional de Antropología e Historia.

Ceja Acosta, Jorge Alejandro 2009. La Huixtocihuatl, Diosa de la Sal o la Problemática de las Identidades Salineras en la Región Sureste Veracruzana en el Posclásico. In *El Alfolí*. 6, 30-39. Guadalajara, Noticiario Salino y Salado de la Asociaciòn de Amigos de las Salinas del Interior.

Dobres, Marcie-Anne 2000. Social Agency and Practice: the Heart and Soul of Technology. In *Technology an Social Agency*, 127-163. London, Blackwell.

Drennan, Robert 1976 Fábrica San José and Middle Formative Society in the Valley of Oaxaca, en Prehistory and Human Ecology of the Valley of Oaxaca. In Kent Flannery (ed.), *Memories of the Museum of Anthropology* 4, 8. University of Michigan, Ann Arbor.

Ewald, Ursula 1997. *La Industria Salinera de México 1560-1994*. México, FCE.

Fournier, Patricia G. 1996. *La Alfarería Tradicional. Resistencia a la Ruptura en Cuerpos Cerámicos*. México, INAH.

Garybay, Ángel María 1975. Historia General de las Cosas de la Nueva España (Bernardino de Sahagun). México, Porrua.

Lemonnier, Pierre 2002. *Technological Choices: Transformation in Material Cultures since the Neolithic* 1-35. London, Routledge.

McKillop, Heather 1995. Underwater Archaeology, Salt Production, and Coastal Maya Trade at Stingray Lagoon Belize. *Latin American Antiquity* 6, 3, 214-228.

McKillop, Heather 2002. *Salt: White Gold of the Ancient Maya*, University Press of Florida, Boca Ratòn.

Navarrete Cáceres, Carlos 2008. El Origen de la Sal en la Tradición Oral de San Mateo Ixtatán, Guatemala y la Peregrinación de los Zapalutas. In Blas Román Castellón Huerta (coord.), *Diario de Campo (Sal y Salinas un Gusto Ancestral)* 51, 143-151.

Parsons, Jeffrey R. 2001. *Last Salt makers of Nexquipayac. México.* Michigan, Ann Harbor.

Pétrequin, Pierre, Anne-Marie Pétrequin and Olivier Weller 2000. Cuire la Pierre et Cuire le Sel en Nouvelle Guinée: des Techniques Actuelles de Régultaion Sociale. In P. Pétrequin, P. Fluzin, J. Thiriot, P. Benoit (eds.), *Arts du feu et productions artisanales. XXe rencontres internationales d'archéologie et d'histoire d'Antibes, octobre 1999,* 545-564. Antibes, APDCA.

Pomeroy, Cheryl 1988. The Salt of Higland of Ecuador: Precious Product of Female Domain. *Ethnohistory* 35, 2, 131-160.

Reina, Ruben and John Monaghan 1981. The Ways of the Maya: Salt Production in Sacapulas Guatemala. *Expedition* 23, 3, 13-33.

Santley, Robert S., and Philip J. Arnold III 1996. Prehispanic Settlement Patterns in the Tuxtla Mountains, Southern Veracruz, Mexico. *Journal of Field Archaeology* 23, 2, 225-249. Boston University, Boston.

Santley, Robert 2004. Prehistoric Salt Production at El Salado, Veracruz, Mexico. *Latin American Antiquity* 15, 2, 199-221.

Shott, Michael J. and Eduardo Williams 2001. Datos Censales sobre la Vida Útil de la Cerámica: Estudio Etnoarqueológico en Michoacán. *Estudios Cerámicos en el Occidente y Norte de México*, 97-124. Michoacán, El Colegio de Michoacán.

Weller, Olivier 2006. Exemples ethnographiques d'organisation du travail: les différentes exploitations de sel en les Hautes Terres de Nouvelle-Guinée. *Techniques et Culture* 46-47, 51-61.

Williams, Eduardo 2003. *La Sal de la Tierra. Etnoarqueología de la Producción Salinera en el Occidente de México*. México, Colegio de Michoacán.

The Saltmakers of Soconusco and Benito Juárez: An Interpretation of Ethnoarchaeological Data from the Perspective of Gender and Identity

María Luisa Martell Contreras
Universidad Veracruzana, Mexico

Abstract
The role played by women in the process of obtaining salt in the communities of Soconusco and Benito Juárez, Acayucan (Veracruz) has developed uniquely from the pre-Hispanic epoch to present day, despite historical modifications introduced by outside groups. Ethnographic information compiled in 2006 during the research project 'Strategies of Salt Procurement in Veracruz, México' has allowed us to observe a few of the social relationships established among the saltmakers, who were typically women, and other members of the community. These data illustrate that hierarchical roles are established for each participant within the saltmaking industry. These roles are intimately bound up with age, experience and the way in which the group perceives the actors. In the same way, younger members of the group are socialized to learn the saltmaking process according to rules, ideas, and representations of 'who they are'. These, in turn, are embedded in the way in which group members perceive gender roles.

This article offers a brief interpretation of the social relations established inside and outside saltmaking contexts, based on gender and identity.

Keywords
saltmaking process, gender, identity, social construction

> '...And whenever April approaches I despair of being here ... for me it would be an honor to die in the saltwater spring ... but as I have already told my family, they should make their farewells here...'

Translation of the words of Mrs. Sebastiana, a saltmaker from Soconusco, Ver.

Introduction

At the present time non-industrial salt-boiling practices can be observed in some parts of the Mexican Republic which adhere to more traditional methods of procurement. Most research into the salt industry in the region describes only the technological processes of its extraction and manufacture. There is very little into the social relationships which surround the procurement and production and salt. The way in which gender and identity affect social interactions is of particular interest as are the social and spatial division of labor, the symbolism of saltmaking activities, and the construction of an 'identidad salinera', or what it means within the group to be a saltmaker (Ceja 2007 personal communication).

All social groups have established behavioral codes which influence how people relate. These codes materialize in the environment and provide visual signals that define 'us' in opposition to 'them' (Gosselain 2000, 108). Gender, in particular, is a social construction that distinguishes the responsibilities of men and women within the group and the spaces where they interact. These gendered differences become part of the 'cultural identities' (Rodríguez-Shadow 2004, 17) which vary according to the group in question and the social and economic contexts in which they develop. Salt procurement consists of a series of activities that take place daily during a short period in late Spring before the rains arrive. The saltmakers of Soconusco and Benito Juárez gather at the salt springs once a year to boil salt brine from the spring and dry it in the sun. The women assume the bulk of the responsibilities to ensure the smooth running of these tasks. Part of that responsibility relates to the ritual appeasement of the spirits inhabiting the spring. The men and children of the saltmakers' families also help by cleaning and preparing the areas around the spring, collecting firewood, and performing other simple tasks delegated by the principal saltmakers. The salt-making activities of the communities of Soconusco and Benito Juárez take place in both public and private places (the saltwater spring and the domestic units within the residential community respectively). As they constantly reinforce the identities of those involved, they give purpose and coherence to their universe.

One of the problems faced by archaeologists is their inability to understand, using the materials they recover, how individuals, and groups of the past, perceived themselves and what type of social identities they constructed among themselves. In this respect, ethnoarchaeology is a very valuable tool. With the information that it generates, archaeologists can attempt to reconstruct non-material aspects that cannot be seen using traditional archaeological methods. Additionally, they learn to observe how to recognize social phenomena that might otherwise go unnoticed in the material record.

The interpretation of information obtained during ethnoarchaeological observation of salt procurement in the Soconusco and Benito Juárez communities in Veracruz[10] seeks to understand what kind of social, spatial and symbolic relationships are established in the saltmaking process. The gender- and age-related

[10] At the present these are the only two communities in the state of Veracruz that boil salt.

construction of space related to salt procurement in these communities provides data that can help interpret archaeological sites where similar activities took place. While the research was not explicitly carried out as a way of understanding archaeological salt- procurement sites, the gendered use of space can serve as a model for future archaeological testing.

Gender, identity and geography of gender

Rodríguez-Shadow (2004, 16) defines gender as follows:

> 'A constitutive element of the social relations based on the differences perceived between the sexes ... from which all social life is constructed, it is a cultural product that changes with time'.

Gender, besides being a social construction, is also a symbolic construction that materializes into a set of practices, ideas, social identities and representations[11]. These are norms of personal conduct attributed to the different sexes. In this way, society constructs the idea of what men and women should do (Lamas 1998, 53 quoted by Campos 2006, 14).

Sabaté *et al.* (1995, 14) emphasize that the division of labor based on gender is one of the most important functional designations in society. Certain tasks are delegated to men and women based on socially structured preconceptions of gender roles. Some gendered division of labor is universal, but the particular way in which it is carried out presents great social and spatial variation across different groups.

Gender is a social discourse that cannot be understood apart from the way in which individuals perceive themselves. The preconceptions and expectations attributed to each gender within a group directly affect the social construction of identity for all men and women. The socialization process begins at a very early age. Just as gender is a social category that structures the identity of different agents within the group, it functions as an analytical category for anthropological study. The boundaries defining masculinity and femininity serve as an axis of comparison among social agents.

Campos (2003, 31) mentions that different social identities form in relation to each other. They are continually negotiated. In the same way, it is important to emphasize that the construction of identity occurs inside spaces that may be sharply delimited (Campos 2003, 32). For example, female saltmakers are the only ones permitted to enter the spaces immediately surrounding the waters of the salt spring. It is taboo for men to enter this space. These limits are established according to the perceptions that the persons have of their universe.

Physical space plays an important role in gender relations, since both social discourse and symbols materialize in the landscape. A landscape is 'a cultural image, a pictorial way of representing, structuring or symbolizing surroundings (Tilley 1994, 24)'. Just as gender forms a major part of the social construction of identity, gender is used to attribute meaning to space. In this respect, the geography of gender, the objective of which is to analyze relationships among space and gender (Sabaté *et al.,* 1995, 14), provides the tools for observing the construction of space with respect to distinctions between the feminine and the masculine.

The geography of gender examines the forms in which:

> 'socioeconomic, political and environmental processes ... reproduce and transform not only the places in which we live but also the social relations among men and women who live there. [It] studies how the relations of gender have impacted the processes mentioned above and in its manifestations in space and in the environment' (Sabaté *et al.,* 1995, 17).

In this regard, gender and identity cannot be completely understood unless the way in which they materialize in space is considered. It is equally important to understand how the socially constructed landscape, in turn, structures the daily practice of gender-related behaviors.

The salt spring and the development of activities related to gender

The activities performed before and during salt procurement in the communities of Benito Juárez and Soconusco take place in concrete spaces and times. Time and space within the natural environment targeted for salt procurement are transformed into a social landscape that takes on social and symbolic meanings for the groups inhabiting it. With regard to this idea, Iwaniszewski (1997, 207) raises the following:

> 'Man uses this arranged space to satisfy his cognitive needs (to shape his vision of the universe) and to establish his social structure (to mark differences of gender, range, age, ethnic origin, etc.)'.

Social relations cannot be understood independently of time and space. Social practices leave a mark within time and space that determines the way in which they materialize in daily life (Sabaté *et al.,* 1995, 288). While there is no set blueprint for any social practice, cultural groups tend to form spatial patterns that correlate in some way with different activities. Saltmakers organize their habitual activities at the beginning of the season and transport their domestic space to the salt spring. In this way, the domestic space maps into the salt-procurement context. However, sometimes saltmakers cannot go to the salt spring as a result of illness, lack of money to purchase the necessary equipment, or lack of help in the case of

[11] '...how groups and individuals construct social categories and the impact this categorization has on the interaction.' Stone 2003, 34.

older salt makers. Because of this, the salt spring context is recreated in their houses (Figures 1-3). They pay someone to transport water from the salt spring. In these cases, space is designated in their backyards to cook salt (personal communication, Ceja 2006). In other words, the salt spring context maps into their own domestic space.

Figure 1. Making salt at a domestic unit (private space), (Photo Ceja 2006).

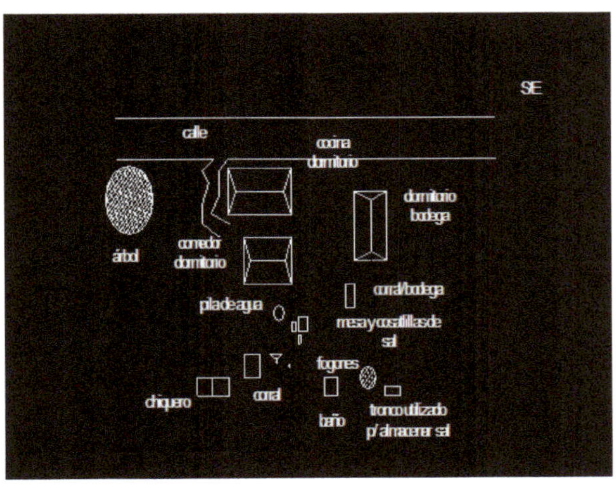

Figure 2. Schematic drawing of the domestic unit (Photo Ceja 2006).

Figure 3. Domestic unit (Photo Ceja 2006).

Soconusco's salt spring (Figures 4-5) lies abandoned throughout most of the year. At the beginning or middle of April (when there is no rain) the community begins local conditioning of the spring. This consists of cutting grass, constructing or remodeling the roofing, removing garbage and cleaning the areas surrounding the salt spring.

These activities are carried out exclusively by men, but the symbolic load that remains in the context is associated with the feminine gender. A very clear example of this is related to the cleanliness of the salt spring. Men are prohibited from entering the salt spring, because 'la chaneca', the female entity that takes care of this space (personal communication Ceja 2006) can 'punish them' through diseases or other maladies. Because of this, men limit themselves to cleaning only the exterior parts of the spring. The participants interact according to pre-established codes in which gender defines the use of space. The symbolic connotations that the settlers attribute to the salt spring are present all the year round, even when the site is abandoned.

As soon as the salt spring is in good condition, families begin to arrive, bringing the necessary tools and supplies to stay there for between five and 20 days. At this point, the participation of men is reduced to the transportation of the firewood in the mornings and the production of a little salt at night, while women rest (personal communication Ceja 2007).

Even the transportation of firewood to cook salt has socially defined rules which relate to gender. In Benito Juárez, for example, men and women transport firewood in different ways. Women carry a roll of firewood on their head or a trunk in their hands while 'men as a rule come with the rolls of firewood on animals' (personal Communication Ceja 2007) (Figure 6). Wiesheu (2005, 11) mentions that: '... in early civilizations demarcated by state development, asymmetrical relations become more sharply defined ... an unequal relationship takes place between gender and technological development ...'. The cases of Sonconusco and Benito Juárez support this assertion. The use of draft animals is confined strictly to the masculine gender. Domestic or traditional technologies are more often assigned to women.

I possess little information at this time to speculate on the causes or results of this technological divide. However, some women in the community have said that, one year, two men obtained more salt than the rest of them. Their access to certain technologies might have influenced these results. Gender limits the resources that both men and women have at their disposal, and also the way in which they perceive themselves.

Figure 4. Salt spring during the year (Photo Ceja 2005).

Figure 5. Salt spring before the beginnig season (Photo Ceja 2006).

Figure 6. Mother and son carrying wood for boiling salt (Photo Ceja 2006).

The saltmakers (who are female) infuse the salt production context with values, symbols, and social and spatial meanings from their domestic contexts. Women recreate their domestic environment at the salt spring, assuming the whole weight of the activity, while men and children only conduct minor activities. Sabaté *et al.* (1995, 289) mentions that a mutual relation exists between the way in which social contexts and spatial behaviors are perceived. The actors' knowledge of space, and the way in which they experience it derive from daily life. In this case, the woman controls the process of salt production by the way in which they control their own homes. The two spaces materialize similarly.

Age determines social status, the order of space, and the transmission of knowledge of the saltmaking process. Older women are in charge of teaching the youngest how to make salt (Figures 7-8). Children help out by checking the levels of water and the stage of 'cooking', controlling the fire and delegating small tasks to other apprentices. Mrs. Isiquia, for example, is a 70-year-old saltmaker who has to ask her 11-year-old granddaughter to add water to the tub because the salt comes out very watery. The status and respect these women garner from their roles as authorities within the community affects their geographical position within the salt spring.

Figure 7. Women taking care of the boiling process (Photo Ceja 2005).

As we mentioned above, adult women are in charge of the general salt-production process. The men who participate are usually husbands, children or grandsons. They are in charge of stirring the boiling salt occasionally, transporting firewood or arranging the logs within the hearth. In some cases men have learned their roles in the process from their mothers or grandmothers since they were children. They do not, however, take charge of the process directly when they come of age. Apparently, both men and women achieve their roles in the saltmaking process in a tacit way. Many adolescents are not interested in taking part in the saltmaking process. They think that old people must carry out the activity. 'They do not want to dirty their hands', or they say that they are frightened by the salt spring. Knowledge of saltmaking is fading away from these communities because of the negative attitudes of adolescents who would ultimately inherit the saltmaking tasks. As the older saltmakers die, their knowledge passes away with them. Shortages of fuelwood have also made traditional means of saltmaking more difficult.

The connotations of 'feminine' and 'masculine' also materialize in the salt itself (Figure 9). Men typically produce salt in block form. Block salt is given the name 'samo', which is exclusively related to the masculine gender. Pink loose-grain salt is associated with the feminine gender.

Gender, age, status and space distribution

Space at the salt spring is divided according to gender, age, and status. Year after year, families are established in places scattered around the salt spring. The only person to establish herself inside the central structure where the spring is located is Mrs. Galdina (Figure 10), a saltmaker who is more than 80 years old. The other women and the rest of the community the respected Mrs. Galdina. She has passed on her experience and knowledge to several generations. Mrs. Galdina occupies the central space at the salt spring because of the status she has acquired over time. In this case, 'acceptance' by 'la chaneca', the feminine entity who takes care of the salt spring is implicit. At the moment, there are no problems boiling the salt. The acceptance of Mrs. Galdina as the highest-status salt maker by both 'la chaneca' and the community can be summarized by the following:

'The places, the landscapes are also constructed facts, not only materially, but also because they are provided with meanings and values' (Sabaté *et al.,* 1995, 294).

Figure 8. Girl helping (part of the learning process), (Photo Ceja 2006).

Gender, age, and status are used to establish spatial order and symbolic meaning within the landscape by those who live in it. Gender influences the way that people reproduce these symbolic charges in the landscape or even in the division of space (Sabaté *et al.,* 1995, 294). The community's tacit acceptance of Mrs. Galdina's central role at the salt spring is reproduced through the ordering of space and a supporting belief system. These are socially constructed facts go unchallenged as long as no problems arise associated with the salt-production process.

Figure 9. Gender embodied within salt, (Ceja and Martell, field Season 2007).

Figure 10. Mrs. Galdina. The oldest saltmaker (Photo Ceja 2007).

Other saltmakers and their families are established in the areas surrounding the salt spring. They organize space around the hearths where they cook salt brine. There are only two 'palapas' or 'ranchos'[1] at the salt spring: one covers the salt spring itself and is occupied by Mrs. Galdina; the other is used by Mrs. Sebastiana. Mrs. Sebastiana, a saltmaker who is aged approximately 55, has been sick. She delegates responsibilities to her grandson, her grandson's wife, and her husband. The other saltmakers remain outdoors.

The domestic context 'moves' to the salt spring during these days. Daily activities, such as cooking the food and washing the clothes, take place at the same time as the preparation of salt. In some families the children, grandsons or teenage nephews are in charge of supplying the foodstuffs, which they bring from the village. Occasionally, relatives or friends 'visit' and bring rice, beans or other types of food with them. Visitors are received with a cup of 'popo'[2] or a glass of soda.

The women of Soconusco and Benito Juárez spend most of the year developing activities inside private spaces (of reproduction) associated with domestic contexts. Al have feminine connotations (Sabaté *et al.,* 1995, 296). However, salt production allows them, in a sense, to take place in public spaces of production, which, in most cases, are associated with the masculine gender.

Gender, space and symbolism

The built environment surrounding the salt springs has strong symbolic connotations reinforced by the actions of the saltmakers. There are spaces that must not be transgressed. For men, the interior of the salt spring and its waters are prohibited. Women, on the other hand, must avoid night activities surrounding the salt spring. Additionally, the salt maker must ask 'permission' from 'la chaneca' to cook the salt.

One of the practices carried out at beginning of the season is the blessing of the cross that has been placed next to the salt spring. This blessing serves two purposes. First the villagers request authorization from 'la chaneca' to work. Secondly, they assure the procurement of a good quantity of salt. The cross is adorned with strips of paper and flowers (Figures 11-12. One woman is in charge of praying and lighting incense. During the ceremony she takes a bottle of liquor and drops some liquid inside the salt spring and on the cross (personal communication Ceja 2007). Some salt makers mention that offerings are also given to the land in the form of food and liquor. These offerings are buried in different areas[3] around the salt spring (Sabaté *et al.,* 1995, 296).

Figure 11. Cross with papers (Photo Ceja 2005).

Day and night are also related to the feminine and the masculine genders. Women are prohibited from being near the spring during night, so they perform their activities during the day. The sacred entities that live in that space prohibit the female presence at night. Men, on the other hand, are not restricted to their use of space during the night.

Finally, it is believed that children who have not been baptized must not be taken to the salt spring. If they are brought to the spring, 'los chaneques' (masculine entities) make them cry or make them ill. In extreme cases, they kidnap them and lose them in the brush.

1 Term that the inhabitants of Soconusco and Benito Juárez use to designate structures with palm roofs.

2 Drink made with cocoa, water and sugar; Typical of the region.

3 This information has not yet been verified.

Figure 12. Praying at the cross (Photo Ceja 2005).

Figure 13. Well and cross (Photo Ceja 2006).

Conclusions

As we can observe through the ethnographic information presented here, gender is used to define many of the actions that the salt makers of Soconusco and Benito Juárez perform during the process of obtaining salt. The social relations that are established in relation to salt production are based on the gendered discourses that define what activities and spaces women and men are responsible for. These relationships permit the cohesion of the group.

Although the principal responsibilities of saltmaking remain with women, men and children conduct some minor activities. Each gender and age group has designated functions that are recognized and accepted by the whole group.

Gender is reflected in the division of physical spaces, as well as in material culture. The socially constructed symbolism attributed to saltmaking landscapes is reinforced through the actions of the members of the salt-making group and of the rest of the community.

The ethnographic observations provide us with very important data, from which it is possible to identify relations between specific activities, objects, spaces and feminine or masculine connotations in the archaeological record. In archaeological cases, a lack of tools often prevents us from associating material culture with social discourses enacted by past groups, especially from the perspective of gender, identity and the use of the physical space. The ways in which gender, age, and status structure the use of space at salt springs in Veracruz, Mexico may serve as models that can be tested at archaeological sites of salt procurement.

Acknowledgements

Dr. Marius Alexianu and Dr. Gabriela Curcă for the invitation to this colloquium, for all their help.
Angeles, Dennisse, Marcela, Mary de Jesús, Rosi and Leti... and the ESAO foundation for their support.
Eduardo Vázquez Yobal for help with the translation.
Dr. Wesley Stone for his time, observations, comments and help with this article.
Jorge, for sharing this experience with me and for his valuable support.
My parents, and, of course, Etienne and Jorge Luis.

References:

Campos Rodríguez, Lilia 2006. *Identidad de Género y Trabajo Femenino: las mujeres ejecutivas y Empresarias*, 10-60. Tesis no publicada. Puebla, Benemérita Universidad Autónoma de Puebla.

Ceja Acosta, Jorge Alejandro 2007. *Estrategias de Obtención de Sal en la Región de Soconusco, Acayucan, Veracruz: un Modelo Etnoarqueológico de Obtención de Sal en Grupos Contemporáneos no Industrializados*. Tesis de Maestría no publicada. México, Escuela Nacional de Antropología e Historia.

Gosselain, Olivier P. 2000. *Materializing Identities: An African Perspective* 187-217. Journal of Archaeological Method and Theory, Vol. 7, No. 3.

Iwaniszewski, Stanislaw 1997. Cotidianidad y cosmología: la representación social del espacio en Otatitlán. *Santuario y región. Imágenes del Cristo negro de Otatitlán* 206-259. Instituto de Investigaciones Histórico-Sociales. Universidad Veracruzana, Xalapa.

Rodríguez-Shadow, María J 2004. *La condición femenina en las comunidades precortesianas.* Diario de Campo. Boletín interno de los investigadores del área de Antropología.

Sabaté Martínez, Ana, Rodríguez Moya, J. and Díaz Muñoz, Mª A 1995. *Mujeres espacio y sociedad. Hacía una geografía del género.* Síntesis.

Stone Tammy 2003. *Social Identity and Ethnic Interaction in the Western Pueblos of the American Southwest* 31-67. Journal of Archaeological Method and Theory, Vol. 10, No. 1.

Tilley Christopher 1994. *A Phenomenology of Landscape Places, Paths and Monuments.* BERG. Oxford/Providence, USA.

Wiesheu, Walburga 2005. *Identidades múltiples en sociedades complejas tempranas: género, elite y especialización artesanal.* México, Escuela Nacional de Antropología e Historia.

Part II. Archaeological Salt Exploitation

Provadia-Solnitsata (NE Bulgaria): A Salt-Producing Center of the 6th and 5th Millennia BC

Vassil Nikolov

National Institute of Archaeology and Museum, Sofia, Bulgaria

Abstract

Tell Provadia-Solnitsata as well as the Chalcolithic production complex nearby both lie upon the only deposit of rock salt in the eastern Balkan peninsula. Salt production in the late Neolithic (5400-5000 BC) was carried out by means of boiling brine from the salt water springs in thin-walled ceramic bowls, which were laid inside massive dome ovens. The ovens are to be found in buildings within the borders of the site. Brine evaporation in ceramic bowls is the earliest recorded case of salt production in Europe using this technology. In the middle Chalcolithic (4600-4500 BC), a big solid salt production center emerged. It went on working until the late Chalcolithic (4500-4200 BC). The dome ovens were replaced by huge open-air features: wide pits, where very deep and voluminous thick-walled bowls were arranged. Brine was evaporated by means of open fire, lit at the bottom of the pit, in the space between the bowls, whose rims were fixed next to each other.

The defense system of the middle Chalcolithic settlement consists of an enclosing ditch and a wall with two oppositely positioned gates. The wall is made up of two linked parts, constructed by using different techniques: wood and clay palisade, and stone bastions. The gates were flanked by two stone bastions each, which collapsed as a result of a strong earthquake in about 4550 BC. New L-shaped stone bastions were constructed behind the previous ones, which also collapsed at the time of a subsequent strong earthquake in about 4500 BC.

Keywords

salt production, brine evaporation, late Neolithic (Karanovo III-IV), middle Chalcolithic (Hamangia IV), defense system, northeast Bulgaria

The excavation at the complex archaeological site of Provadia-Solnitsata started in 2005. It is located in the valley of the Provadiyska River, near Lake Varna, about 45km from the Black Sea coast. Until recently, we used the term 'salt-producing center' to designate solely the prehistoric tell, upon which, and partially by means of whose cultural layer, a Thracian tumulus was piled up much later. The excavation of the prehistoric layers of the tell (late Neolithic and Chalcolithic) has been carried out for four seasons now, providing results of exceptional interest, some of which bear relation to the 'salt issue' discussed here (Николов 2008). The tell exploration up to now covers an area of about 700m²; however, the main results have been secured by a sector of approximately 600m² in its southeast periphery, beyond the borders of the tumulus. Two phases have been identified so far in the prehistoric layer: late Neolithic (Karanovo III-IV period) and middle Chalcolithic (Hamangia IV period), having an average thickness of 1.20m each. The late Chalcolithic layer (Varna period), which existed in this part of the settlement, was used as a source of building material for erecting the Thracian tumulus and thus destroyed.

During the last 2008 season, an exceptionally large Chalcolithic salt-producing complex was also recorded in the immediate vicinity of the tell, extending the archaeological site both territorially and structurally, and corroborating my suggestions as to the role of the Provadia-Solnitsata prehistoric salt-producing center in the Neolithic and Chalcolithic.

The tell, which used to have a cultural layer about 6m thick (preserved in some areas beneath the tumulus) before being partially destroyed for the erection of the Thracian (burial?) tumulus, and featuring a diameter of 105m, as well as the Chalcolithic production center next to it both lie upon the huge truncated cone of the sole East Balkan rock salt deposit. Brine springs with an almost maximum salt concentration (310g/l) used to gush out from the 9 to 20m deep 'salt mirror'.

Field observations and the technology of salt production

At the time of the late Neolithic Karanovo III-IV culture, which spread all over Thrace in the period 5400-5200 BC, a group of its bearers crossed the Balkan Mountains and settled close to the brine springs near today's town of Provadia, where they started extracting salt.

The late Neolithic salt production technology at Provadia-Solnitsata involved boiling the spring brine in specially designed thin-walled ceramic bowls, which were put into specially constructed massive dome ovens. The ovens, whose productivity amounted to about 10 tons of solid rock salt per year, were located within the settlement. It turns out that boiling brine in ceramic bowls at Provadia-Solnitsata is the earliest recorded use of this salt production technology in Europe.

The remains of a late Neolithic two-storey building with a floorage of about 55-60m² were revealed in 2005-2007. A big dome-shaped clay installation designed for boiling brine was found on the ground floor (see Николов 2008a). It has four bulging sidewalls with rounded corners (Figure 1). Its dimensions are 1.70 x 1.50m along the two axes. The installation comprises a massive dome and a thick inside bottom, but lacks the variably thick base, elevated above the room floor and typical of the domestic ovens. The accumulating function in this case was transferred exclusively on the massive dome, suggesting a purpose, different from that of the ordinary ovens.

The dome is a massive clay structure. Its walls near the base are about 25cm thick and get thinner with height, reaching 13-14cm thickness at 40cm; this thickness is likely to have been the same for the entire upper section. Judging by the dome walls, preserved to our day in their original condition, its maximum inside and outside height was respectively 50 and 65cm.

Figure 1. Provadia-Solnitsata prehistoric salt producing center. Southeast area. Remains of a late Neolithic dome oven for brine evaporation.

The installation used to have two insertion openings. They were shaped in the eastern and southern sides of the dome. The eastern one is 26cm wide and its preserved height is 25cm. A low platform is shaped in front of it. The second opening was positioned in the southern wall and was considerably larger, most probably about 60cm wide, but it is in poor condition. It also had a platform in front.

The presence of two openings in the dome of a late Neolithic oven is quite unusual. The most likely reason for the existence of the small side opening is the technological requirement for a particular temperature regime during the brine evaporation process and the salt crystallization, as well as securing draught to lead away the steam. All this was achievable only by means of regulated air access into the installation. The big opening served as an inlet for the ceramic bowls, which were also taken out through it at the end of the cycle, with the solidified salt inside.

The installation floor, taking up the entire area below the dome, was made of tamped clay. Its surface is about 1.35 m².

The above mentioned ceramic bowls, designed for producing pure salt by boiling of salt solution (brine) represent a specific ware type, which has been discovered for the first time in the late prehistory of Europe. The vessels in question are thin-walled, deep ceramic bowls with special function (Figure 2) (see Стоянова 2008).

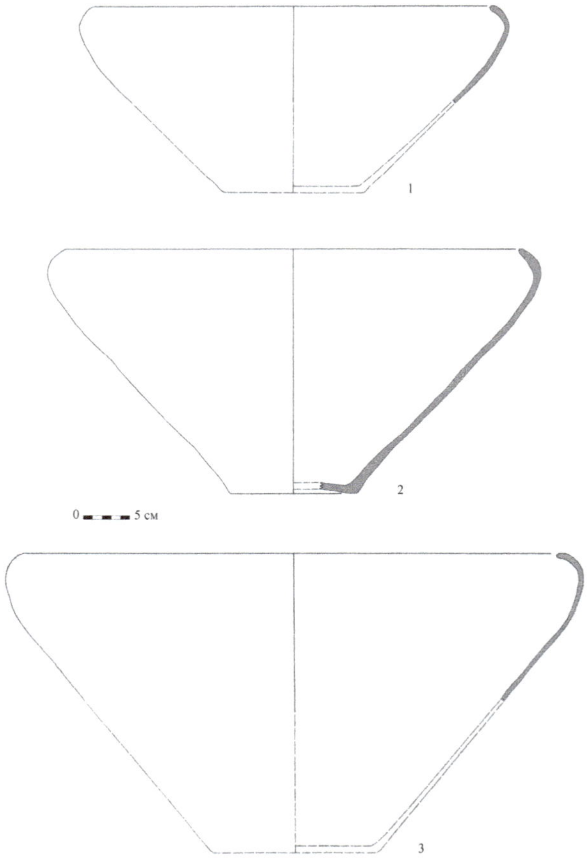

Figure 2. Provadia-Solnitsata prehistoric salt producing center. Southeast area. Late Neolithic ceramic bowls for brine evaporation in an oven (after П. Стоянова).

The surface of the production bowls is roughly smoothed, and their usage has contributed for the accumulation of a thick whitish build-up. Only sherds of them can be found due to their thin walls. The thickness of the walls varies between three or four to five or six mm, reaching 10mm at the mouth. The thin walls and the high porosity of these specially made bowls both facilitated brine evaporation.

The production bowls are wide open shapes with deep carinated bodies, with the carination located at two-thirds of their height. The bottom diameter is 11 to 18cm, and that of the mouth is 32 to 56cm. The mouth rim is rounded and slightly thickened on the inside.

The inverted rim prevented brine spilling at the time of inserting the bowl into the oven, as well as unwanted seething during the evaporation. The maximum temperature during this boiling process had to be lower than the boiling point of brine with a concentration of sodium chloride around 310g/l (the natural brine concentration of Provadia springs), which is 105°C. Bearing in mind the surface area under the dome of the oven, as well as the volume of the bowls (from 5 to 36l when filled up to their maximum diameter), one may presume that, at an optimal arrangement of bowls with various diameters, an average of 90l brine could have been evaporated at one load of the installation. Therefore, given the respective sodium chloride concentration of the brine and at one production cycle per day (however, two and possibly three are theoretically plausible), the process will result in 26-28kg of purified solid salt, i.e. up to 10 tons annual production in a single oven. And should the probability be accepted for a preliminary brine concentration by way of sun exposure (during the warmer months), the quantity produced could have been much more.

Boiling in brines was a normal practice through the late Neolithic, i.e. until 5th millennium BC, but there is no evidence of early Chalcolithic production as yet (such a layer has not been revealed in the excavated area of Tell Provadia-Solnitsata).

Life at the tell continued in the middle Chalcolithic as well, i.e. between 4600 and 4500 BC. It was precisely in the middle Chalcolithic that a big production center for solid salt extraction originated near the settlement. Probably, it continued its work through the late Chalcolitic, i.e. as late as the third quarter of the 5th millennium BC. It owed its existence to the obvious necessity to increase salt production, which was carried out by technological modification.

At the time of the middle Chalcolithic or even earlier the dome ovens at Provadia-Solnitsata were replaced by large open-air features: wide pits, where ceramic bowls of a new ware type - very deep thick-walled bowls whose capacity was much bigger than that of the late Neolithic ones - were tightly arranged, with their rims touching each other. Brine evaporation was carried out then upon an open fire, lit at the bottom of the pit in the space between the bowls.

The Chalcolithic production center was revealed at the end of the 2008 season. So far, its area seems to come up to at least fifty ares, but it may turn out to be larger in reality. It is located immediately north-northeast of the tell.

A small amount of data was collected from a narrow trench in one of the middle Chalcolithic production pits, whose diameter has not yet been precisely identified, but is certain to exceed 3m. Its depth is about 1.60m and there is another cylindrical pit in the middle (?) of the bottom, at least 30-40cm deep and 1.50m in diameter. The bottom of the pit at 1.60m is thick, probably tamped, and red in color, with clear indications of being exposed to fire. The walls of the pit were not revealed by the trench; however, judging by the found fragments of fired clay 'wall' with holes from burnt vertical sticks within, the pit walls were probably plastered.

The pit was densely packed with predominantly large sherds of very deep and wide thick-walled ceramic bowls, and a great quantity of wood ash (Figure 3). The mouth diameter of the bowls normally exceeds 50cm, and that of the bottom is ca. 18-20cm, and the likely height is 60-70cm. The walls are ca. 1.5-2.5cm thick. The outer surface is rusticated, with vertical pairs of big conic buttons below the mouth. The inner surface of the bowls is smoothed. The sherds testify to a secondary firing. Only a few thin-walled sherds have been found in the pit, allowing for dating the feature to the end of the middle Chalcolithic or the beginning of the late Chalcolithic in this region.

Figure 3. Provadia-Solnitsata prehistoric salt producing center. Wall of a trench made in a middle Chalcolithic negative feature for brine evaporation.

The shift in the brine evaporation technology is obvious: the ovens were replaced by much more efficient negative features. I suggest that the deep ceramic bowls were laid on the bottom of the pit in such a way that their rims were tightly pressed onto each other, while the bowls in the periphery touched the pit wall. Bearing in mind their height and their inverted cone shape with the mouth diameter considerably exceeding that of the bottom one, it is obvious that fairly large spaces, widest at the bottom, were left between the lower parts of the bowls. It was evidently a purposeful solution. These spaces were probably filled in with firewood. Throughout the evaporation process, mostly the upper layer of the brine was laterally heated by the fire; its level fell with the advance of evaporation, and so did the flames' height outside. Temperature went down simultaneously with the subsiding fire, creating thus conditions for crystallization of the salt. Hard, cone-shaped salt ingots remained in the vessels, which made their transportation convenient even over long distances.

The Chalcolithic salt producers near today's Provadia invented a faultless technique for a lot more high-speed extraction of considerable for that time amounts of salt that can rightly be labeled 'industrial'. The details of this novel technology will be clarified by the future field work and laboratory tests. This also applies to the role of the negative middle part of the bottom of the production pit.

The recorded shift in the salt extraction technology, which led to a serious increase of the production capacity of the 'factory' near today's Provadia in the middle and late Chalcolithic, lends strong support to the presumed connection between the salt production and respectively salt trade there on the one hand, and the amazing abundance of prestige items at the late Chalcolithic 'gold' Varna cemetery in the same region (Николов 2005) on the other hand. The long-sought-for clue to the existence of such a unique concentration of prestige items at a cemetery from the second half of the 5th millennium BC seems to be finally solved.

Defense system. Most probably, the late Neolithic salt producers satisfied their own needs for salt by using the spring brine itself, whereas solid salt was exported south of the Balkan Mountains. Salt production obviously amounted to quantities that might be called 'industrial' with regard to that time. Solid salt, which played the role of money at the time, is likely to have turned into an equivalent in a large-scale trade with neighboring regions. The riches, amassed by the salt producers, had to be defended; this is why the tell settlement was fortified by means of a strong fortification system dating back to the middle Chalcolithic.

The defense system comprises an enclosing ditch and a wall close behind, with two diametrically opposed gates. According to the archaeological evidence and geophysical surveys, the ditch has the shape of an irregular circle. The main source of information is the Southeast area of the tell, where the ditch has been uncovered over the entire range of the excavated area, while its south and southwest parts have been recorded by means of several trenches. Remains of the fortification wall and the southeast gate have been uncovered only in the Southeast area.

The presumed diameters of the inner side of the ditch are as follows: about 102m in the north-south direction and about 98m from east to west. It was dug out in the virgin yellow soil, at the beginning of the late Chalcolithic Hamangia IV period, when a late Neolithic cultural layer, at least 1m thick, had already existed along the periphery of the natural elevation. Since in the excavated area, this layer was partially destroyed during the erection of the Thracian tumulus, the width of the ditch in its upper part can only be subject to surmise. The width of the enclosing ditch on the level of the virgin soil varies from 1.15 to 1.95m, but its prevailing width is ca. 1.40m. At these measurements, the actual width of the ditch in its upper part must have been 2m at the least, extending to 3m at certain places. The depth of that part of the ditch that was dug into the virgin soil varies from 1.20 to 2.70m, measured from the higher inner side, but is 1.60-1,80m at most places. Bearing in mind also the thickness of the late Neolithic layer, I presume that the real depth of the feature was between 2.20 and 3.30m. The ditch features a markedly asymmetric and tapering down trapezoidal profile.

The defense wall, excavated in the Southeast area, consists of two joined parts, constructed by use of different techniques: a palisade of wood and clay, and stone bastions (Figure 4). Almost only their substructure has been preserved. The palisade was located at the distance of 3-4m from the ditch, while the lower level of its preserved superstructure used to stand higher than the upper part of the higher inner wall of the ditch.

The palisade consists of a massive upright wooden structure and a thick clay plastering. The wooden posts were at least 20cm in diameter. In some of the sectors, where the Neolithic cultural layer was thinner, they were rammed straight into the virgin soil to a depth of 70cm. At other locations with burnt ruins, the depth of ramming into the natural terrain is smaller, but the clay substructure compensated for it, functioning to a great extent as a virgin soil. The distance between the wall posts is 30 to 50cm. The width of the palisade was about 80cm. The wall was constructed of well-tramped yellow clay. So far as its height is concerned, the matter is quite complicated; however, I would assume a minimum of about 3m.

Figure 4. Provadia-Solnitsata prehistoric salt producing center. Southeast area. A panoramic view of the middle Chalcolithic defense system.

The palisade type defense wall probably also fenced in the entire southern part of the settlement. Northeast of the excavated palisade section, within the Southeast area, the defense structure continues in the form of a stone wall. Its length to the gate is 8.60m. Only the substructure has been preserved *in situ*, comprising eight big adjacent limestone rocks with irregular shape and a maximum size between 1.05 and 1.40m.

The back (interior) side of the stone wall base is located about 3m from the former inner edge of the ditch. A few more large and medium-sized rocks of the same kind lie in front of the northeast half of the stone wall. They come from the base of a massive stone bastion, filling the space to the former inner edge of the ditch at this particular location. They outline an impressive solid rock structure with dimensions about 4.50 x 3.00 to 3.30m.

This structure, also including the main wall, must have been tall enough for a defense facility. One should bear in mind the important fact that at least 22 big and medium-sized rocks have been found during the excavation of the ditch (the big ones among them are 11); hundreds of smaller ones should also be added. Besides, part of the stones in this area were taken out and used for the then newly-constructed Thracian stone feature at the base of the tumulus.

The stone bastion and the adjoining short stone wall fell down as a result of a strong earthquake (Figure 5), the shock having a north-northeast to south-southeast direction, which is evidenced by the position of some stones in the ditch. They were thrown diagonally towards southeast. According to the seismologist, B. Ranguelov, the earthquake had a magnitude of 7.5-8 and was related to Shabla-Kavarna seismic focus. Having in mind that this happened in the Hamangia period and that life in the settlement went on in the same period, I suppose the earthquake took place approximately around 4550 BC.

Figure 5. Provadia-Solnitsata prehistoric salt producing center. Southeast area. Remains of the southeast bastion (the large rocks) and the southeast L-shaped structure (in the rear). Middle Chalcolithic Hamangia IV period.

A second bastion, probably built in a similar way (Figure 6), existed north-northeast of the one discussed above. Only a few big stones have been preserved from it. One can think of several reasons for this poor condition, but I will not discuss them here.

The two stone bastions flank the gate of the fortification. The gate is about 2.40m wide; as wide as the street, starting from there. It leads northwest, heading precisely for the settlement's center. It is clear that the street extends to the opposite side of the fortification, where a piling of big and medium-sized rocks was recorded through trench excavation. These can be the remains of the second stone gate.

The two bastions, flanking the southeast gate, were at least 3m high, but most probably slightly more than that. Their upper leveled surface served for deploying fortress defenders, where they had the advantage of a better strategic position against invaders. The special attention paid to the most vulnerable spots in the fortification (the gates) along with enhancing their defense by flanking bastions, is perhaps the earliest known case of such fortification works. One should also add here the ditch, particularly deep and wide at this place, as well as the purposeful expressive drop in elevation between the lower level of the gate and the bastions, respectively the inner edge of the ditch, on the one hand, and the then outside edge of the ditch, on the other hand – ca. 2.50m! If we take for granted that the bastions were more than 3m high, it means that the defenders had an elevation advantage of about 6m, which rendered the gate practically impregnable. There must have been an at least partially movable wooden draw-bridge across the ditch, but as yet there is no conclusive evidence to confirm its existence.

Figure 6. Provadia-Solnitsata prehistoric salt producing center. Southeast area. Remains of the northeast bastion (the large rocks) and the northeast L-shaped structure (in the rear). Middle Chalcolithic Hamangia IV period.

I will summarize the above information. The defense system of the first middle Chalcolithic settlement at Tell Provadia-Solnitsata consisted of a wood and clay palisade, crossed at two places by a stone wall with bastions, flanking both gates of the fortification; the gates in the defense wall are opposite each other, with the settlement's main street connecting them. The defense wall is probably an almost regular circle with a diameter of 92-95m, while the inner diameter of the ditch is 98-102m. The average height of the defense wall may have been ca. 3m; the thickness of the palisade was ca. 80cm, the thickness of the stone wall exceeded one meter; the bastions protruded over 2m before the wall line; the gates' width was ca. 2.40m. This source data means that the overall length of the defense wall was approximately 295m, about 40 of which were taken by the two stone walls with the gates, and about 255m, by the two arcs of the palisade. The enclosed area amounted to some 70 ares.

The defense system of Tell Provadia-Solnitsata was badly damaged by the earthquake, which occurred around 4550 BC. This is certainly true for the two stone bastions, almost entirely demolished by it. Obviously, it was not possible to lift the enormous rocks back into place or bring new ones in because of the already existing settlement and the enclosing ditch. The fastest and most plausible option to close the disrupted fortification wall was to erect two L-shaped walls from smaller stones immediately behind the demolished bastions. Most probably, the palisade's resilient structure was not damaged beyond repair. However, a new connection with the newly built stone walls had to be made. The ditch in its part around the southeast gate was filled in by the huge bastion stones and hundreds of smaller ones as well. So it turned out to be easier to dig a new by-pass ditch immediately in front of the clogged one, which had to connect the still functional neighboring parts of the enclosure ditch. The length of the bypass is 24m.

The newly-built L-shaped stone structure behind the demolished southwest bastion (Figure 5) was so positioned that its short wall (about 4m) was constructed as an extension of the northeast side of the base of the bastion in front of it. The long wall of the new structure (about 9m) was almost parallel to the base of the bastion's northwest side. The walls of the new stone structure were about 1.20m thick. The southwest end of the long wall was connected with the then existing or rebuilt palisade. The preserved height of the walls, established at the excavations, reaches 45-50cm, but the interior space of the facility is filled in with a much higher layer of crushed stones, used as a building material.

The newly-built L-shaped stone structure behind the demolished northeast bastion (Figure 6) was so positioned that its short wall (about 4m) was erected upon the main street, thus narrowing it from 2.40 to about 1.30m. The short walls of the two symmetrical facilities were parallel to each other, forming a gate, about 1.30m wide. The long wall of the northeast L-shaped structure was about 15m. The walls of the second stone structure were also about 1.20m thick and their preserved height at the excavations is more than 1m. It has not yet been clarified whether the long stone wall of the structure goes northwest in the form of a stone wall or palisade; excavation will continue in this direction.

The new defense facilities were built by crushed stones without any mortar to hold them together. They were no less than 3m tall. The period, during which both new 'bastions' were in use is not likely to have lasted long, either. They were demolished by a subsequent strong earthquake, probably at the end of the middle Chalcolithic Hamangia IV period (ca. 4500 BC). The seismic wave, however, came then from either east or southeast; the walls collapsed inside, towards the fortification. Due to the presence of red clay among the remains of the northeast L-shaped structure, the stones have a specific color shade; this allowed for measuring their total volume and establishing the above cited height of the structure. It stands to reason that the neighboring symmetrical structure was equally tall, i.e. at least 3m. The total length of the two structures with the street between was about 25m. Together with the southeast palisade, the length of the uncovered sector of the fortification is about 40m.

Several fortified tells have been explored in northeast Bulgaria. In this context, the fortification system of the middle Chalcolithic settlement at Tell Provadia-Solnitsata has a few peculiarities. The space, encompassed by a defense wall, is considerably larger compared with that of the excavated early and middle Chalcolithic settlements in the region. The combination of defense wall and ditch has been recorded at one more site, but at Tell Provadia a section of the defense wall, which is near the gates at that, was built by stone, and the gates themselves were flanked by stone bastions, to be replaced by L-shaped stone structures at a later stage. This is the first case of this type on the Balkans.

The research at the Provadia-Solnitsata prehistoric salt producing center is going on.

References:

Николов, В. 2005. Първи свидетелства за най-ранното солодобиване в Европа. *Археология* 1-4, 109-117.

Николов В. (ed.) 2008. *Праисторически солодобивен център Провадия-Солницата. Разкопки 2005-2007 г.* София, НАИМ-БАН.

Николов В. 2008a. Постройка 5: архитектура и съоръжения. In В. Николов (ed.), *Праисторически солодобивен център Провадия-Солницата. Разкопки 2005-2007 г.*, 87-115. София, НАИМ-БАН.

Стоянова П. 2008. Постройка 5: керамични съдове за производство на сол. In В. Николов (ed.), *Праисторически солодобивен център Провадия-Солницата. Разкопки 2005-2007 г.*, 135-154. София, НАИМ-БАН.

Tell Provadia-Solnitsata (Bulgaria): Data on Chalcolithic Salt Extraction

Viktoria Petrova
National Institute of Archaeology and Museum, Sofia, Bulgaria

Abstract
Buildings 1 and 2 at Tell Provadia-Solnitsata probably refer to two successive stages in the development of Hamangia IV culture. A completely preserved vessel and numerous sherds, whose surface is coated with an additional clay layer with lots of organic temper, have been found in the earlier Building 2. An exceptionally small number of coated ceramic sherds have been found, however, in the ceramic assemblage of the later Building 1. Their presence testifies to the existence of a technology for the manufacturing of specific purpose vessels. Apart from the outer clay coating, they also feature uniform conic shapes and smoothed inner surface. These vessels have no parallel in either earlier or simultaneous ceramic material; they are likely to be related to salt extraction in the middle Chalcolithic.

Keywords
salt exploitation, brine, middle Chalcolithic (Hamangia IV), late Chalcolithic (Varna), northeast Bulgaria, briquetage

Tell Provadia-Solnitsata is located 5km south of the town of Provadia, upon the salt mirror of a rock salt deposit. It features an approximately circular shape with a base diameter of ca. 105m and total height of 20m. The site was excavated 2005-2008 by a team directed by Professor Vassil Nikolov from the National Institute of Archaeology and Museum, Sofia.

The periods of the late Neolithic, middle Chalcolithic, and late Chalcolithic have been distinguished in the general periodization and chronology of the cultural layers.

A convincing case for the salt production in the late Neolithic Karanovo III-IV period is presented by the numerous briquetage sherds and a dome oven revealed in a building at the site. It has been presumed that the brine was boiled in the oven (Николов 2008, 87-115). The long-lasting existence of salt springs near the tell and the rich ceramic assemblages of the Hamangia IV culture lead to the conclusion that salt extraction and production technologies were known in the middle Chalcolithic as well.

This report considers certain features and elements of the ceramic assemblages of the middle Chalcolithic Buildings 1 and 2, which suggest that some wares had a special function and might possibly be linked to brine boiling during the Hamangia IV culture.

Building 2
Building 2 refers to the earlier middle Chalcolithic horizon of the Hamangia IV culture. It was destroyed by a strong fire. Its uncovered southeastern part has a total area of ca. 18m². The building was made of wattle and daub; complete or fragmented pots have been found on its floor, next to the southeast wall. One of them is a conic bowl, found in a reclining position (Figure 1). Its outer surface is coated with clay, containing a large quantity of plant temper such as hulled wheat grains, as well as small snail shells. The clay coating on the bowl walls is up to 6mm thick, whereas that on the bottom reaches 18mm. Opposite each other on the well smoothed inner surface of the bowl, there are a zigzag groove and an oval concavity. The base is flat, but subsequently made round by means of applying the coating, so that the bowl itself becomes conic. There is a small horizontal groove, 6cm long and 1.2cm wide, upon the rounded base. It probably served for securing the vessel at the time of its use.

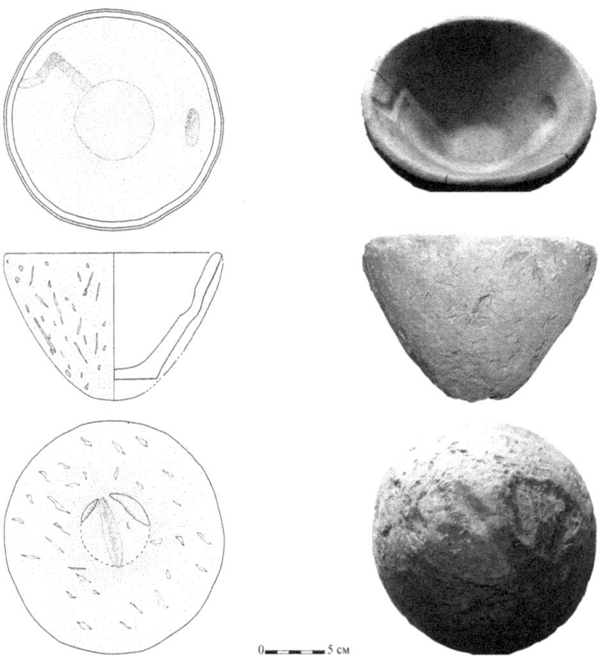

Figure 1. Provadia-Solnitsata prehistoric salt producing center. Southeast area. Building 2. A conic bowl with coated outside surface. Middle Chalcolithic Hamangia IV period.

Coating of the outer surfaces has been recorded on a significant number of sherds from the ceramic assemblage of Building 2 (Figure 2). The surface under the coating is rough and often covered by barbotine. The inner surface is well smoothed and there are oval concavities upon some of the bowls, similar to that of the complete vessel (Figure 2 *6*). The external coating of the bowls starts from the outer mouth rim and is up to 6mm thick. As with the coating of the complete bowl, there is a great deal of vegetable temper in it as well. The thickness of the bowl walls varies from 8 to 13mm, reaching 18mm above the base. These are mainly conic bowls, rarely deeper ones, with diameters ranging between 22 and 38cm (Figure 2 *5-6*). The bases are massive and feature a specific profile; they often bear mat imprints (Figure 2 *4*). A conic bulge has been recorded upon the inner surface of one of them.

The ceramic assemblage of Building 2 includes elements that can be referred to the early Chalcolithic in the region, but at the same time it has lots of peculiarities characterizing the middle Chalcolithic Hamangia IV culture. Most common are the stamped and incised and white incrusted patterns. The stamped patterns include rectilinear and curvilinear motifs and are often combined with dotted patterns (Figure 2 *1-3*). Similar patterns and motifs can also be found in the ceramic assemblage of Cucueti–Slatina veche (Munteanu *et al.*, 2007, Figure 7).

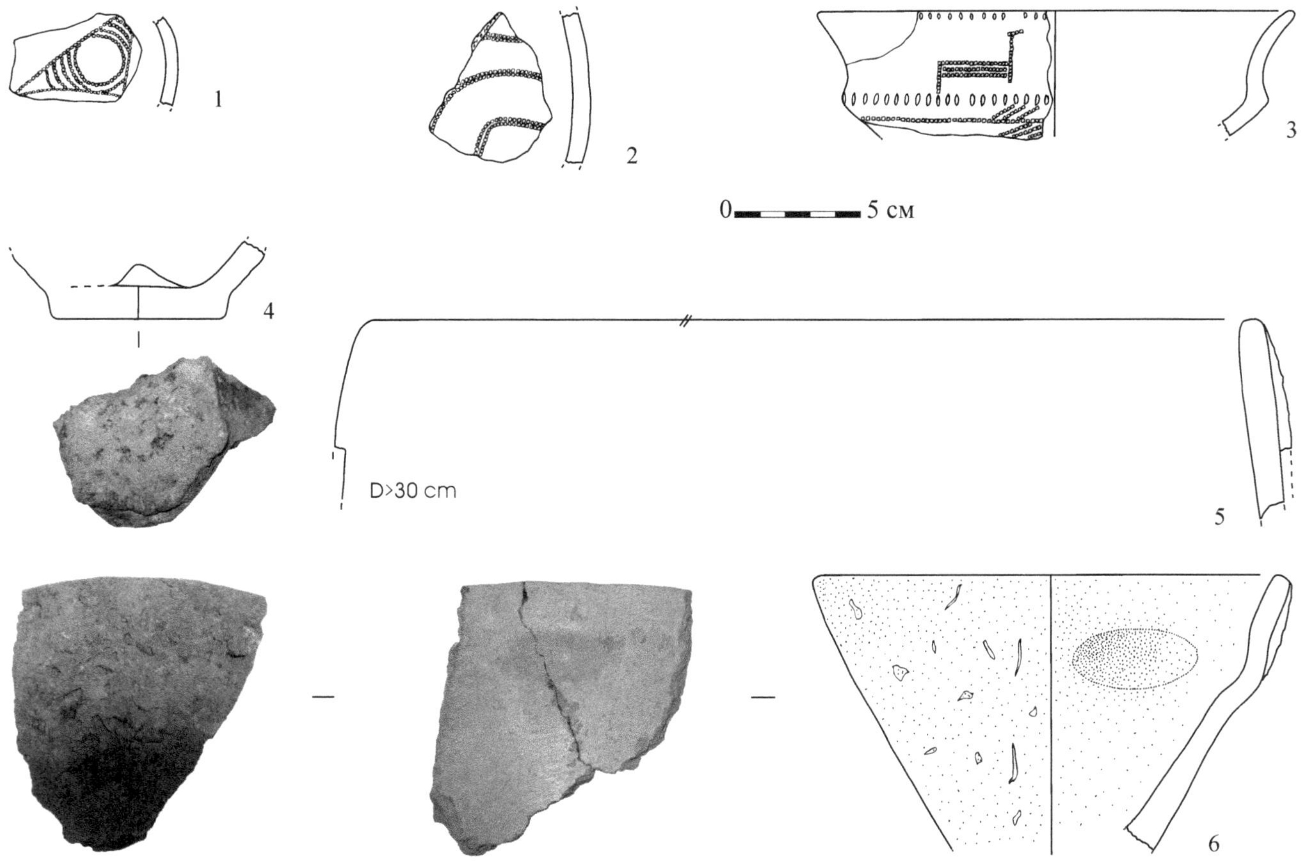

Figure 2. Provadia-Solnitsata prehistoric salt producing center. Southeast area. Building 2. Ceramic vessels. Middle Chalcolithic Hamangia IV period.

Building 1

The upper middle Chalcolithic horizon of Tell Provadia-Solnitsata presents a later phase in the development of the Hamangia IV culture, to which Building 1 refers. It has been recorded over a total area of about 180 m². It had two floors and at least two rooms, where the remains of a hearth and numerous preserved or fragmented ceramic pots have been discovered. Coarse ware prevails in this rich and varied ceramic assemblage. As opposed to Building 2, sherds with outer surface coating are rare here (Figure 3 *1*). The material from Building 1 has its best parallel in Tell Durankulak's horizon VII and the adjoining cemetery (Slavčev 2003, 145-175; Todorova 2002). Some channeling and impressed patterns – horizontal lines in the upper parts of the vessels, a triangular motif of impressed lines near the base (Figure 3 *2, 4*) – relate this assemblage to some ceramic sherds and vessels from Cucueti–Slatina veche and Tîrpeşti, which have been referred to Precucuteni culture, as well as from Turnu Măgurele, referred to the Hamangia IV/Varna culture (Munteanu *et al.*, 2007, Figure 7; Marinescu-Bîlcu 1983, Figure 10 2, 8; Voinea 2005, pl. 72 3). A good parallel of the flat lid (Figure 3 *3*) has been found in the Solacolu collection, dated to the Hamangia IV/Gumelniţa A1 culture (Voinea 2005, pl. 19 2).

Conclusion

Buildings 2 and 1 at Tell Provadia-Solnitsata probably represent two successive stages in the development of Hamangia IV culture. The presence of coated ceramic sherds in their assemblages testifies to the existence of a technology for manufacturing special purpose vessels. The outer surface clay coating, the uniform conic shapes, as well as the concavities on the inner surface, are all part of this technology. One hypothesis about the function of the coating is that it helped preserve longer the heat in the vessel. It is interesting to note that regardless of the larger excavated area of Building 1, an exceptionally small number of coated surface sherds have been found there as opposed to Building 2, where they are abundant. On the one hand, this might be related to the different degree of usage of this vessel type in the two chronological stages of the Hamangia culture. On the other hand, it raises the question of the function of the buildings, where they have been found.

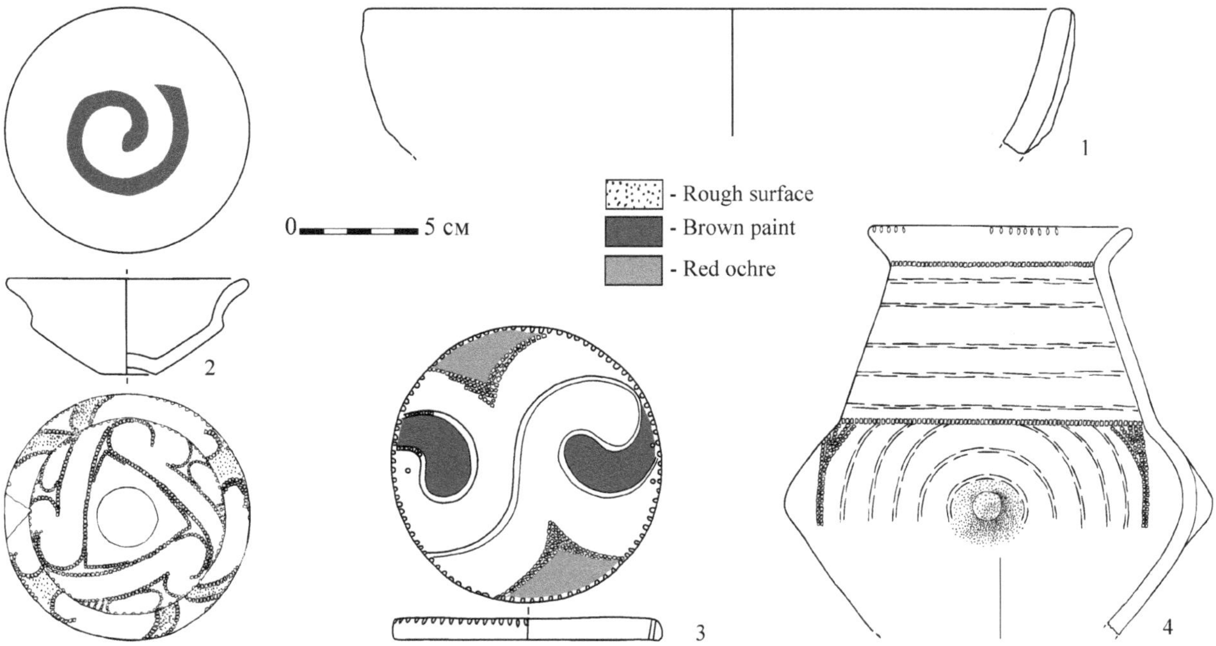

Figure 3. Provadia-Solnitsata prehistoric salt producing center. Southeast area. Building 1. Ceramic vessels. Middle Chalcolithic Hamangia IV period.

Obviously, the technology of salt extraction and production underwent changes through the middle Chalcolithic at Tell Provadia-Solnitsata. Most probably, the coated surface vessels from Buildings 1 and 2 were related to it. A proof thereof are the ceramic sherds with coated outer surface discovered at the Chalcolithic salt production center, which was recorded in the immediate vicinity northeast of Tell Provadia-Solnitsata in 2008. This vessel type lacks parallels in either earlier or simultaneous ceramic material. Moreover, they also differ from the published briquetage from Solca, which have been referred to the Cucuteni culture (Nicola *et al*., 2007, figs. 14 and 15). The forthcoming research at Tell Provadia-Solnitsata, as well as other sites that are related to the salt exploitation, provide an opportunity to further our knowledge on the prehistoric technology of salt extraction and production.

References:

Николов, В. 2008. Пласт 5: постройка 5. Късен неолит. In В. Николов (ред.), *Праисторически солодобивен център Провадия-Солницата. Разкопки 2005-2007 г.*, 87-115. София, НАИМ-БАН.

Marinescu-Bîlcu, S. 1983. Le début et les étapes de la culture Précucuteni ainsi que ses relations avec la culture de Tripolje. *Thracia praehistorica, Supplementum Pulpudeva* 3, 23-44.

Munteanu, R., Garvăn, D., Nicola, D., Preoteasa, C. and Dumitroaia, Gh. 2007. Cucuieţi-Slatina Veche (Romania). Prehistoric Exploitation of a Salt Resource. In D. Monah, Gh. Dumitroaia, O. Weller and J. Chapman (eds.), *L'exploitation du sel à travers le temps*. Bibliotheca Memoriae Antiquitatis XVIII, 57-70. Piatra Neamţ, Constantin Matasă.

Nicola, D., Munteanu, R., Garvăn, D., Preoteasa, C. and Dumitroaia, Gh. 2007. Solca – Slatina Mare (Roumanie). Preuves archéologiques de l'exploitation du sel en préhistoire. In D. Monah, Gh. Dumitroaia, O. Weller and J. Chapman (eds.), *L'exploitation du sel à travers le temps*. Bibliotheca Memoriae Antiquitatis XVIII, 35-56. Piatra Neamţ, Constantin Matasă.

Slavčev, V. 2003. Charakteristik der Keramik von Schicht VII aus dem Tell 'Die Große Insel' bei Durankulak, Nordostbulgarien. *Добруджа* 21, 145-175.

Todorova, H. (Hrsg.) 2002. *Durankulak, Band II. Die prähistorischen Gräberfelder von Durankulak*, Teil 1-2. Sofia, Publishing House Anubis Ltd.

Voinea, V. 2005. *Ceramica complexului cultural Gumelniţa – Karanovo VI. Fazele A1 şi A2*. Constanţa, Muzeul de istorie naţională şi arheologie.

Spatial Analysis of Prehistoric Salt Exploitation in Eastern Carpathians (Romania)

Olivier Weller
CNRS, Laboratoire de Protohistoire européenne, UMR 7041-ArScAn
Maison de l'Archéologie et de l'Ethnologie, Paris, France

Robin Brigand
Université de Franche-Comté, UMR 6249
Laboratoire de Chrono-Environnement, Besançon, France

Laure Nuninger
Université de Franche-Comté, UMR 6249
Laboratoire de Chrono-Environnement, Besançon, France

Gheorghe Dumitroaia
Complexul Muzeal Judeţean Neamţ, Romania

Abstract
This paper presents the first results of our project, an interdisciplinary Franco-Romanian collaboration focused on the dynamics of salt spring exploitation in the longue durée. In this analysis spatial and statistical measures, notably kernel densities, are used to investigate the relationship between salt spring exploitation and settlement dynamics from the Neolithic to Chalcolithic (6000-3500 BC). Generalising the results, natural and anthropogenic factors that could shape the observed patterns are proposed.

Keywords
settlement patterns, salt springs, Kernel densities, Neolithic-Chalcolithic, Cucuteni, Romania, Moldavia

In Moldavia, at the base of the Carpathian Mountains, salt springs are plentiful. One of the region's most important resources, they have been the subject of archaeological and ethnographic research since the 1980s. In particular, attention has been focused on the *Poiana Slatinei*-Lunca, where archaeological evidence shows exploitation from the Early Neolithic (early 6th millennium BC), intensifying through the 5th and 4th millennia BC (Dumitroaia 1987, 1994a; Weller and Dumitroaia 2005). Research on this site and its landscape has been led by the Museum of Piatra Neamţ (Alexianu *et al.,* 1992; Monah and Dumitroaia 2007). In 2003, after a first collaboration in 1995 in the framework of a PhD (O. W.), our French-Rumanian team began collaborating under the aegis of several interdisciplinary programmes supported by the CNRS, the French Foreign Affairs and the Romanian CNCSIS (Weller *et al.,* 2007a; Alexianu and Weller 2009; Alexianu *et al.* in this volume).

To investigate the impact of salt springs on settlement patterns, land use and landscape perception in the long term, an interdisciplinary approach, including archaeology, ethnology, paleoenvironment and geomatics, to the material was devised. One aspect of this approach is the exploration of socio-economic and environmental factors in the spatial organisation of the landscape, as analysed through GIS-based statistical analyses. The objective is to test the hypothesis that the pattern of salt springs was the driving factor, structuring the settlement pattern. The research questions, protocols and database design were first presented at the Antibes and Piatra Neamţ colloquium in 2004 (Weller and Nuninger 2005; Weller *et al.,* 2007a). After some difficulties in obtaining cartographic informations, and building a Geographic Information System (GIS), the first results have emerged, and provided an overview of population dynamics at micro-regional scale (Weller *et al.,* 2007). In this paper, we begin to apply spatial analysis tools to investigate the relationship between the two patterns, salt and settlement.

1. GIS and spatial archaeology in sub-Carpathian Moldavia

1.1. The use of GIS in archaeology: short historiographic follow-up

A GIS, is a system designed for the acquisition, storage, management, analysis and modelling of the spatial distribution of different phenomena: for example, environmental, economic, and social patterns. It is an organised set of instruments (software, computers), persons, and data registered in space and structured in order to extract and analyse the geographic information (Collet 1992, Didier 1990, Tourneux 2000). The interest in GIS applied to archaeology is not recent. The first archaeological examples, published in the United States starting in the 80s, used predictive models to locate areas likely to contain archaeological sites (Kvamme 1983a, 1983b, Kohler 1988, Allen *et al.*, 1990). Soon after, GIS appeared in European archaeology - first in Great Britain, then in the Netherlands, Denmark and France. In the European context, the pioneering work on the island of Hvar in Croatia (Gaffney and Stancic 1991) is particularly notable for its application of GIS to the interpretation of field walking data. Several recent handbooks present concepts and methods for the study of the spatial organisation of landscapes through GIS (Wheatley and Gillings 2002, Conolly and Lake 2006). The introduction of GIS to French archaeology was motivated by the heritage management sector and by the research sector, to study long-term landscape and settlement pattern changes and human-environment relationships. It is this context that we should understand the experience of the European programme *Archaeomedes,* which developed methodologies and applications specific to archaeology for the first time: multi-scale studies, integration of spatial and temporal dimensions for defining the studied objects, and socio-environmental relationships (van der Leeuw 1995, 1998, Durand-Dastès *et al.*, 1998, Favory *et al.,* 1999, van der Leeuw *et al.*, 2003). The work of the *Archaeomedes* project marks the beginning of French spatial archaeology, which was then structured within the ISA (Spatial Information in Archaeology) network (Berger *et al.*, 2005). In 2004, the *Archaedyn* project was begun, pursuing a related issue: the understanding of territorial dynamics (Nuninger and Favory 2006, Nuninger *et al.,* 2008). The spatial analysis component developed within the interdisciplinary project studying Romanian Moldavia addresses the same themes as the previous project, focusing on the role of salt springs.

1.2. Application to Sub-Carpathian Moldavia

In order to understand the Neolithic and Chalcolithic populations dynamics (6000-3500 cal BC) related to the use and exploitation of salt water, we take a multi-scalar approach strictly depending upon the quality of available archaeological information (Figure 1).

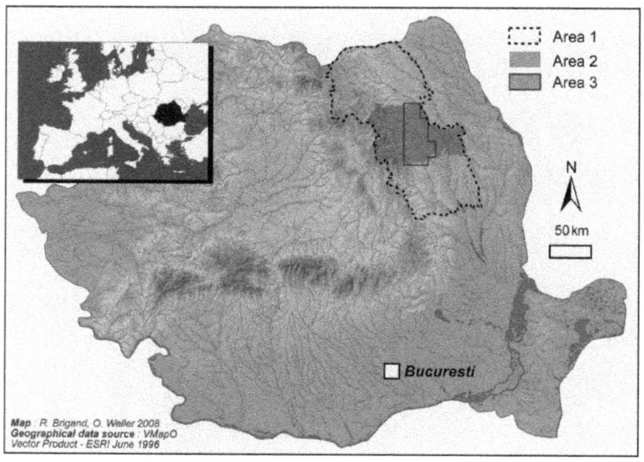

Figure 1. Study areas in Romania.

Area 1, approximately 20 000 km^2 and covering the Sub-Carpathian depression, is the object of mapping showing the distribution of salt springs and archaeological sites. GPS campaigns have started in the departments of Suceava and Bacau but a full inventory is only available for the Neamţ department (area 2).

Area 2, approximately 6 000 km^2, covers the current department of Neamţ. It is the only sector where the archaeological data could be checked in the field and organised within a database (245 settlements). Field survey and GPS data were collected for all mineral sources in the area (83 springs, out of which 63 sodium chlorides). This area is the subject of statistic analyses to look at variation between communes. This documentation was generously put at our disposal by M. Cosinchi's team, from the University of Lausanne.

Area 3, approximately 2 000 km^2, is the central area of the Neamţ department and corresponds to the Piedmonts of the Oriental Carpathians. In this sector, all archaeological sites were geo-referenced (202 confirmed settlements) and several had later assessments in the field (41 settlements). Based on the geographic location of known archaeological sites and mineral springs, occupation densities were evaluated. In this area, the study of the area surrounding the archaeological sites, the control and accessibility of salt resources, were accomplished using a high resolution (25m) Digital Elevation Model (DEM) processed by our colleagues from the Institute of Anthropological and Spatial Studies, ZRC-SAZU of Ljubljana (Slovenia) derived from ERS radar images.

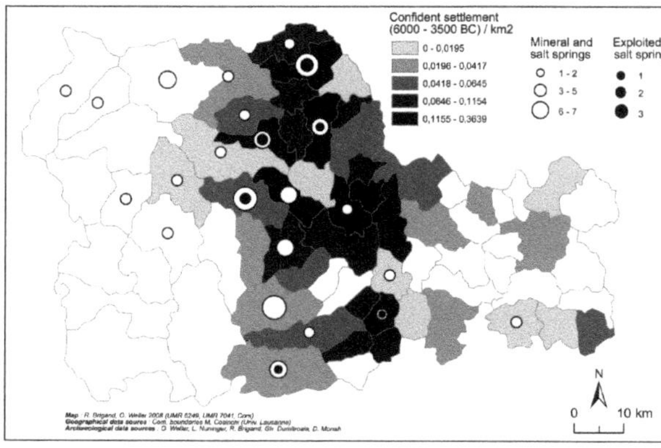

Figure 2. Settlement density and salt springs with or without remains of exploitation.

In order to build an efficient GIS, exhaustive databases, for both archaeological and salt springs sites are required. In each DB the following are recorded: the quality of the archaeological information (confirmed or uncertain settlement), the spatial precision (GPS coordinates, topographic location at 1:25 000, imprecise or undetermined location), the intensity of investigation (isolated discovery, field survey, exploratory trench or excavation), the nature of the settlement (undetermined occupation, salt exploitation site, simple settlement, hilltop or fortified hilltop settlement), its chronology and its extent. The 'DB salt' database takes the inventory of all mineral springs recorded within the territory of the current department of Neamţ which were all the object of systematic field survey (Weller *et al.*, 2007b). For each spring the following criteria were documented: access, rate of flow, pH, known uses, geographic coordinates and chemical composition of the water, spring type (non-salted and mineral, slightly salted to very very salty, undetermined), the mean of collection (pit, tree trunk wells, wooden square wells, stone wells, complex composite wells, concrete wells, cistern wells or undetermined), the depth of the well (inf. at 1m, between 1 and 3m, sup. at 3m or undetermined), the rate of flow (weak, medium, strong or undetermined) and the current use (none, animal husbandry, hunting, forage, food conservation, cheese industry and curing, crystallised salt exploitation, therapeutic use, milk curdling or undetermined).

Both database were linked using the geo-referenced data as shown in following both examples for the Neamţ department (area 2). Figure 2 emphasises the population trends during the entire studied period (6000-3500 BC). In general, we see a greater occupation density in proximity to the Carpathian Piedmonts, which are rich in salt springs, compared to two other geographic units (alluvial plains towards the East, mountain massif towards the West). Although there are unexploited salt springs where there is no population, such as at the North-East of the department, it is important to point out that all salt springs where we find archaeological evidence of exploitation are correlated with relatively high population density. We should also underline, as shown in Figure 3 that all salt springs with evidence of exploitation from the Neolithic and Chalcolithic times are situated in areas where there are salt springs which have high rates of flow and salinity. Both criteria explain modern day exploitation for crystallised salt (*husca*) production and visits to them justify long-distance travel from several surrounding communes (see Alexianu *et al.*, in this volume). The ethnological evidence has interesting implications for our interpretation of archaeological patterns.

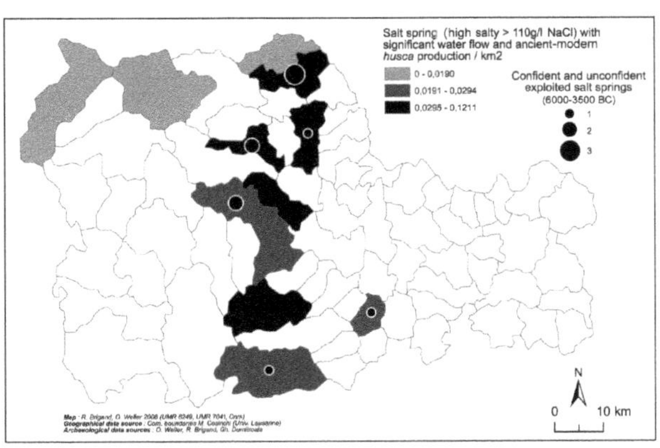

Figure 3. Modern and Prehistoric salt exploitation. Comparison of the densities of salt springs with high salinity and significant flow rate used for modern crystallised salt production (*husca*) and the distribution of salt springs exploited during recent Prehistory (6000-3500 BC).

These statistical maps, succinctly presented, were already discussed (Weller *et al.*, 2007a, 2008a). They allowed the validation of a spatial relation between salt springs and Prehistoric population dynamics. These maps also represent an important step of the analysis; they illustrate the state of the documentation registered within the inventories, which were realised according to an administrative logic. Nevertheless, these statistical maps are biased in two ways: first, modern boundaries between communes introduce 'breaklines' in the map that are unrelated to the archaeological data; second, the calculated site densities are affected by the sizes of the communes, which vary widely. To improve our understanding of the settlement pattern and use of salt springs on the regional scale, we selected a case study (area 3) where the locations of sites were verified, geo-referenced and recorded in the GIS (Figure 4).

2. Population density and diachronic evolution

For the initial case study we defined an area (area 3) with a high site density, allowing us to make reliable spatial analyses (Figure 5, datasets 1 and 2). The first analysis counted the number of sites per period within 5km of each salt spring. As a control, a random point set was created, and the number of points from the random set found within 5km of each salt spring was also counted. (Figure 5, dataset 3). The distance of 5km was used because according to site catchment theory (Higgs 1975) 5km is the maximum distance that can be travelled (there and back) in a single day, and therefore most resources used regularly by an agricultural community should be within this distance. The result of the analysis shows that 50 % of all sites are located within the 5km radius. However, a comparison with the random sample shows the same distribution: 50 % of the random sites are also located within 5km of a salt spring (Figure 5, dataset 4). Therefore this distribution does not illustrate an active preference for locations within 5km of salt springs. In the first two periods, and in Cucuteni AB and B, there appears to be a slightly higher probability for a site to be located within 5km of a salt spring, but the numbers involved in the analysis for the Criş and Linear Pottery Periods are not large enough to be statistically significant. With the exception of the last period (and possibly of the Early Neolithic) the presence of a salt spring within 5km does not seem to be a determining criterion for the choice of settlement location. Salt exploitation is well attested during the period from Criş to Cucuteni, and its importance to the development of the region is not in doubt. Therefore, it is unclear why the salt springs themselves do not appear to play a significant role in the choice of settlement locations. To understand why salt spring locations are not influential, we must look at each site within its local socio-economic network in relation to the locations of other nearby settlements. We propose that the choice of settlement location depends not only upon the raw material resource (the salt spring), but also upon the whole system contributing to its exploitation (from salt water supply to crystallised salt diffusion).

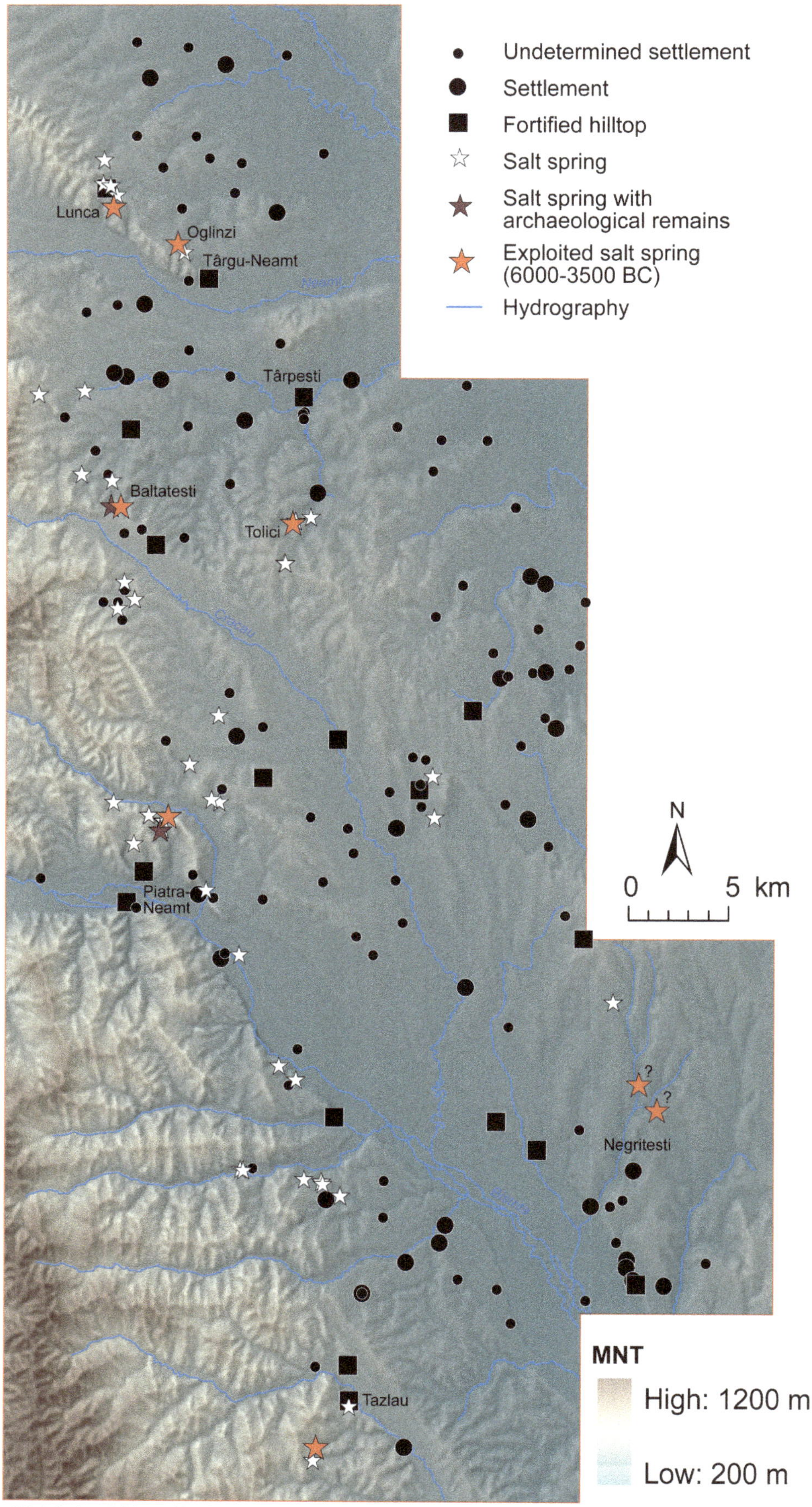

GIS : L. Nuninger, R. Brigand ; Map : R. Brigand, Chrono-Environnement UMR 6249 - LEA ModelTer ;
***Archaeological data sources** : O. Weller, L. Nuninger, R. Brigand, Gh. Dumitroaia, D. Monah ;*
***Geographical data sources** : Topo 1/25000, DEM ModelTER K. Ostir*

Figure 4. Fortified hilltop settlements, settlements and salt springs with or without remains of exploitation in the area 3 (6000-3500 BC).

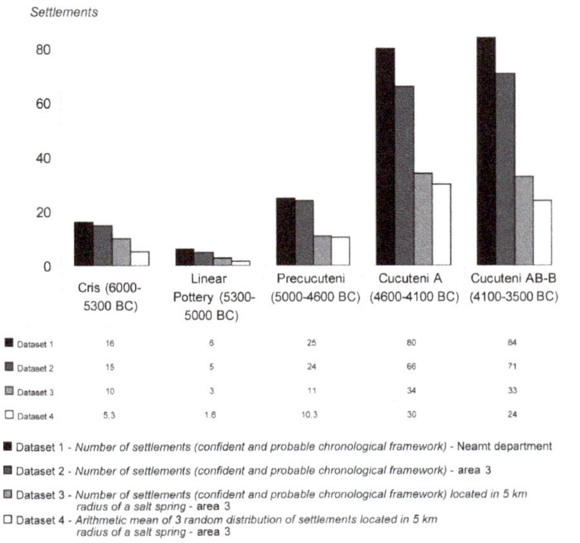

Figure 5. Statistical distribution of the population by cultural period (6000-3500 BC) in the department of Neamț and in the sub-Carpathian area (area 3) in relation to a salt spring (less than 5 km).

2.1. GIS density estimations

In order to understand the settlement pattern in the context of salt exploitation, we must go beyond looking at the settlement pattern as a point distribution; we must treat each spring and the sites around it as a coherent unit (a populated area). To this end, a series of analyses, beginning in 2007 (Weller *et al.*, 2008), were undertaken using *kernel density estimation* (KDE). The KDE method provides an estimation of the site density, defined by a moving window. The density value obtained takes into account the size of the neighbourhood. The weight assigned decreases proportionally with the distance from the centre of the window, following functions defined in the kernel density model. This method is well known (Silverman 1978 and 1986, Wand and Jones 1995, Zaninetti 2005) and has been used for archaeological applications, in particular for intra-site analysis (Baxter *et al.*, 1997, Beardah 1999). The method is more traditionally used by geographers (Grasland *et al.*, 2000). The density estimation obtained using KDE is dependent on two main parameters: 1) *k* the kernel function chosen, 2) *h* the radius[1] chosen.

As a first step, we used the kernel function implemented in the ArcGIS. The kernel function used by ArcGIS is based on a quadratic kernel function (Silverman 1986) and there is no way to change it (more open software should be used to provide 'controlled' results). Based on the assumption that the result of the analysis is not strongly influenced by the kernel function as long as the function is symmetrical (Silverman 1978), we saw no need for additional tests to be made using different *k* at present.

[1] Kernel functions originate in signal processing, where they are used in the frequency domain. In the literature (and software) the term bandwidth is used to specify the "window" of density estimation (bandwidth is the radius of frequencies or wavelengths) both in frequency and space. However, in this paper we will use the term radius consistently to highlight the spatial application of the method.

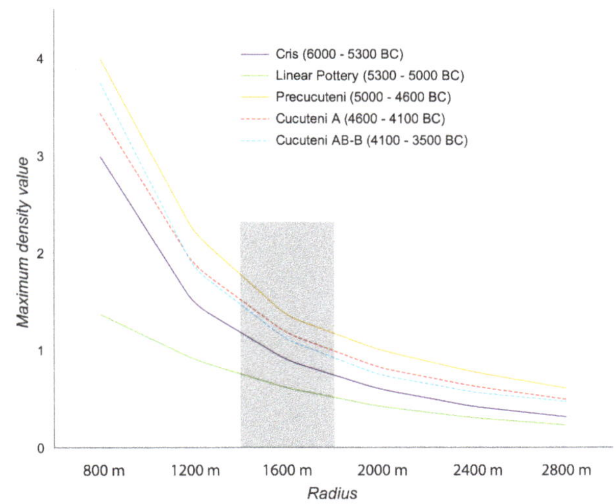

Figure 6. Determination of the optimal radius. It corresponds to the inflection point of the maximal density curves, meaning around 1600 m.

In archaeological analyses, the choice of the radius (*h*) is the most important parameter (Silverman 1978) because *h* determines the degree of smoothing. There are many ways to define the optimal value of *h*, which can be fixed for the entire study area or adjustable. Statistical indicators based on the data distribution can help choose *h* (Baxter *et al.*, 1997, Zaninetti 2005) but there is no explicit determination method and the user has to make an arbitrary decision. In general, using too small radius will produce an irregular surface that is problematic, especially when the total number of points (*i*) is relatively small. On the contrary, too large a radius will result in a loss of precision, favouring general trends. For this project, based on the Archaedyn experience (Nuninger *et al.*, 2008), a simple graphical approach was used to define the optimal value of *h*. Based on this method, several radii were tested, at 400 meter intervals, to define the range of optimal values (Figure 6). Very high values are found to give too much importance to isolated points, whereas the lowest values provide almost no information. The optimal radius for density estimation corresponds to the inflection point of the distribution, in this case at 1600m. This radius was consistently optimal across all periods.

2.2. Data classification and organisation into a hierarchy

In practical terms, the density calculations pose two serious problems: 1) when we calculate the density per period, sites with uncertain poorly calibrated dating are not taken into account; 2) all sites are treated equally in the model regardless of size or importance, the smallest salt exploitation site is the equal of the largest fortified settlement. Clearly the size and importance of each site should be taken into account when performing the density calculation, because the number of sites is not directly indicative of population size.

In order to overcome the first difficulty, sites with uncertain dates have been distributed over each period.

In the database, a site definitely belonging to a period has a value of 1, and a site definitely not belonging to a period has a value of 0. An example of the table showing chronological attributes is given below (Figure 7). The example is given for a random site with an uncertain date within the Chalcolithic.

Detailed chronology			Large chronology (1)			Large chronology (2)		
Criş	6000/5300 BC	0	Neolithic	6000/5000 BC	0	Neolithic	6000/5000BC	0
Linear Pottery	5300/5000 BC	0						
Precucuteni	5000/4600 BC	0	Precucuteni	5000/4600 BC	0	Chalcolithic	5000/3500 BC	1
Cucuteni A	4600/4100 BC	0	Cucuteni	4600/3500 BC	0			
Cucuteni AB and B	4100/3500 BC	0						

Figure 7. The database structure of the 'chronology' attribute of a random site dating to the Chalcolithic period.

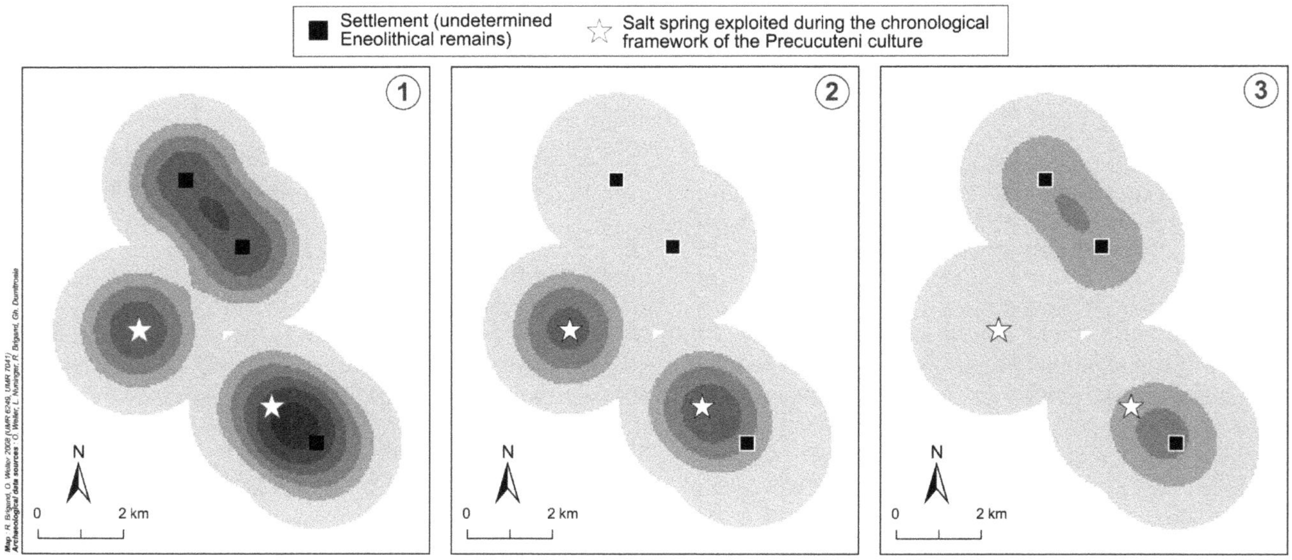

Figure 8. Estimate of the Precucuteni settlements density in the area of salt springs at Oglinzi-Răuceşti. 1: without weighting, 2: chronological weighting, 3: weighting concerning the nature of the settlement.

In this example, since the Chalcolithic covers the 'Precucuteni', 'Cucuteni A' and 'Cucuteni AB and B' periods, we have decided to give a weighting to the site in accordance with the duration of each cultural period, which gives the following breakdown:

Estimated detailed chronology		
Neolithic		
Criş	6000/5300 BC	0
Linear Pottery	5300/5000 BC	0
Chalcolithic		
Precucuteni	5000/4600 BC	0.3
Cucuteni A	4600/4100 BC	0.35
Cucuteni AB and B	4100/3500 BC	0.35

We used the same weighting procedure for the other periods, for which the dates are uncertain: 'Neolithic' thus becomes 0.7 for 'Criş' and 0.3 for 'Linear Pottery'; 'Cucuteni' becomes 0.45 for 'Cucuteni A' and 0.55 for 'Cucuteni AB and B'. The undeniable advantage of this method is that it allows us to consider the sites that were previously excluded from the analyses by focusing upon precise chronological periods. The attribution of a lesser weight for these sites makes the density vary (Figure 8, 1 and 2), giving a more realistic image, whilst avoiding insofar as possible, the spatial gaps due to dating issues.

Weighting in relation to the nature of the site	
Possible occupation	0.5
Confirmed occupation	1
Confirmed settlement	2
Fortified or non-fortified hilltop settlement	3
Salt exploitation site	0.5

In order to refine the result, we have included a second weighting to attenuate the second bias. The weighting given of each site is no longer automatically equivalent to 1, but has been redefined, insofar as possible, by taking into account the nature of the archaeological sites (Figure 8, 3). Since the project aims at underlining the importance of the salt resource as regards the manner in which the population was structured within the territorial organisation, it seemed appropriate to emphasise the distribution of known settlements, whilst moderating the importance of salt exploitation sites; the objective being not to give too much importance to the areas of production, which were probably less important in terms of population and power. The small settlements – for the most part identified by surface surveys – and the salt spring exploitation structures have been attributed a reduced value of 0.5, whilst the settlements and fortified sites have a standard value of 1 progressing to 2 or 3 (see below).

We have chosen to present here the site density maps with the weightings according to their nature, insofar as they are visually the more telling (Figure 9, top). The combination of the two weighting factors would be fairer, but we need to undertake more robust tests, which are currently under way.

In order to apprehend the general framework of the population dynamics between 6000 and 3500 BC, the site density maps have been associated with 'transition' maps obtained by subtraction of the site density maps weighted for the nature of the site (Figure 9, bottom). Thus, for the period 6000-5000 BC, we have eliminated the sites relating to the Linear Pottery period (5300-5000 BC) from the site density map of the Criş period (6000-5300 BC). The negative values correspond to the Criş archaeological sites that were abandoned during the following period. Conversely, the positive values correspond to new sites, which did not exist previously. Finally, the median values emphasise the overall stability of the population in a given locality, if it is locally evidenced by the abandonment of a site and the creation of a new one nearby. We should also mention that stability may express itself by the absence of occupation from one period to another.

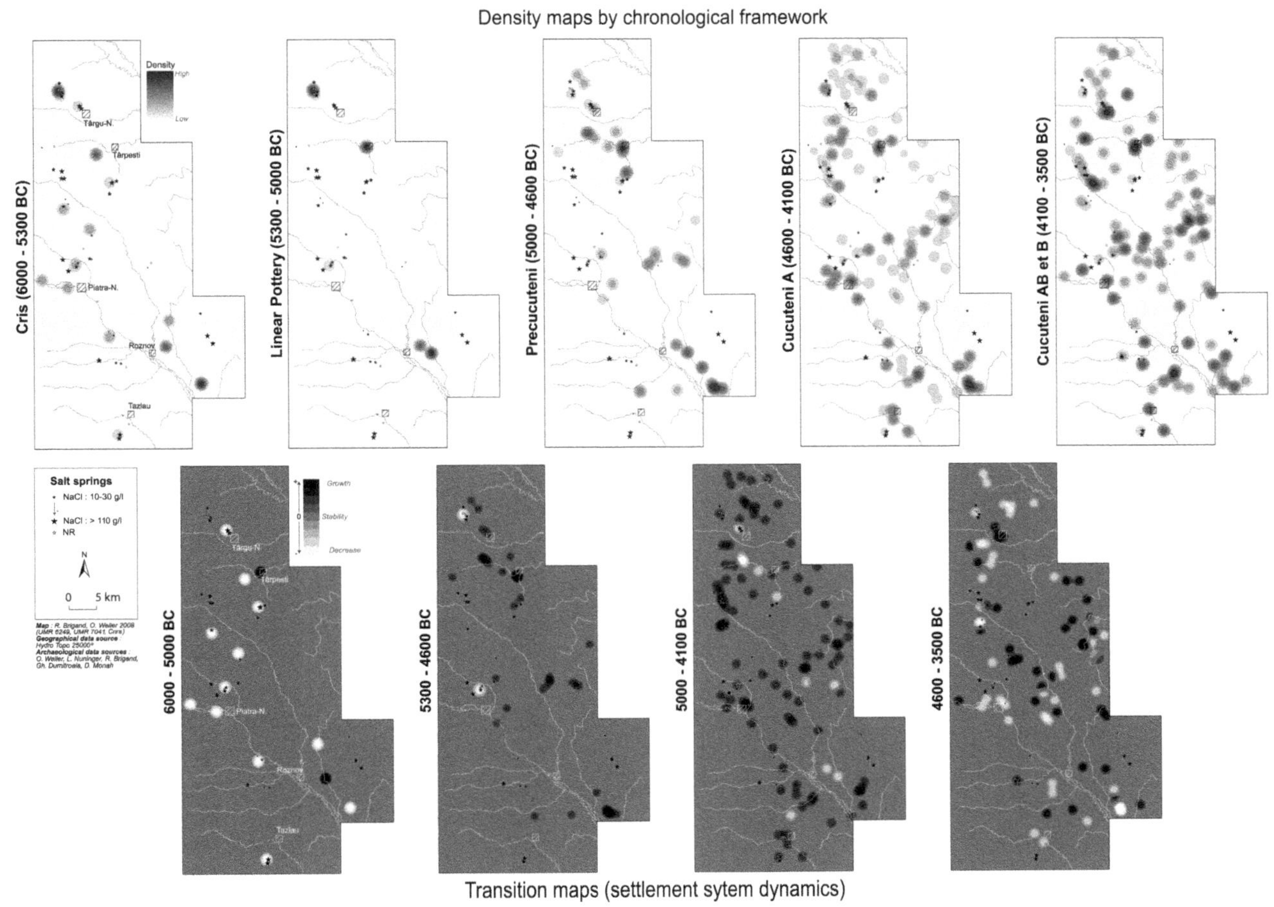

Figure 9. Settlement densities (*at the top*) and settlement dynamics (*below*) between 6000-3500 BC in relation to salt springs salinity.

The instability index, whether it be negative (abandonment) or positive (creation/development), is finally confronted to the salt springs distribution, which is itself static. The springs are classified according to their degree of salinity, which allows their appeal to be evaluated according to the periods analysed (Figure 9).

2.3. Dynamics and evolution of the Neolithic and Chalcolithic population

Overall, the site density maps by period (Figure 9, density maps) show an increase in the dynamics of occupation between the Early Neolithic and Cucuteni. There is a marked increase in the number of settlements during the Cucuteni A period (4600-4100 BC) compared to previous periods. Nevertheless, when mapped to the location of salt springs, it would appear that in general, whatever the period, neither the presence nor the high salinity of a spring constitutes a determinant factor for the installation of a settlement. In fact, if the salt springs appear to have been more heavily exploited during the Cucuteni AB and B period, a comparable, if not a larger number of sites were settled at more than 5km from a salt spring and oriented towards the valley terraces.

The transition maps (Figure 9, transition maps) suggest new interpretations in terms of population stability and instability:

– Between Criş and Linear Pottery periods (6000-5000 BC), there is a marked general decrease in the number of sites, including close to salt springs, which is in large part due to the scarcity of sites attributable to Linear Pottery culture. This situation seems to be primarily due to research conditions, since the sites are more difficult to identify through field surveys.

– Between the Linear Pottery and Precucuteni periods (5300-4600 BC), we witness the establishment of sites which are generally at some distance from salt springs, other than the ones associated with those exploited of Lunca-Oglinzi and Ţolici.

– Between the Precucuteni and Cucuteni A periods (5000-4100 BC), whilst some sites are abandoned in secondary valley sectors, many new sites are created, covering almost the entire area, be it near salt springs (Bălţăteşti, Piatra Neamţ, Tazlău) or along important river systems (such as the Bistriţa and Cracău) and their direct tributaries.

– Between the Cucuteni A and Cucuteni AB-B periods (4600-3500 BC), numerous sites are abandoned, particularly in a number micro-regions sometimes rich is salt springs (Piatra Neamţ, Bălţăteşti, Lunca), but there are also many new settlements, which are often in adjacent sectors, as if we were witnessing the displacement and resettlement of the population. We should also mention that these population dynamics during the Cucuteni AB-B period concern essentially the occupations and half of the confirmed settlements or hilltop settlements, but to a lesser extent the fortified hilltop settlements. In fact, 7 of the hilltop settlements continue in occupation, whilst 5 from the Cucuteni A period are abandoned and 5 new Cucuteni B sites are established. As regards the fortified hilltop settlements, 5 remain in occupation, whilst 6 sites are deserted and only 1 new one is created during the Cucuteni B period (*Pometea* – Târgu Neamţ), probably in relation with the exploitation of the salt spring of Oglinzi (Dumitroaia 2004). At the same time, the majority of salt springs exploitations from the Cucuteni A period continues into the Cucuteni B.

When we consider the population dynamics around salt springs, we can see that several of them, whether with a higher or lower degree of salinity, are occupied since the Early Neolithic (Criş), although they do not enjoy the same destiny. Only two areas have proven to be almost continuously occupied, close to salt springs with a high degree of salinity, which have yielded clear evidence of exploitation, the region of Lunca in the north and, to a lesser extent, the area to the north of Piatra Neamţ:

– Human occupation developed progressively around the group of springs at Lunca-Oglinzi, whether by the creation of new sites or by site expansion. If, generally speaking, there is demographic growth in the area, we need to point out an alternating rhythm of occupation between the group of salt springs at Lunca, and that at Oglinzi (Figure 9, transition maps). When one of these groups of salt springs is less densely occupied, the other compensates with a higher level of occupation, as though there existed a sort pendulum marking the rhythm of salt exploitation between the two areas.

– The area to the North of Piatra Neamţ shows a similar development, albeit with a lower overall density of occupation; there is a marked decrease in the level of occupation from the Linear Pottery to the Precucuteni periods, before witnessing a increased resurgence during the Cucuteni A.

The other areas occupied during the Criş period are abandoned during the Precucuteni but sometimes recolonised during the Cucuteni A, when almost all the areas around salt springs are densely occupied. Under these conditions, prior zones of occupation are no longer a determining factor in the choice of settlement site, which lends credence to the idea of new populations exercising specific choices, as was the case for the communities of the Linear Pottery and the Precucuteni groups.

Finally, in the whole study-area, certain springs appear to be attractive despite their low level of salinity, having a relatively high population density, attested at least during Cucuteni period. We nevertheless need to point out that these springs are invariably located less than 15km from an important salt spring with very high salinity with either gouged-out tree trunk wells, or those, like tanks, made with assembled squared timbers, to capture the brine.

Some areas located more than 5km from salt springs are also occupied beginning in the Criş cultural period, with a growing density of occupation throughout the period (excepting during the Linear pottery). Most of these areas correspond to the mouths of side valleys, and a few others to the confluence of river systems. From a chronological standpoint, these settlements become generalised during the Cucuteni A period, but we need to particularly mention the existence of a specific area, heavily occupied from the Linear Pottery period. It is the important settlement of *Râpa lui Bodai* – Târpeşti (Marinescu-Bîlcu 1981), south-east of present town of Târgu Neamţ, which continued in occupation until the Cucuteni B period. In this particular case, the 1500 years long, and early occupation of this village at the confluence of two rivers, is certainly to be contextualised in relation to the contemporary and continuous exploitation of the salt spring of *Hălăbutoaia* at Ţolici, 6km to the south (Weller *et al.*, 2007b, Dumitroaia *et al.*, 2008). During Precucuteni period, a hilltop settlement named *Şipot-Mohorâtu,* situated beneath the Ţolici cemetery, was installed upstream from the confluence between the small valley in which the salt spring is to be found, and which is almost visible 2km upstream, and the Ţolici Valley, which exits 4km downstream at *Râpa lui Bodai.* This hilltop settlement appears to have functioned as a relay station between the village of *Râpa lui Bodai* and the exploited salt spring.

Another interesting case is worth mentioning. It concerns a group of salt springs south-east of the area studied, which, despite having a high degree of salinity, remained isolated, with sites situated at a distance of 3 to 8km, regardless of the period. The question arises as to why these salt springs remained so isolated in relation to the settlements. This is of particular interest since area 3 has been well surveyed (and that the salt springs are more than 5km from its limits), so that it is unlikely that the apparent vacuum is due to a lack of research.

From a geographical standpoint, there are no special circumstances that would justify the situation. Access is no more difficult than in the other areas, and the type of soil, with cambic horizons, is similar to that of the other areas of occupation (Weller dir. 2007). Besides, the ethnological survey in the area shows that today, one of the salt springs, *Slatina Mare* – Negriteşti, provides a supply on a supracommunal scale, being one of the largest of the area, together with *Poiana Slatinei* at Lunca (Alexianu *et al.*, 2008; see Alexianu *et al.* in this volume). According to their current characteristics, these two springs have a similar profile, with a very high degree of salinity, even if not as high as that of *Slatina Mare* (150g/l NaCl compared to 165g/l as maximum measurement for *Poiana Slatinei*). The use of natural brine appears to be less diversified for this spring, as it only used for food conservation and animal feed.

From an archaeological standpoint, only the salt spring of *Poiana Slatinei* is documented as being exploited from the beginning of the 6th millennium (Weller and Dumitroaia 2005). The salt spring of *Slatina Mare* had only yielded Dacian evidence of exploitation and material from the 4th-3rd centuries BC (Dumitroaia 1992). However, as recently as July 2009, one of us (O.W.) re-examined the archaeological material for the Cucuteni hilltop settlement of *Movila Flocoasa*-Negriteşti (Dumitroaia 1994b), located 3km south of the salt spring at *Slatina Mare*, and discovered a fragment of briquetage, or salt mould from the Cucuteni period, fragments of Cucuteni C ceramic style, and even a pottery sherd that can certainly be attributed to the steppe area, as we had already identified at the *Hălăbutoaia* – Ţolici salt spring (Dumitroaia *et al.*, 2008, Figure 10). It is thus very probable that the salt spring of *Slatina Mare*, or the one nearby of *Slatina mica* – Hoiseşti, was exploited during the Cucuteni period, even if archaeological proof has yet to be discovered in their vicinity. How can we then explain their relative isolation during this period ? The previous example of *Hălăbutoaia* – Ţolici leads us to propose, for Cucuteni, not an attraction to establish a settlement near the exploited springs, but rather to control the access to the springs with a settlement downstream, at the mouth of the valleys and often near confluences. Thus, in the case of the Negriteşti-Hoiseşti salt springs, there are no less than 4 Cucuteni hilltop settlements which control, downstream, the small valley of Verdele before its confluence with the River Bistriţa (see Figure 4).

This settlement model, not only in relation to the important circulation corridors, but also to the salt springs exploited upstream in secondary valleys, has already been noted in area 3 by mapping only Cucuteni hilltop settlements, whether fortified or not, with the salt springs on our high resolution DEM (Weller *et al.*, 2007). In fact, at this scale, we could determine that hilltop settlements are in many instances tied to the control of the valleys giving access to salt springs. There is no direct visual control, but by and large, the ensemble of hilltop settlements are distributed along the Pre-Carpathian area, particularly at the exit of valleys giving access to salt resources or, more generally, along the main river corridors. Thus, the Cucuteni fortified settlements are generally established on the edges of river terraces, for an optimal control of the communication corridors, as well as access to salt resources. All this suggests that salt spring exploitation, and probably also the circulation of salt, were part of a dense territorial network, in which visual control seems to be essential, and where certain Cucuteni sites play a central role (see the first visual control analyses in Weller *et al.*, 2007). Here (Figure 9), with a dynamic analysis of the global population, we see the importance of the confluences in the choice of settlement sites starting with the Cucuteni A period, choices sometimes revised during the Cucuteni AB-B period. Prior to that, these locations did not appear to be of particular interest, other than the special case of the village of *Râpa lui Bodai* – Târpeşti which, since the Linear Pottery period, is durably installed, most probably in relation to the salt exploitation upstream, at *Hălăbutoaia* – Ţolici.

Finally, except for the continuous occupation of some rare areas (Lunca-Oglinzi, Piatra-Neamţ and Târpeşti) and the probable lack of occupation in the area of Negritesti, the territorial occupation seems to be rather dynamic. Thus, as shown in the transition maps (Figure 9), particularly as concerns the last three periods (from the Precucuteni to the Cucuteni B), the settlements development phases alternate with periods of abandonment. We need to point out, that the fragmented appearance of these maps, where site-development and abandonment form a mosaic, suggests the hypothesis of a progressive displacement of the settlements. It is important to note the development of stable areas of occupation during the Cucuteni A period, more rarely during the Precucuteni. We should however push the analysis, to better characterise those areas which today cannot be distinguished from a stability expressed by the absence of occupation from one period to another.

Conclusion

These pioneering approaches to spatial analysis help to characterise the population dynamics during Neolithic and Chalcolithic times in relation to the distribution and exploitation of salt springs of greater or lesser degrees of salinity. The cartographic analysis of the results, far from emphasizing a simple and monolithic evolution, underlines the complexity of the Prehistoric population system. Within the same studied area, salt

springs do not all have the same powers of attraction for the establishment of settlements, and their direct exploitation does not constitute the sole reason for the development of the settlements. Furthermore, seemingly contradictory situations can appear in the use of groups of salt springs, which are the most productive to this day (*Poiana Slatinei* – Lunca, *Hălăbutoaia* – Ţolici, *Slatina Mare* – Negriteşti). The results raise more questions than they answer, but they do show that we cannot understand the issues relating to salt exploitation on the unique basis of the relationship between the settlement and salt spring.
In fact, as shown by the ethnological survey (Alexianu *et al*., 2007), the analysis should be included within a more complex population system, based upon settlement networks, taking into account not only the issues of supply from the salt spring, but also those relating to the redistribution of this resource, which indubitably plays a important role in the development of certain sites, such as *Râpa lui Bodai* – Târpeşti. We should also enlarge area 3, currently very much limited to the region of salt springs, and in particular towards the east and the plains of the Siret River. The future analyses should also take into consideration the departments adjacent to that of Neamţ, particularly Bacau and Suceava, for which the inventories are almost completed, in order to determine the rhythms of occupation, the complementarities between the sites, and the settlement strategies for the entire sub-Carpathian Moldavian area.

It will also prove useful to analyse the high instability observed between the Cucuteni A and Cucuteni AB-B periods, by testing the role of river corridors, that of the soil quality, as well as the importance of the territorial organisation in terms of visual control, in respect to which certain Cucuteni fortified hilltop settlements seem to play a central role, particularly during the Cucuteni A period. The issues of territorial control and the hierarchical importance of settlements are key, if we want to understand the settlement system dynamics in terms of hierarchic networks and the control of human circulation as well as that of products such as salt.

Acknowledgements
Since 2003, this project is supported by the CNRS (Humanities department), the French Foreign office (DGMDP-DPMA) and Franche-Comté University, and is developed in association with the History and Archaeology Museum in Piatra Neamţ (International Centre of Research on Cucuteni Culture), the Al. I. Cuza University and the Archaeological Institute in Iasi (Romania). We also want to thank M. Cosinschi from the University of Lausanne for giving us the geo-referenced communal corpus for the departments of Moldavia realized in 1998 with the University of Iasi, the ArchaeDyn program (dir. F. Favory and L. Nuninger, UMR 6249 - MSH Ledoux) for the methodological contributions (KDE analysis) and J. Mudry (UMR 6249 - Franche-Comté University) for the chemical analysis of all the salt springs.

References:

Alexianu, M., Dumitroaia, Gh. and Monah, D. 1992. Exploatarea surselor de apă sărată din Moldova: o abordare etnoarheologică. *Thraco-Dacica* 13/1-2, 159-167.

Alexianu, M., Weller, O. and Brigand, R. 2007. *Izvoarele de apă sărată din Moldova subcarpatică. Cercetări etnoarheologice*. Iaşi, Casa Editorială Demiurg.

Alexianu, M., Weller, O. and Brigand, R. 2008. Usages et enjeux autour des sources salées de Moldavie précarpatique, Roumanie In O. Weller, A. Dufraisse and P. Pétrequin (eds.) *Sel, eau et forêt: hier et aujourd'hui*, 49-72. Cahiers de la MSH Ledoux 12 (coll. Homme et environnement, 1). Besançon, Presses Universitaires de Franche-Comté.

Alexianu, M. and Weller, O. 2009. The *Ethnosal* project. Ethnoarchaeological investigation at the Moldavian salt springs. *Antiquity* 83 (321). http://antiquity.ac.uk/projgall/weller321/index.html

Allen, K. M. S., Green, S. and Zubrow, E. B. W. (eds.), 1990. *Interpreting space: GIS and archaeology, Applications of Geographic Information Systems*. London, Taylor and Francis.

Barge, O., Rodier, X., Davtian, G. and Saligny, L. 2004. L'utilisation des Systèmes d'Information Géographique appliquée à l'archéologie française. *Revue d'archéométrie* 28, 15-24.

Baxter, M. J., Beardah, C. C. and Wright, R. V. S. 1997. Some archaeological applications of kernel density estimates. *Journal of archaeological science* 24, 347-354.

Beardah, C. 1999. Uses of multivariate kernel density estimates. In L. Dingwall, S. Exon, C.F. Gaffney, S. Laflin and M. van Leusen (eds.), *Archaeology in the Age of the Internet: Computer Applications and Quantitative Methods in Archaeology 1997*, Oxford, BAR Publishing, British Archaelogical Reports, International Series, 750.

Berger, J.-F., Bertoncello, F., Braemer, F., Davtian, G. and Gazenbeek, M. (eds.) 2005. *Temps et espaces de l'homme en société, analyses et modèles spatiaux en archéologie*. XXVe rencontres internationales d'archéologie et d'histoire d'Antibes, 21-23 octobre 2004, 534 p. Antibes, APDCA.

Collet, C. 1992. *Système d'Information Géographique en mode image. Gérer l'environnement*. Lausanne, Presses Polytechniques et Universitaires Romandes.

Conolly, J. and Lake, M. 2006. *Geographical Information Systems in Archaeology*. Cambridge Manuals in Archaeology, Cambridge, Cambridge University Press.

Didier, M. 1990. *Utilité et valeur de l'information géographique*. Paris, Economica.

Dumitroaia, Gh. 1987. La station archéologique de Lunca-Poiana Slatinei. In M. Petrescu-Dîmbovita, N. Ursulescu, D. Monah and V. Chirica (eds.), *La civilisation de Cucuteni en contexte européen*, 253-258. Iasi, Bibliotheca Archaeologica Iassiensis, I.

Dumitroaia, Gh. 1992. *Săpături şi cercetări arheologice de suprafaţă în jud. Neamţ (1987-1991). Memoria Antiquitatis* 18, 287-297.

Dumitroaia, Gh. 1994a. Depunerile neo-eneolitice de la Lunca şi Oglinzi, Judeţul Neamţ. *Memoria Antiquitatis* 19, 7-82.

Dumitroaia, Gh. 1994b. Săpături şi cercetări arheologice de suprafaţă în jud. Neamţ (1992-1993). *Memoria Antiquitatis* 19, 515-517.

Dumitroaia, Gh. 2004. Săpături arheologice sistematice şi periegheze din judeţele Neamţ, Bacău şi Suceava (2002-2003). *Memoria Antiquitatis* 23, 431-449.

Dumitroaia, Gh., Munteanu, R., Weller, O., Garvan, D., Diaconu, V. and Brigand, R., 2008. Un nou punct de exploatare a apei sărate în preistorie: Ţolici-*Hălăbutoaia*, jud. Neamţ. In D. Monah, Gh. Dumitroaia and D. Garvăn (eds.), *Sarea. De la prezent la trecut*, 203-224. Bibliotheca Memoriae Antiquitatis, XX. Piatra Neamţ, Constantin Matasă.

Durand-Dastès, F., Favory, F., Fiches, J.-L., Mathian, H., Pumain, D., Raynaud, C., Sanders L. and van der Leeuw, S. 1998. *The ARCHAEOMEDES Project. Understanding the natural and anthropogenic causes of land degradation and desertification in the Mediterranean basin. Research results.* Luxembourg, European Communities.

Favory, F., Girardot, J.-J., Nuninger, L. and Tourneux, F.-P. 1999. Archaeomedes II. Une étude de la dynamique de l'habitat rural en France méridionale, dans la longue durée (800 av. J.-C. – 1600 ap. J.-C.). *AGER* 9, http://ager2.site.voila.fr/

Gaffney, V. and Stancic, Z. 1991. *GIS approaches to regional analysis: a case study of the island of Hvar.* Ljubljana, Filozofska fakulteta.

Grasland, C., Mathian, H. and Vincent, J.M. 2000. Multiscalar Analysis and map generalisation of discrete social phenomena: Statistical problems and political consequences. *Statistical Journal of the United Nations ECE*, 17, IOS Press.

Higgs, E.S. 1975. *Palaeoeconomy.* Cambridge University Press.

Kohler, T.A. 1988. Predictive locational modelling: history and current practice. In W. J. Judge and L. Sebastian (eds.), *Quantifying the present, predicting the past*, 19-59. Denver, US DIBLM.

Kvamme, K.L. 1983a. Computer processing techniques for regional modelling of archaeological site locations. *Advances in Computer Archaeology* 1, 26-52.

Kvamme, K.L. 1983b. *A manual for predictive site location models: examples from the Grand Junction District, Colorado.* Colorado, Bureau of Land Management, Grand Junction District.

Marinescu-Bîlcu, S. 1981. *Tirpeşti. From Prehistory to History in Eastern Romania.* Oxford, British Archaelogical Reports, International Series, 107.

Monah, D. and Dumitroaia, Gh. 2007. Recherches sur l'exploitation préhistorique du sel en Roumanie. In D. Monah, Gh. Dumitroaia, O. Weller and J. Chapman (eds.), *L'exploitation du sel à travers le temps.* Bibliotheca Memoriae Antiquitatis XVIII, 13-34. Piatra Neamţ, Constantin Matasă.

Nuninger, L. and Favory, F. 2006. Archaedyn. Dynamique spatiale du population et ressources naturelles: vers une analyse intégrée dans le long terme, de la Préhistoire au Moyen Age. *AGER* 16, 7-11. http://ager2.site.voila.fr/

Nuninger, L., Ostir, K., Saligny, L., Poirier, N., Fovet, E., Gandini, C., Gauthier, E. Kokalj, Z., Tolle, F. and Archaedyn's team 2008. Models and tools for territorial dynamic studies (ArchaeDyn Project). In F. Favory and L. Nuninger (eds.), *7 millennia of territorial dynamics settlement pattern, production and trades from Neolithic to Middle Ages. ARCHAEDYN ACI "Spaces and territories" 2005-2007 Final conference.* Dijon, 23-25 June 2008, 25-44. Dijon, Pre-proceedings.

Silverman, B. W. 1978. Choosing the window width when estimating a density. *Biometrika* 65/1, 1-11.

Silverman, B. W. 1986. *Density estimation for statistics and data analysis.* London, Chapman and Hall.

Tourneux, F.-P. 2000. *Modes de représentation des paysages.* Unpublished PhD thesis, Besançon, Université de Franche-Comté, Géographie.

van der Leeuw, S. (ed.) 1995. *Volume 3: dégradation et impact humain dans la moyenne et basse vallée du Rhône dans l'Antiquité, partie II:la mobilité de l'habitat rural dans la moyenne et basse vallée du Rhône dans l'Antiquité. The Archaeomedes project:understanding the natural and anthropogenic causes of soil degradation in the mediterranean basin.* Bruxelles, European Communities, unpublished.

van der Leeuw, S. (ed.) 1998. *The Archaeomedes project, understanding the natural and anthropogenic causes of land degradation and desertification in the Mediterranean basin*. Bruxelles, Science Research Development, European Communities.

van der Leeuw, S., Favory, F. and Fiches, J.-L. (eds.) 2003. *Archéologie et systèmes socio-environmentaux: études multiscalaires sur la vallée du Rhône dans le programme ARCHAEOMEDES*. Paris, CNRS (Monographies du CRA, 27).

Wand, M.P. and Jones, M.C. 1995. *Kernel Smoothing, Monographs on Statistics and Applied Probability*. London, Chapman and Hall.

Weller, O. and Dumitroaia, Gh. 2005. The earliest salt production in the world: an early Neolithic exploitation in *Poiana Slatinei*-Lunca, Romania. *Antiquity* 79 (306). www.antiquity.ac.uk/ProjGall/weller/index.html

Weller, O. and Nuninger, L. 2005. Les eaux salées de Moldavie roumaine: étude interdisciplinaire autour d'une ressource structurante du territoire. In J.-F. Berger, F. Bertoncello, F. Braemer, G. Davtian, M. Gazenbeek (eds.), *Temps et espaces de l'homme en société. Analyses et modèles spatiaux en archéologie*, 511-516. Antibes, APDCA.

Weller, O. (dir.) 2007. *Exploitations pré- et protohistoriques des sources salées de Lunca et ses environs (dép. Neamţ, Roumanie). 4e année: analyses et synthèse sur Lunca et le Neamţ*. Rapport de mission archéologique CNRS, multigraphié. Paris, Ministère des Affaires Etrangères et Européennes.

Weller, O., Dumitroaia, Gh., Monah, D. and Nuninger, L. 2007a. L'exploitation des sources salées en Moldavie: un exemple de ressource structurante du territoire depuis le Néolithique. In D. Monah, Gh. Dumitroaia, O. Weller and J. Chapman (eds.), *L'exploitation du sel à travers le temps*. Bibliotheca Memoriae Antiquitatis XVIII, 99-113. Piatra Neamţ, Constantin Matasă.

Weller, O., Brigand, R. and Alexianu, M. 2007b. Cercetări sistematice asupra izvoarelor de apă sărată din Moldova. Bilanţul explorărilor din anii 2004-2007 efectuate în special în judeţul Neamţ. *Memoria Antiquitatis* 24, 121-190.

Weller, O., Brigand, R., Nuninger, L., Dumitroaia, Gh. and Monah, D. 2007. Analyses et dynamiques spatiales autour des sources salées de Moldavie sous-carpatique durant la Préhistoire roumaine. In N. Morère Molinero (ed.), *Las salinas y la sal de interior en la historia: economía, medio ambiente y sociedad. Inland saltworks and salt history: economy, environment and society*, 165-185. Madrid, Dykinson.

Weller, O., Brigand, R. and Nuninger, L. 2008. Spatial analysis of salt exploitation in Moldavian Pre-Carpatic Prehistory (Romania). In F. Favory and L. Nuninger (eds.), *7 millenia of territorial dynamics: settlement pattern, production and trades from Neolithic to Middle Ages. ARCHAEDYN ACI "Spaces and territories" 2005-2007 Final conference*. Dijon, 23-25 June 2008, 215-220. Dijon, Pre-proceedings.

Wheatley, D. and Gillings, M. 2002. *Spatial technology and archaeology. The archaeological applications of GIS*. London, Taylor and Francis.

Zaninetti, J.-M. 2005. *Statistiques spatiales, méthodes et applications géomatiques*. Paris, Lavoisier.

The Cucuteni C Pottery near the Moldavian Salt Springs

Roxana Munteanu
Complexul Muzeal Județean Neamț, Romania

Daniel Garvăn
Complexul Muzeal Județean Neamț, Romania

Abstract
The archaeological investigations undertaken in the field of salt archaeology in eastern Romania revealed the great extent of salt exploitation in these territories during the Neolithic and the Chalcolithic. In several sites involved in salt water evaporation an important quantity of shell-tempered pottery – the so-called Cucuteni C type – was unearthed. This type is seen in the literature either as a result of trade between the steppe communities and local population, either as an indicator of the actual presence of individuals of eastern origin within the Cucuteni tribes.

In the following lines Cucuteni C vessels found near salt springs are analyzed based on morphological and decorative patterns. Although paste and decorations sign a ceramic style completely different from those of Cucuteni A and B, we shall notice the existence of vessels decorated according to criteria of the Cucuteni C style but with a typically Cucuteni (A or B) paste. The opposite is also shown since they sometimes find in 'classical' ceramics of the Cucuteni culture modelled in shell-tempered paste.

The authors believe that the Cucuteni C pottery is a category made by the local communities for specific activities, such as salt water evaporation.

Keywords
chalcolithic, shell-tempered pottery, Cucuteni C, salt water

The next lines concern a ceramics species, which appears in the Chalcolithic sites for exploring the salted waters. We are talking about the Cucuteni C type of ceramics, discovered in the archaeological deposits of Solca-*Slatina Mare*, Lunca-*Poiana Slatinei*, Cacica and Ţolici-*Hălăbutoaia*.

The context of the discoveries

In the first work dedicated to the discoveries of Solca-*Slatina Mare* in 1968 there were discovered, in the level between 0 and 0.30m, ceramic artefacts '... dating from several periods: Criş, Precucuteni, Cucuteni B (in association with the C type)' (Ursulescu 1977, 311). The elements composing the Cucuteni B level, identified between 0.30 and 0.60m are not specified, and in the following papers published is cited just one shell-tempered pottery fragment with pieces of shreds and gravels added from Solca, as well as some vase fragments modelled in a clay paste with sand and limestone granulites (Cucoş 1985, 69). Recently, the statement below was reiterated in a study regarding the technology and raw material used to prepare the Cucuteni C ceramic clay paste (Cotoi 2007, 157), without mentioning the results of the more recent researches, which emphasized the existence in this site of an important Chalcolithic layer. From the consistent material that was found at Slatina Mare during the 2003 excavations, the Cucuteni C species reaches 40% (Dumitroaia *et al.*, 2004b, 314-315; Nicola *et al.*, 2007).

Identified in 1980 and archeologically investigated since 1983, the Lunca-*Poiana Slatinei* site is, taking into account the deposit, the stratigraphy, and the materials discovered, one of the most important sites regarding the prehistoric exploitation of salt in the Sub-Carpathian Moldavian region (Dumitroaia 1985, 735; 1987, 255; 1994, *passim*). In the anthropic deposits of *Poiana Slatinei*, only in the B zone there have been discoveries of Cucuteni A and Cucuteni B ceramic type, briquetage and Cucuteni C pottery. The shell-tempered category appears only in the Cucuteni B layers and is representing 33% of the entire ceramic lot of that particular level (Dumitroaia 1994, 58).

Another site related to the exploitation of salt is that of Cacica (dep. of Suceava). After the surface researches and the inquiries during the 90s of the past century, there have been many briquetage discoveries (Şandru 1952; Andronic 1989; 2006; Andronic and Mareş 1996, 21). We should also mention the presence of the Cucuteni C pottery, but we do not know its percentage within the Chalcolithic discoveries. A 2006 field research with one of the authors of the inquiries during the 90s allowed us to appreciate (on basis of surface discoveries) that the Cucuteni C ceramic is indeed numerous, probably as numerous as that of Solca.

In 2007 began the systematic researches of Ţolici-*Hălăbutoaia*, in the dep. of Neamţ (Dumitroaia *et al.*, 2008). On the superior side of the anthropic deposit, in the level belonging to the Cucuteni B phase, there was a great amount of shell-tempered pottery. The Cucuteni C ceramics of Ţolici (Figure 7) seems to be less numerous and more fragmented than that of Solca, Lunca or Cacica. We can estimate (as a preliminary result) that the Cucuteni C pottery of Ţolici represents around 10% of the whole Cucuteni B layer.

Besides the four aforementioned sites, there were also discoveries regarding the exploitation of salt of the salt springs at Oglinzi-*Băi*, the dep. of Neamţ (Dumitroaia 1994, 75-78) and Cucuieţi-*Slatina Veche*, the dep. of Bacău (Dumitroaia *et al.*, 2004a; Munteanu and Garvăn 2007, 526-527; Munteanu *et al.*, 2007). The shell-tempered category is not documented in any of these sites, the Cucuteni materials being – certainly at Cucuieţi, probably at Oglinzi – older than the appearance of the Cucuteni C pottery.

So far there have been studies of such materials at Lunca (Dumitroaia 1994) and Solca (Nicola *et al.*, 2007), the last constituting the core of our analysis. The Chalcolithic ceramics of Ţolici – the most recent resort for the exploitation of salt (Dumitroaia *et al.*, 2008) – is currently being studied, which allows us to have only preliminary conclusions on the report. The lot of shell-tempered pottery from this site was rather reduced (70-80 fragments). As regards the Chalcolithic pottery from Cacica, the focus was on the briquetage category, as the Cucuteni C ceramics was not presented (Andronic 1989). For this reason, for Cacica, for a comparative analysis we used the shell-tempered shreds (30 fragments, probably from 30 recipients), gathered during a surface research, in a deranged area in the mine yard.

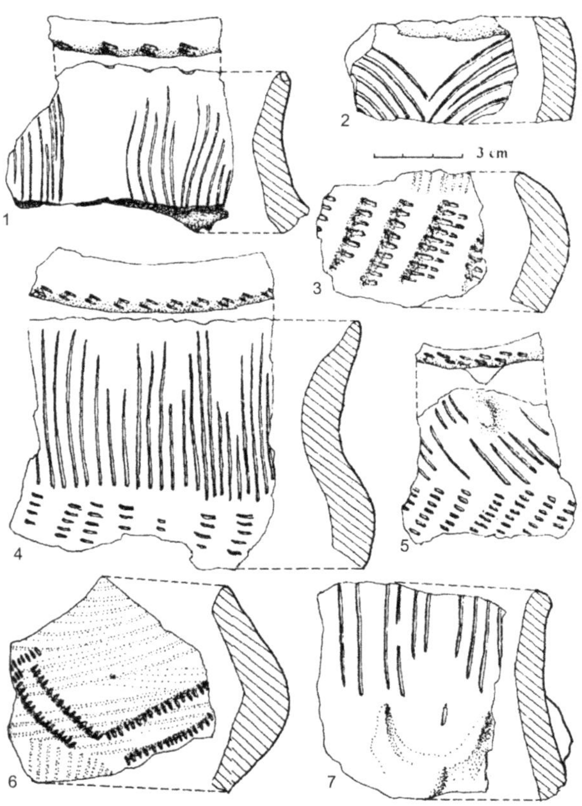

Figure 1. Cucuteni C pottery from Lunca-*Poiana Slatinei* (after Dumitroaia 1994).

The ceramic analysis

In all Cucuteni B levels identified in the deposits around the salted areas there was a great amount of Cucuteni C broken recipients discovered, which does not seem to occur in the usual contemporary settlements. The analysis of this ceramic species starts from several general observations: the shell-tempered pottery is obviously different from the Cucuteni specific one, being stylistically and technologically uniform, which goes for the materials in all the seasonal salt exploiting places researched so far.

The Cucuteni C ceramics is distinct through the specific degreaser (hashed shells), and through the decorative motifs. Both characteristics are usually present.

The paste for the Cucuteni C recipients, discovered at Lunca, Solca, Ţolici and Cacica is homogenous, and the colour varies from yellowish to bricky and brown-grey. The most often used degreaser is that of hashed shells. We cannot tell for sure if they are shells of snail shells collected by the Chalcolithic population or if there had been any additions of bioclastic sands. Without an analysis of the clay paste, we believe that the specific degreaser for the C type of ceramic within these sites might be a secondary product of a gathering activity. In other words, for the supposed scenario of the Chalcolithic salt exploitations – with a seasonal frequency – and without other anthropic deposits with rich fauna remains, the gathered items must have constituted an important part of the nurture. The great amount of snail volutes discovered at Ţolici-*Hălăbutoaia*, in a level that does not contain any pottery with a shell-mixed paste, sustains our hypothesis.

Figure 2. Cucuteni C pottery from Solca-*Slatina Mare*.

The *shell*-tempered pottery of all four sites is unitary as regards the quantity of the degreasing substance. The greatest part of the ceramic fragments analyzed has many shells, which are generally well hashed, forming a homogeneous mix, with just a bit of sand.

Besides the category with the shell in the paste, the most common one, there are also fragments coming from recipients modeled in a paste more specific to the Cucuteni cultural environment, with sand and broken shreds. This kind of ceramics represents less than 10% of all the categories. This category is slightly better represented at Ţolici than at Solca, Lunca and Cacica. The recipients modelled from the paste mentioned before are decorated in the same manner as the ceramics with shell degreaser. We note three ceramic fragments at Solca, one of them containing broken shell in the paste (Figure 2/2), and the two other having no such specific degreaser (Figure 2/1, 3), whose surface, less homogenous, has traces of vegetal materials.

Figure 3. Cucuteni C pottery from Solca-*Slatina Mare.*

Another category, which contains limestone granulites, was discovered only at Solca and Cacica and is represented by two recipient fragments (Figure 2/4).

As regards the form of the recipient found near salt springs, we do not have much to say, as the pieces are too fragmented. Given the thickness of the fragments (0.8-1.5cm) it seems that the most common ones are the medium-sized recipients, large and flat-based, specific to Cucuteni C category.

*

According to the rendering method, we find the following decorating styles:

1. striations

- organized, vertically on the neck – the so-called combed decoration (Figure 1/3, 4, 5, 6, 7, 8; 2/1; 5/3, 5, 6, 7), or with a veiled decoration on the shoulder (3/1; 5/4; 6/1). Technically, this decoration pattern comprises:

a. deep striations– probably made with a sharper instrument; the lines are clear and the distance between is 1-2 mm (Figure 5/5, 7; 7); they are closer to the incisions;

b. superficial striations – closer and less visible, sometimes intercut; they seem to be made with a more rounder-edged instrument (Figure 2/1, 4); they are closer to imprints than to the incision technique;

- unorganized, on the recipients, mostly on the outside and more rarely on the inside.

a. continuous, it covers the whole recipients or at least the inferior side (Figure 3/3; 5/3); it consists on superficial striations, made probably through an imprinting method;

b. groups of short striations with a random disposal; usually thin, barely visible (Figure 3/4, 6/2).

The striated decoration is the most common one within the analysed lot, being present on almost all decorated pottery in a large variety of associations (a combination between the two categories or one type of striations with corded ornaments, incisions, alveoles or prominences).

Figure 4. Cucuteni C pottery from Solca-*Slatina Mare.*

2. incisions (rafters – Figure 3/7, lines in wave – Figure 6/4; 3/5, rows of short, vertical incisions – Figure 2/2, 7); they are disposed at the top of the recipients (short, deep incisions, associated with striations on the neck of the vessels) or on the shoulder (zigzag incisions associated with organized and unorganized striations,

bands of veiled lines). On a fragment at Cacica, the zigzag decoration which covers the shoulder of the recipient seems to have been rather scratched then incised (Figure 5/4);

3. imprinted: they appear as rows of horizontal (Figure 3/5) or rectangular imprinted triangles (Figure 2/3), arch motifs (Figure 3/6) or line decorations. The last one has two variants:

- segments of 1-5cm of deep, long, imprinted rows, forming the so-called wrapped cord motif, which appears on the rim (Figure 2/5), on the shoulder and neck of the recipients, oblique (Figure 2/1, 5; 4/1), horizontal (Figure 4/2) or vertical (Figure 2/6; 3/6); at least for one fragment at Cacica (Figure 5/7) it is obvious that the corded motif on the shoulder was realized with the same instrument as the striations on the neck of the vessels;
- twisted cord – oblique or right imprinted lines, long, disposed on several short, parallel rows, on the shoulder of the recipients (Figure 4/4, 5, 6, 7); segments of transversal imprinted lines on the top (Figure 2/6).

Figure 5. Cucuteni C pottery (1, 2 – Solca; 3-7 - Cacica).

4. alveoles:

- rows of circular alveoles, realized by removing the paste (Figure 4/8);
- irregular, obtained by laterally pushing the paste with the finger or by pinching (Figure 2/4).

They usually appear at the upper part of the vessels – right under the rim or on the neck.

5. prominences:

- applied – conic (Figure 5/7; 6/3), long (Figure 5/1), cylindrical (Figure 5/2), as an non-perforated tubular handle (Figure 4/8), as a bucrane (Figure 1/7) or a small, flat disk (Figure 3/6); mostly at the top and associated with other decorating elements (lines, striations, prints);
- pushed from the interior of the recipient, the empty side being filled and levelled up (Figure 4/4).

These ornament patterns may be seen equally on the shell-tempered pottery as well as on vessels made of other types of paste.

Observations

As regards the Cucuteni C pottery, the main contribution is that of the studies of Ann Dodd-Oprițescu (1980; 1981; 1982; 1992), which discuss the opinions up to the 70s-80s of the past century referring to this subject and in which, on the basis of several materials analyzed, there is an attempt to clear the origin and the cultural significance of the Cucuteni C shell-tempered pottery. We should also mention the studies elaborated by I.T. Dragomir (1982) and Șt. Cucoș (1985), seen as compulsory reference in the next papers on the subject, when analyzing or just mentioning the ceramics pieces with such a degreaser in the newly studied Chalcolithic sites.

Figure 6. Cucuteni C pottery from Cacica.

The C type is said to have an origin different from Cucuteni, in the northern side of the Cucuteni-Tripolie world (Dodd-Oprițescu 1980, 554; 1982, 79), or the north-pontic steppes (Dragomir 1982; Sorochin 2002, 180, 184). The conclusions of the aforementioned studies have a tendency towards generalizing the results of studies on more or less material. This is why we do not know whether to give credit to the theory according to which there was a *cucutenization* of the C ceramics paste, after which it transformed into a Cucuteni type (Nestor and Zaharia 1968, 27; Dodd-Oprițescu 1980, 549-550). This statement implies the foreign origin of people who modelled this ceramic type or the existence of certain relationships (regardless of their nature – product exchanges or living together) between these population groups and the local communities, in the same parameters, for the whole Cucuteni A_3-B_2 interval. Not even for the Tripolie settlements which are geographically closer to the steppe and the forestry steppe (if this is the origin for the influences or the individuals that created the C ceramic type – (Sorochin 2002, 180-184) we do not have documents of such intense cultural contacts (Tsvek and Rassamakin 2005). Moreover, we do not believe that this is the reason of the appearance of the C ceramics in the Cucuteni settlements in the Moldavian Sub-carpathians as the older references (Dodd-Oprițescu 1980; 1982), and the more recent works (Rassamakin 2004; Anthony 2007, 225-262) do not find reasonable common points between the steppe ceramics (Sredni Stog or Skelya) and the decorating motifs and forms of the shell-tempered Cucuteni C type. There is also the possibility that these C-type recipients travelled on long distances within exchanges, but we do not believe that we may attribute to a foreign population the whole C-type ceramics in the Cucuteni Moldavian areas.

Figure 7. Cucuteni C pottery from Ţolici-*Hălăbutoaia.*

Even more, the idea of the *cucutenization* of the C-type ceramic paste, as the whole analysis for this ceramic is based upon stylistic criteria – according to which the shell-tempered pottery is inferior to the Cucuteni painted category. The aesthetic interpretations are not relevant for the real interest of Cucuteni people. The petrographic studies insist upon the dominant rapport between the composition of the paste and the functioning of the final product, stressing that the recipients for different purposes were made from different compositions, that the main preoccupation was for the recipient to have the standards required – parameters regarding the endurance to thermal or mechanic shocks, porosity (Stoltman 2001, 309-310).

The shell used as degreaser for most of the Cucuteni C recipients offers, apparently, an increased endurance and the capacity to evaporate the liquids faster (Stoltman 2001, 310; Rowlett and Shaw 2005, 165; Anthony 2007) and maybe this can explain the presence of the Cucuteni C ceramics in the salt exploitation seasonal camps.

Ann Dodd-Oprițescu speaks of an evolution in the forms, decorations and paste of the shell-tempered category during Cucuteni A_3-B_2 and it is precisely during Cucuteni B_1 when the depersonalization of the C ceramic type took place (Dodd-Oprițescu 1980, 549-550), by giving up the specific degreaser and adopting the Cucuteni motifs. This idea is sustained by other authors, also (Cotoi 2007, 155), without performing an analysis on several Cucuteni C materials. The generalizations regarding the Cucuteni C ceramic type cannot be made only by agreeing with the other studies, without verifying the archaeological materials.

Yet, as we presented here, far from being reduced as number, the Cucuteni C pottery is frequent in Cucuteni B_1 layers in the specific sites used to exploit salt water. Most of this type we recognise it because of the shell-tempered, which is well represented.

Also, the statement according to which the cord decorations are the most frequent during Cucuteni B in comparison to the other C-type ornaments (Bem 2007, 60), is not validated by the ceramics at Solca, Lunca, Ţolici and Cacica, where most of the recipients were decorated with striations.

Figure 8. Fragments of shell-tempered briquetages.

Upon the observations made while analysing the C-type ceramics within the sites close to salt springs, we believe that this ceramic species belongs to Cucuteni culture. Even though at the beginning it might have been a 'technologic import', the Cucuteni population adopted it and it became a part of their repertoire. This repertoire is in fact far more complex than that of the painted ceramics; it's comprising, for example, during the B phase of the culture, a type of recipient closely related to the salt exploitation – the tronconic glass, rudimentarily modelled, named briquetage – which does not have, taking into account the paste, burning and shape, many common points with the 'classic' Cucuteni ceramics. Likewise the briquetage recipients, we believe that the C-type ceramics too are modelled for clear functional purposes. In this situation, the variation in the paste reflects the various functionalities. In a lesser extent this might be the result of certain experiments, as the two shell-tempered briquetage fragments at Solca-*Slatina Mare* could be interpreted (Figure 8/1, 2) (Nicola *et al.*, 2007, 47).

As regards the mechanism of exploiting the salt from the salt water in Moldavian Sub-carpathian during Chalcolithic, the opinions expressed until now are very different. We shall mention here two of them which annul each other. According to the first (Cotoi 2007, 159), these salt springs would have been directly exploited during Cucuteni B by groups of Eastern populations that could have exerted some kind of a temporary *condominium*, alongside the local population. This C-type ceramics would exist here because of these Eastern populations that had come to the Cucuteni area. This scenario reflects, we believe, the lack of knowledge in the archaeological realities, on the one side, as pottery alone could not have been the evidence of a migration, and, on the other side, because the deposition at Lunca-*Poiana Slatinei* (see Dumitroaia 1994, 58-62), in which the Cucuteni B and the Cucuteni C material appear in the same layer, could have formed only if the local and the Eastern populations, each of them using their own recipients, would have exploit the salted source at the same time with an equal intensity.

According to the second opinion (Monah 1991; 2002), the Cucuteni population is the one that exploits the salt springs, and the product – the salt – is being traded at long distances (directly or through individuals attracted to this area); as a result, the C-type pottery arrives in the Cucuteni area.

If the C-type ceramics is an 'import' product, we are still to explain its massive presence among the remains of the seasonal camps near the salt springs and its poor evidence within the remains of the bigger and longer inhabited settlements.

As local product, yet, its existence is justified in the same manner as the briquetage or other Cucuteni recipients near the salt springs, meaning the exploitation of the salt by the local population in a manner which is not that far from the current one.

References:

Andronic, M. 1989. Cacica – un nou punct neolitic de exploatare a sării. *Studii şi Cercetări de Istorie Veche şi Arheologie* 40/2, 171-177.

Andronic, M. 2006. Cacica (jud. Suceava). In V. Cavruc and A. Chiricescu (ed.), *Sarea, Timpul şi Omul*, 60-61. Sfîntu Gheorghe, Angustia.

Andronic, M. and Mareş I. 1996. Cacica, jud. Suceava. *Cronica Cercetărilor Arheologice. Campania 1995*, 21.

Anthony, D.W. 2007. *The Horse, the Wheel and Language: How Bronze-Age Riders from the Eurasian Steppes Shaped the Modern World.* New Jersey, Princeton University Press.

Bem, C. 2007. *Traian Dealul Fântânilor: fenomenul Cucuteni A-B*. Târgovişte, Cetatea de Scaun.

Cotoi, O. 2007. Observaţii privitoare la tehnologia şi materiile prime utilizate la prepararea pastei ceramicii Cucuteni C. *Annales Universitatis Apulensis, Series Historica* 11/I, 153-160.

Cucoş, Şt. 1985. Ceramica de 'tip C' din aria culturii Cucuteni. *Memoria Antiquitatis* IX-XI, 63-92.

Dodd-Opriţescu, A. 1980. Consideraţii asupra ceramicii Cucuteni C. *Studii şi Cercetări de Istorie Veche şi Arheologie* 31/4, 547-557.

Dodd-Opriţescu, A. 1981. Ceramica ornamentată cu şnurul din aria culturilor Cucuteni şi Cernavoda I. *Studii şi Cercetări de Istorie Veche şi Arheologie* 32/4, 511-528.

Dodd-Opriţescu, A. 1982. La céramique Cucuteni C. Son origine. Sa signification historico-culturelle. *Thracia Praehistorica. Supplementum Pulpudeva* 3, 70-79.

Dodd-Opriţescu, A. 1992. Die schnurverzierte Keramik in der Cucuteni- und Cernavodă I –Kultur. In P.I. Roman, A. Dodd-Opriţescu and J. Pál, *Beiträge zur Problematik der Schnurverzierten Keramik Südosteuropas,* 38-44. Mainz am Rhein, Verlag Philipp von Zabern.

Dragomir, I.T. 1982. Elemente stepice „Cucuteni C", descoperite la Bereşti (jud. Galaţi). *Studii şi Cercetări de Istorie Veche şi Arheologie* 33/4, 422-429.

Dumitroaia, Gh. 1985. Săpăturile arheologice din judeţul Neamţ 1977-1983. *Memoria Antiquitatis* IX-XI, 733-740.

Dumitroaia, Gh. 1987. La station archéologique de Lunca-Poiana Slatinei. In M. Petrescu-Dîmboviţa, N. Ursulescu, D. Monah, V. Chirica (eds.), *La Civilisation de Cucuteni en Contexte Européen*, 253-258. Iaşi, Bibliotheca Archaeologica Iassiensis, I.

Dumitroaia, Gh. 1994. Depunerile neo-eneolitice de la Lunca şi Oglinzi, judeţul Neamţ. *Memoria Antiquitatis* XIX, 7-82.

Dumitroaia, Gh., Munteanu, R., Nicola, D., Preoteasa, C. and Monah, D. 2004a. Cucuieţi, com. Solonţ, jud. Bacău. Punct: Slatina Veche. *Cronica Cercetărilor Arheologice din România. Campania 2003*, 110-111.

Dumitroaia, Gh., Nicola, D., Munteanu, R., Preoteasa, C., Monah, D., Boghian, D. and Ignătescu, S. 2004b. Solca, com. Solca, jud. Suceava. Punct: Slatina Mare. *Cronica Cercetărilor Arheologice din România. Campania 2003*, 314-315.

Dumitroaia, Gh., Munteanu, R., Garvăn, D., Weller, O. and Brigand, R. 2008. Ţolici, com. Petricani, jud. Neamţ. Punct: Hălăbutoaia. *Cronica Cercetărilor Arheologice din România. Campania 2007*, 325-326.

Monah, D. 1991. L'exploitation du sel dans les Carpates orientales et ses rapports avec la Culture de Cucuteni-Tripolye. In V. Chirica and D. Monah (eds.), *Le Paléolithique et le Néolithique de la Roumanie en contexte européen*, 387-400. Iaşi, Bibliotheca Archaeologica Iassiensis, IV.

Monah, D. 2002. L'exploitation préhistorique du sel dans les Carpates orientales, in O. Weller (ed.), *Archéologie du sel: Techniques et sociétés dans la Pré-et Protohistoire européenne*, Actes du colloque 12.2 du XIVe Congres de l'UISPP, 2001, 135-146.

Munteanu, R. and Garvăn, D. 2007. Cercetări arheologice în judeţele Neamţ şi Bacău (2004-2006). *Memoria Antiquitatis* XXIV, 525-534.

Munteanu, R., Garvăn, D., Nicola, D., Preoteasa, C. and Dumitroaia, Gh. 2007. Cucuieţi-Slatina Veche (Romania). Prehistoric exploitation of a salt resource. In D. Monah, Gh. Dumitroaia, O. Weller and J. Chapman (eds.), *L'exploitation du sel à travers le temps*. Bibliotheca Memoriae Antiquitatis XVIII, 57-70. Piatra-Neamţ, Constantin Matasă.

Nestor, I. and Zaharia, E. 1968. Sur la période de transition du néolithique à l'age du bronze dans l'aire des civilisations de Cucuteni et de Gumelniţa. *Dacia*, N.S. 12, 17-43.

Nicola, D., Munteanu, R., Garvăn, D., Preoteasa, C. and Dumitroaia, Gh. 2007. Solca-Slatina Mare (Roumanie). Preuves archéologiques de l'exploitation du sel en préhistoire. In D. Monah, Gh. Dumitroaia, O. Weller and J. Chapman (eds.), *L'exploitation du sel à travers le temps*. Bibliotheca Memoriae Antiquitatis XVIII, 35-56. Piatra-Neamţ, Constantin Matasă.

Rassamakin, J. Ja. 2004. *Die nordpontische Steppe in der Kupferzeit. Gräber aus der Mitte des 5. Jts. bis Ende des 4. Jts. v. Chr.*, Teil I – Text. Archäologie in Eurasien 17. Mainz, Verlag Philipp von Zabern.

Rowlett, R. M. and Shaw, M. 2005. Shell tempered pottery in the Cucuteni-Tripolye areal culture. In Gh. Dumitroaia, J. Chapman, O. Weller, C. Preoteasa, R. Munteanu, D. Nicola and D. Monah (eds.), *Cucuteni. 120 ans de recherches. Le temps du bilan/120 Years of Research. Time to sum up*, 157-171. Bibliotheca Memoriae Antiquitatis, XVI. Piatra-Neamţ, Constantin Matasă.

Sorochin, V. 2002. *Aspectul regional cucutenian Drăguşeni-Jura*. Bibliotheca Memoriae Antiquitatis, XI. Piatra-Neamţ, Constantin Matasă.

Stoltman, J. B. 2001. The Role of Petrography in the Study of Archaeological Ceramics. In P. Goldberg, V. T. Holliday and C. R. Ferring (eds.), *Earth Sciences and Archaeology*, 297-326. Kluwer Academic/Plenum Publishers. New York.

Şandru I. 1952. Contribuţii geografico-economice asupra exploatării slatinelor în Bucovina de Sud. *Studii şi Cercetări Ştiinţifice* 3, 407-422.

Tsvek, E. V. and Rassamakin, I. Ia. 2005. The Interactions between the Eastern Tripolye Culture and the Pontic Steppe Area: Some Aspects of the Problem. In Gh. Dumitroaia, J. Chapman, O. Weller, C. Preoteasa, R. Munteanu, D. Nicola and D. Monah (eds.), *Cucuteni. 120 ans de recherches. Le temps du bilan/120 Years of Research. Time to sum up*, 173-192. Bibliotheca Memoriae Antiquitatis, XVI. Piatra-Neamţ, Constantin Matasă.

Ursulescu, N. 1977. Exploatarea sării din saramură în neoliticul timpuriu, în lumina descoperirilor de la Solca (jud. Suceava). *Studii şi Cercetări de Istorie Veche şi Arheologie* 28/3, 307-317.

Some Salt Sources in Transylvania and their Connections with the Archaeological Sites in the Area

Gheorghe Lazarovici

Universitatea "Lucian Blaga" Sibiu, Romania

Cornelia-Magda Lazarovici

Institutul de Arheologie, Iaşi, Romania

Abstract

Salt has been playing important role in the evolution of society, because of its being an indispensable nourishment for man and animals. Additionally, salt has also a therapeutic role, being used for this purpose by traditionalist societies, but even today. In that respect it plays an important role as well. The Neolithization process started with animal breading (sheep/goats, cows) and garden agriculture. The oldest Neolithic sites in the Intra-Carpathian area (Gura Baciului, Ocna Sibiului) or in the Extra-Carpathian one (Govora – *Valea Răii* / Copăcele, Lunca – *Poiana Slatinii*) are related with salt springs / lakes, currently salt bathes (Miercurea Sibiului, Ocna Sibiului, Cojocna), areas with salt, areas with previous salt springs. Some disappeared today but are present in the local toponimy, such as Şeuşa (Early Neolithic), Bădeni (Bronze Age, Hallstatt (*So* = salt or *sosviz* = salt water: Hungarian language). In our opinion not only the salt itself is important to be studied, but also the roads to the salt areas are also important also to be studied and understood. We intent to present some case studies that may be more important as a general presentation. These are related to our older or newer special research in regard to the salt sources. In the paper we present some archaeological areas located in the vicinity of salt sources such as: Tureni, Buneşti, Cojocna - Căian Vamă, etc. Some other examples, situations or ethno-archaeological researches related to salt area also presented: use, transportation and its role in historical processes (especially the Neolithization). The salt, obtained from different sources (sea, lakes, salt springs, salt mines) because of its qualities and taste determine terms of comparison or mythological elements that benefit of a large symbolism. Many proverbs or sayings express elements that draw our attention: *...beloved as salt in the food, evil as salt in the eyes...* or *to put salt on the wound...*, etc.

Keywords

salt sources, Transylvania

Among recent interdisciplinary research in Transylvania we wish to highlight that related to the exploitation of gold and salt (Rusu 1977 and the appendix). The discovery of a workshop for jewels led to investigations connected with the exploitation and processing of gold (Lazarovici and Lazarovici 2008).

Figure 1. Figa, Beclean town (district BN) old bathes (after Harding, Kavruk).

Figure 2. Figa, Beclean town (district BN) salt water basins (after Harding, Kavruk).

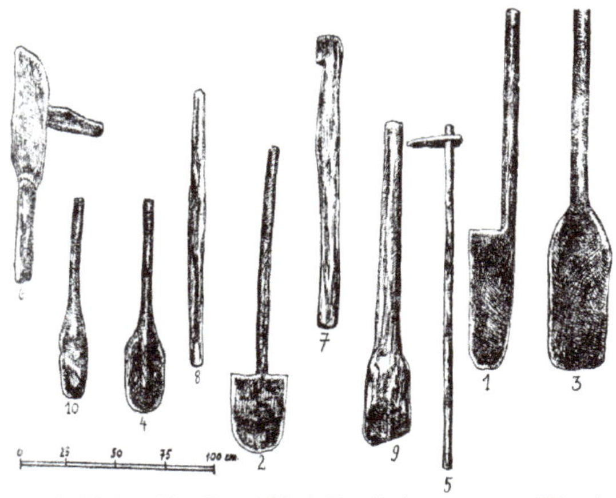

Figure 3. Valea Florilor (Cluj district), storage of Dacian wooden tools from the mine.

Salt exploitation and his use are in connection with older[1] projects of the Romanian archaeology and the last cooperation with an English team at Figa (Figures 1-2, 5)[2].

History of the research

Earlier, V. Wollmann (1996) had analyzed the problems of the exploitation and marketing of salt in Roman times. His starting point was the Roman salt exploitation at *Potaissa* (Turda), *Salinae* (Ocna Mureşului), Sânpaul Homorodului, Domneşti (Bistriţa-Năsăud district, hereafter BN), and *Apulum* (Alba Iulia), where there is information about the so-called *conductores*, *conductores salinarum* (at Porolissum), and *conductores pascui et salinarum* (Rusu 1977; Wollmann 1996, 240 ff.).

Our research is connected to salt springs in the area of some Neolithic sites. They were excavated and the archeological area was also studied, including Miceşti (related to the archaeological research and ethnoarchaeological investigations from the Petrind Mountains and Feleac Hills), the well at Buneşti (related to research into the Transylvanian Plain from Iclod, the Roman road from Gherla and Ţaga), and Valea Florilor – Cojocna – Căian (research into special surfaces).

The idea of this analysis was necessary when a Neolithic complex with Precucuteni pottery was discovered at Mintiu Gherlei. Similar materials were found at several other Neolithic sites (Ţaga, Iclod, Fundătura). The western limit of Precucuteni culture lies about 200km distant, connected by the Ţaga - Fizeş Valley between Someşul Mic River basin and the eastern part of Transylvania. It starts from the border of Mintiu Gherlei River, passing through Ţaga and reaching Eastern Transylvania through Reghin, or Central Transylvania through Zau de Câmpie (from here to Târgu Mureş). Another road goes north, to Bistriţa. Zau de Câmpie (Zau for short) was a cultural and economic center, one of the biggest Neolithic sites, the eponymous site of Zau culture. Recently, cultural groups from Central, Northern and Eastern Transylvania have been linked to Zau, as well as those from Eastern Crişana, the Northern Apuseni Mountains, and the Western Maramureş area, especially Oaş - Satu Mare (Lazarovici, 2008, 2009).

Museums in Moldavia and east Transylvania, such as Piatra – Neamţ, and the Eastern Carpathians Museum at Sfântu Gheorghe together with Bistriţa - Năsăud County museum have initiated similar projects, which include the development of research into salt in Transylvania; the most important research relates to Figa. Transylvania is very rich in salt (Figures 6-7) and its exploitation is mentioned from the 18th century onwards. The first objects relating to the exploitation of salt from the Roman or Dacian periods (Figures 3-4) (Maxim, 1971; Rusu, 1977; Wollmann, 1996, 246) were discovered at the Valea Florilor railway station (Rusu, 1977, on the Cluj – Apahida – Câmpia Turzii railway. Mircea Rusu mentions 62 localities with salt sources in Transylvania (Rusu, 1977). We have visited more then 20. There are many Neolithic sites in the area. Here we will mention only those, which have been intensively inhabited.

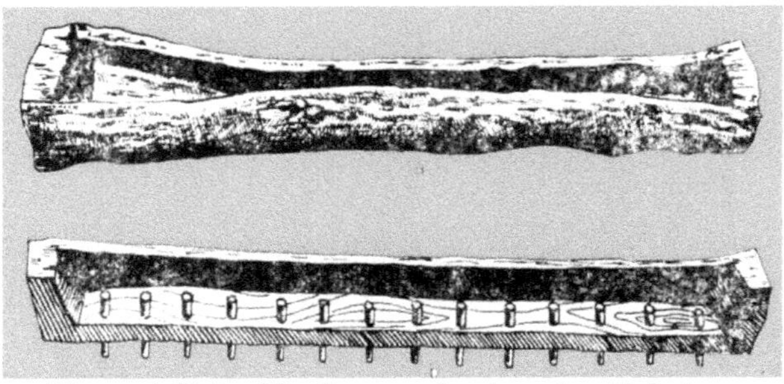

Figure 4. Valea Florilor, wooden trough with holes.

Figure 5. Wooden through with holes at Figa - Beclean.

Many wooden tools (Figure 3), as well as two Dacian grinding stones, were discovered in 1938, when the railway station at Valea Florilor was extended. The objects were found at a depth of 8-10m at the bottom of a well covered with wooden branches.

The wooden pieces made of hornbeam (*carpinus betulus*) are better preserved, some of them having been discovered in similar conditions as those from the recent research at Beclean - Figa[3]. It is the same type of piece; some corks with holes (Figure 4) are absolutely identical to those described by A. I. Maxim. Professor A. Harding, with our colleagues from Sfântu Gheorghe Museum, investigated the Figa site and at the same time created the first school for modern research in this field of excavation, and in the preservation and interpretation of organic materials, especially those which are wooden. This can open new perspectives, including the calibration of the radiocarbon data.

[1] Ethnoarchaeological research in the Petrind Mountains, in cooperation with J. Nandriş in 1982-1986, *Salt programm*, unpublished.

[2] Piatra Neamţ - Sfântu Gheorghe, project *Salt 2006*, exhibition exposed in several museums in Transylvania.

[3] Comparing situations at Figa with those at Valea Florilor, we believe that even here was not a well. It was a former underground gallery in which the pieces and the installations mentioned were used. Even it is a chronological postponement (dendrocronological and radiocarbon data for Figa are still being processed), the pieces and the functionality of the installation prove their long lifetime. The active principle was kept for very long time (such as the axe with handle, used for over 7 millennia); *Sarea, Timpul şi Omul*, catalogue of the exhibition, Ed. Angustia, Sfântu Gheorghe, 2006. This was our inspiration.

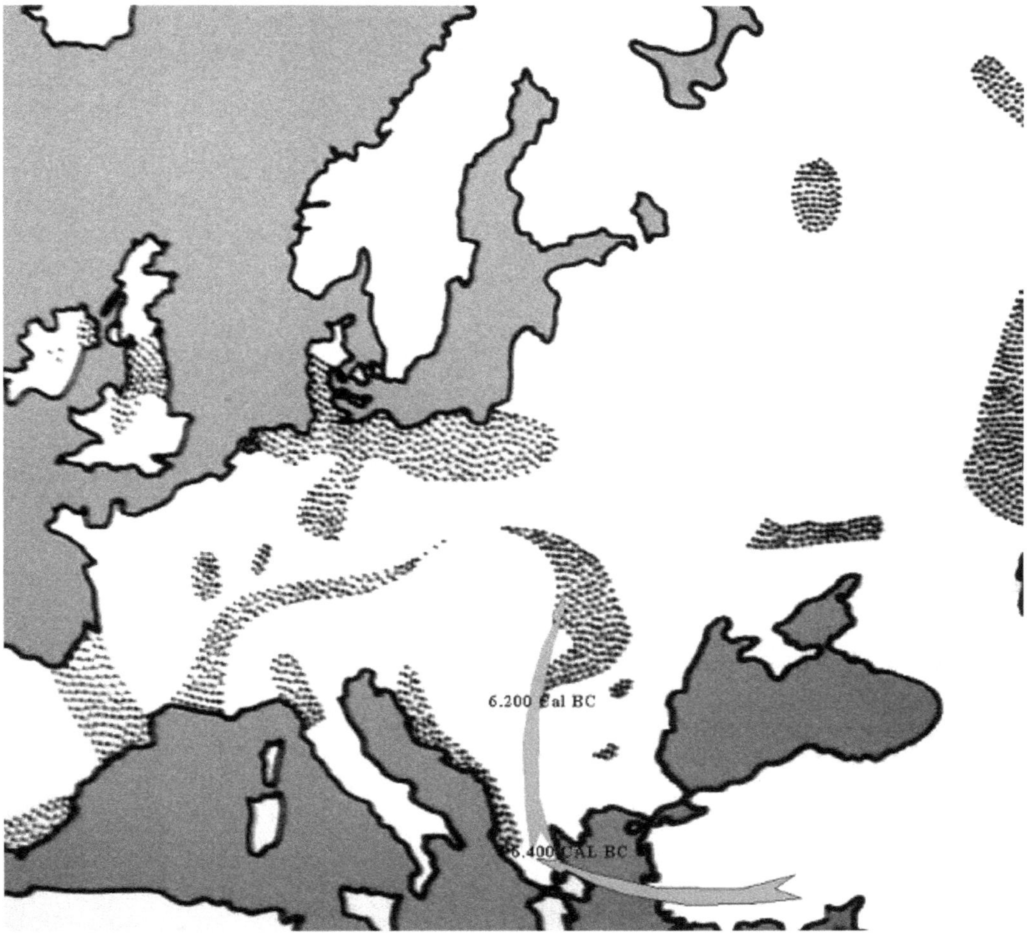

Figure 6. Salt areas in Europe and the first migrations in Neolithic.

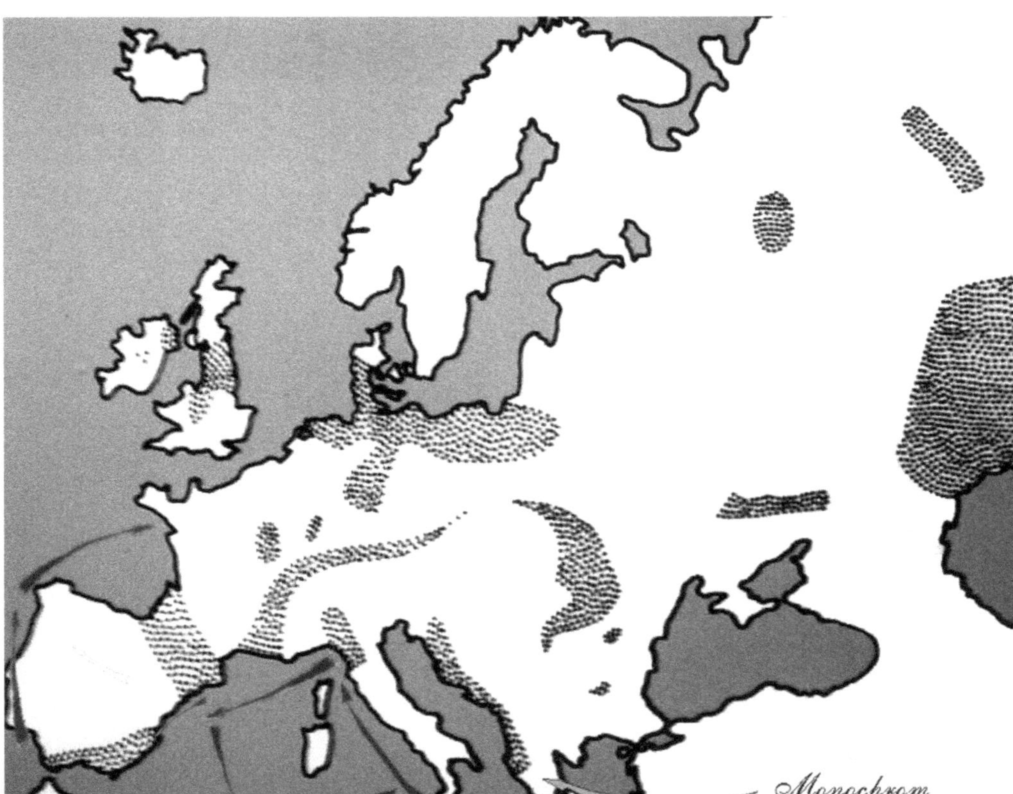

Figure 7. Salt areas in Europe and migrations in Neolithic: FNT, Cardial - impresso.

Salt and the Neolithization process

We believe that the need for salt, especially for animals and the conservation of products, drove the Neolithization process from Anatolia to Thessaly, Macedonia and from there to the Danube area, and the salt areas in Transylvania. (Figures 6-7).

Many studies reflect the salt sources, the moment of use, the necessity of salt as a good foodstuff and for the good health of domestic animals, especially of sheep (Weller and Dumitroaia 2005; *Sarea* 2006).

In the Olt Valley, from the south to the north, in the area of major salt sources, there are several early Neolithic sites (Starčevo - Criş, SC for short), followed by phases, according to our chronological system[1]), such as:

- **Copăcelu** ex Valea Răii, near Băile Govora, SC IC and IIIB-IVA (Petre-Govora);
- **Ocna Sibiului** (Figure 8), near baths on a high terrace that dominates the lakes, the river and the salt springs, SC IB/IC – IVA (Paul 1981, 1992, 1995; Rusu 1977; *Sarea* 2006; Figure 4);
- **Miercurea Sibiului** – Pietriş (Figure 8), at about 1km distant from salt baths at Miercurea Sibiului, SC IA – III, Vinča A-B, Zau culture (the former cultural complex Lumea Nouă – CCCTLNZIS) (Luca *et al.*, 2006; Luca, Pinter and Georgescu 2003, and bibliography.);
- **Şeuşa** (Figure 8), SC IC-IIA; the name of the village derives from the Hungarian language, from *só* (= salt (Ciută *et al.*, 2000, 2001, 2005; Ciută 2005, etc.), *sósviz* = salt water, the name of the spring in Bădeni village; during the Hallstatt period, bronze deposits and a large site);
- **Ciunga – Uioara de Sus and Uioara,** included in Ocna Mureşului, areas with major salt exploitations from SC IC/IIA until now (Lazarovici and Cristea 1979).
- **Gura Baciului,** SC I-IV, lies about three hours' walk away from the salt deposits from Apahida – Cojocna. At 6km the Chinteni Valea Fânaţelor/Valea Caldă, has a well which is salted periodically (surface investigations) (Figure 6);
- **Cluj – Napoca,** SC II-IV, the biggest site of the developed Neolithic (Zau culture, phases II-IV) in Transylvania (an area between 30-70ha with over 4m thick Neolithic levels) lies at the margin of the salt mountains, Turda – Apahida (Rusu 1977, 211; see bellow);
- In Moldavia the most important site lies at **Lunca – Poiana Slatinii** (Figure 8), related to the oldest salt exploitations from the Early Neolithic period, in connection with the third migration. We believe that there are also workshops for pottery firing here (Monah 1991; Weller and Dumitroaia 2005; *Sarea* 2006);
- **Iclod - *Doroaie*** and **Livada**, SC IC/IIA, secondary site, at about 4-6km far from the salt well at Buneşti (Figure 8);
- **Ţaga**, SC IC - IIA (local bittern) from the source at Buneşti, situated at about 20km distant.

These sites have played an important role in the Neolithization processes. All have extended ethno-cultural connections with Transylvania, central (Szarvas – Ob. 23) and the southern part of the Panonian Plain (Donja Branjevina), farther to the south with Serbia (Grivac, Divostin, Pavlovac, Dobanovć, etc.), with Macedonia through Vardar – Axios (Anzabegovo I – Vršnik I), and through such sites as Cârcea, Grădinile with the Early Neolithic civilizations in Bulgaria (Ruse, Gradešnica, Vaksevo, Slatina, Gălâbnik, Kovačevo I, etc.), to Macedonia through the Struma Valley (Lazarovici 1996, 1998, 2006, etc.).

Because salt sources have been very weakly investigated, we believe it is preferable to present some case studies, even if we make generalizations.

Figure 8. Cheile Turzii area with the salt source at Miceşti, general view.

Figure 9. Cheile Turzii area with the salt source at Miceşti, detail.

[1] Conclusions about the connections with Anzabegovo for stage SC IC are the salt sources, Lazarovici 2006, 140-141.

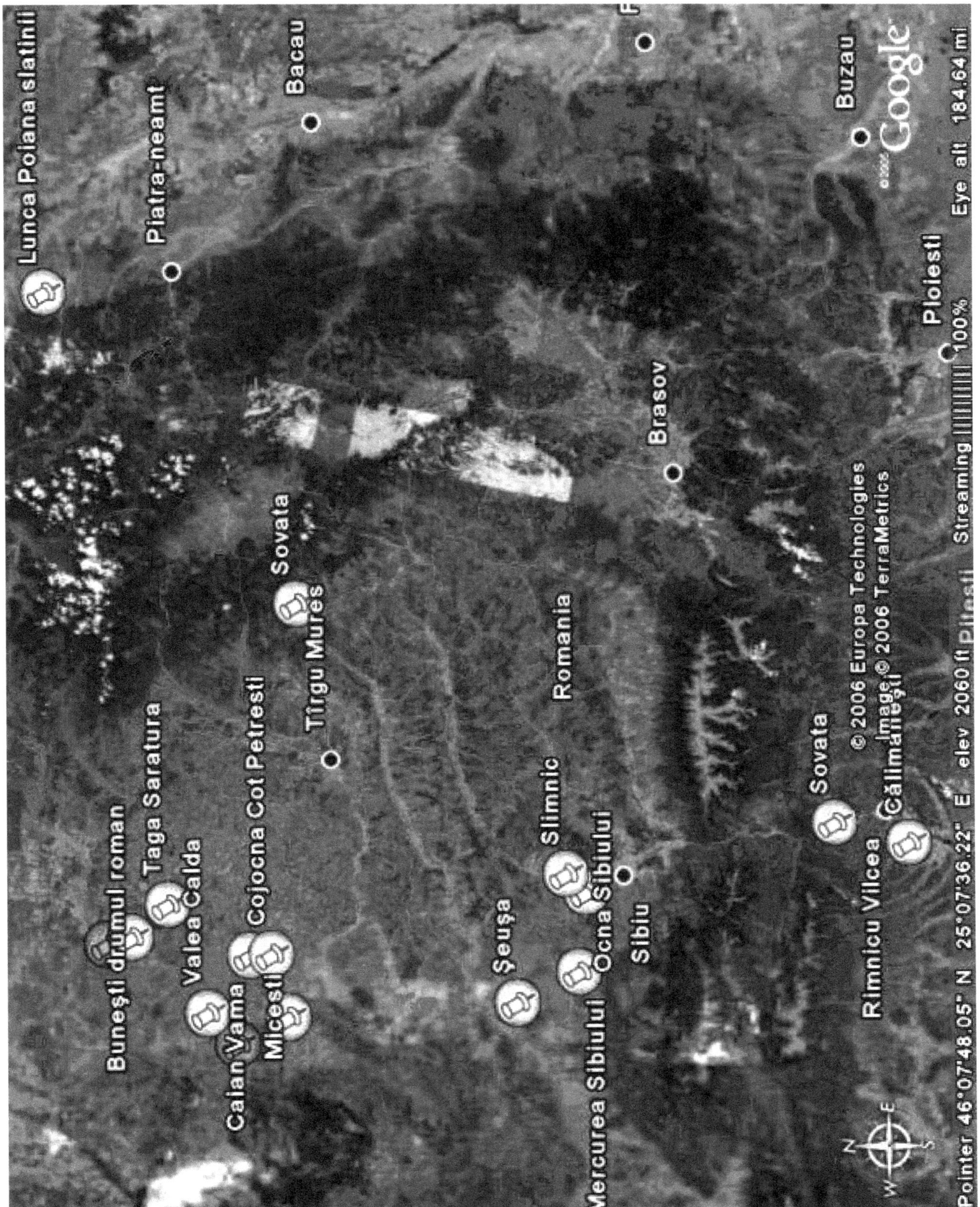

Figure 10. Salt sources in central Transylvania and the Neolithic sites in the area.

Figure 11. Miceşti, typical salt vegetation.

Figure 12. Miceşti, the salt well and modern exploitation.

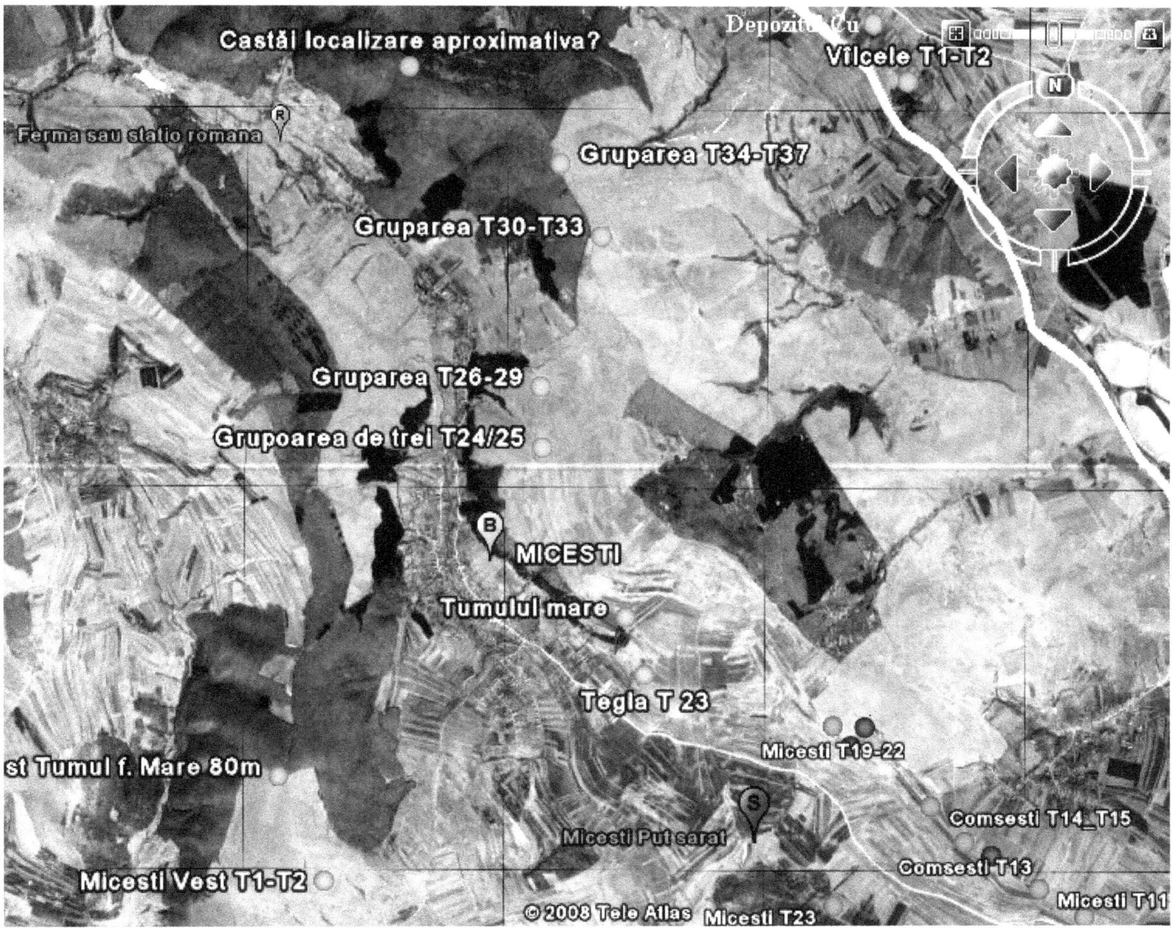

Figure 13. Miceşti – Puţul Sărat and the north archaeological area: green circles indicate tumuli.

Figure 14. Miceşti – Puţul Sărat, salt well.

The Miceşti well

As we have seen, in the area of Turda town, there are old salt exploitations related to the Dacian and Roman eras. Neolithic materials (pottery and flint tools, Kovrig collection) were found in a location in the same area, at the Carolina lake (Crişan *et al.*, 1992, 403, nr. 52). With all these proofs, some specialists in the Roman era are skeptical about the Roman exploitations (Crişan *et al.*, 1992, 403, no. 51). The well is located in the large valley of Miceşti, its opening orientated to Hăşdate Valley, across Cheile Turzii (Figures 9-10).

We visited the spring in November (Figures 11-12, 14) before the slaughter of the pig. On this occasion we did not observe any sort of installation at the wells. For one hour, the duration of our investigation, we saw two farmers, one with a wagon (Figure 14) the other with a tractor with attachment, which collected salt water for several families (Figure 11). Each carried about 100-200l of salt water. Digging in the salt-water trench leads to the fresh water (Figures 10), but both ends of the valley are full of salt, proof that the water is rising. The entire large valley is very damp (Figures 12). In all this area of salt water, salty vegetation can also be observed (Figure 12), present in other places too, such as Buneşti or elsewhere in Transylvania.

Sheep from a nearby sheepfold, 400-500m distant, eat this vegetation, selecting some plants. In particular, they lick the salty earth. We collected information from the farmers about the former exploitation of the well (This is more related to an ethnographic or ethno-archeological study). The archeological area was little investigated. Many Roman period materials were discovered across 19-20 centuries. Up from the village, on *Pe Cărămidă* point, Roman bricks and tiles were discovered. Investigations have proved the existence of a medieval habitation and a grave that dates from 12-15 centuries ago.

Figure 15. Archaeological area to the south and west (Cheile Turenilor and Cheile Turzii - aquamarine big circles; tumuli - green small circles).

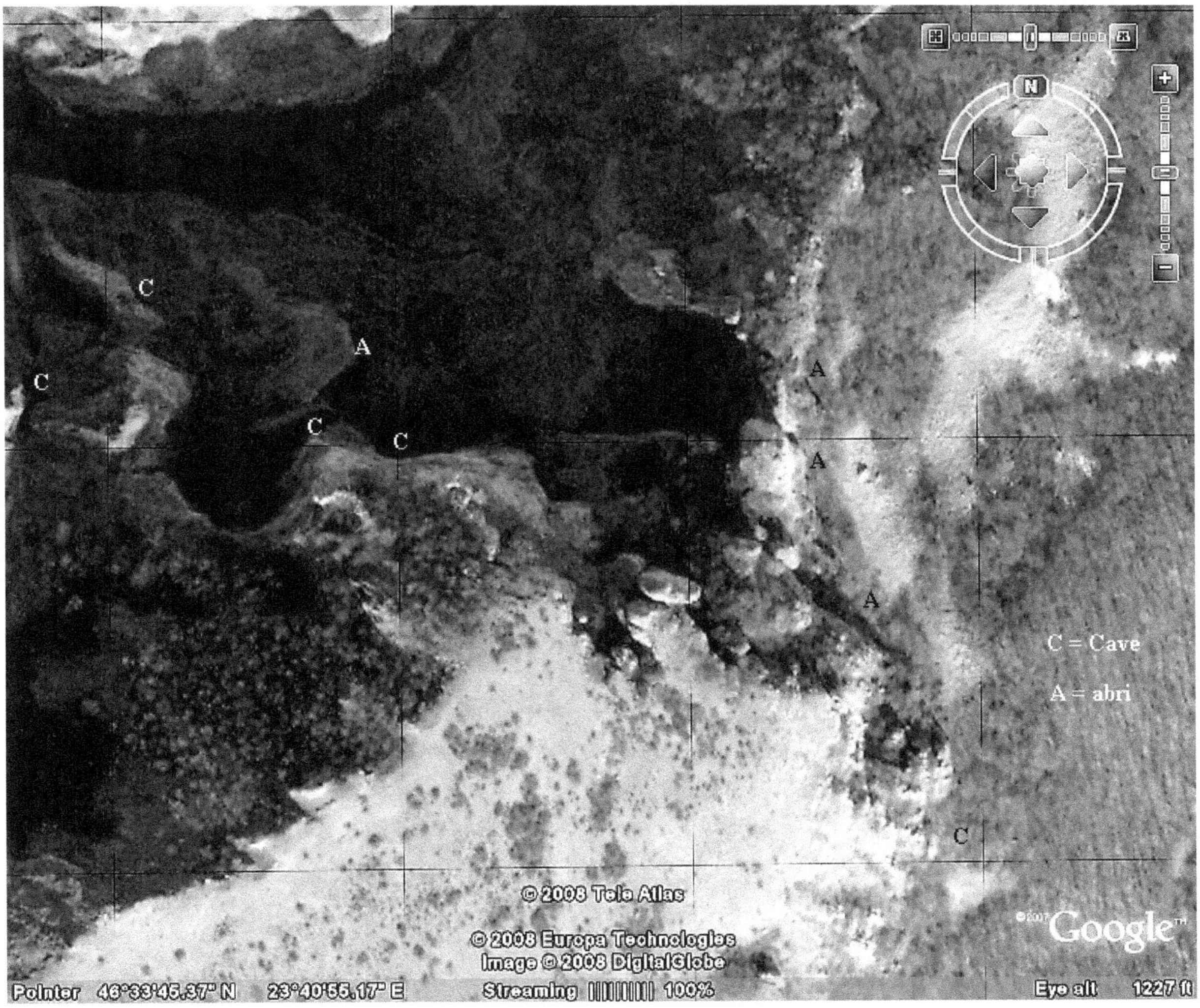

Figure 16. Cheile Turzii canyon, Peretele Vulturilor in the center, neighboring caves (C) and shelters (A).

Roman graves and sarcophagi mentioned at Miceşti can be related to a Roman road from Napoca (Cluj-Napoca) to Turda (*municipium* and Roman *castrum*). There is a definite connection between the Roman road and the salt spring. At Gherla, meanwhile, salt springs are located about 200-400m distant from the Roman road, which is not by chance. During the Copper Age near the Miceşti salt spring we observe the largest tumuli concentration from the Cluj district (Figure 15). The shepherds used the ridge roads close by, avoiding the heights of the Feleac hill. In the area between the salty spring and the forest located to the north, over 40 tumuli were discovered (Lazarovici and Kalmar/Maxim 1987-1988). This connects us with the 39-43 copper axes found in the deposit at *Banyabic* and the tumuli mentioned above. We investigated some of them (Crişan *et al.*, 1992, 429-431).

Furthermore, south of Miceşti, in the Cheile Turenilor area and more to the south, the tumuli line continues. At Câmpia Turzii a large tumulus and another three were located at the mouth of Valea Florilor in the direction of the salt sources. Valea Florilor separates the salt areas from Turda from those at Cojocna (Crişan *et al.*, 1992, 116-117).

Many permanent Neolithic sites are located in the southern area, at 4-12km distances from the salt spring (over the hill the road is even shorter). In this area there are materials related to Starčevo - Criş IIA (at Copăceni on Valea Turenilor), sites belonging to Zau culture and others, until the end of Copper Age (when we believe that the tumuli stopped, because the Bronze Age began). The oldest tumuli are related to Coţofeni culture, the end of the fourth millennium and the first half of the three millennia following (calibrated data) (Popa 2009, 1008-1010).

According to their dimensions and other features, such as a main chamber covered with wooden girders or other wooden installations (but without inventory, or with just a flint blade), some tumuli are related to the phenomena in the east (Câmpia Turzii, Petreştii de Jos – *Continit*, excavations made by Z. Maxim, M. Wietemberger), that breakthrough in Transylvania. Other older tumuli have pottery from the Coţofeni culture era up until the Early Bronze Age. Several discoveries related to the Neolithic and Copper Ages were discovered in the Cheile Tureni area; they are located at about 3-4km, as the crow flies, from the salt spring. The archeological area is very rich, with over 40 archeological points, with habitations from the Neolithic until the Early Medieval period. The most intensive habitation is related to the Tiszapolgár culture, followed by Dacian pit houses (with grinding stones inside), Medieval pit houses with large stone slates, similar to those discovered at Iclod.

Figure 17. Cheile Turzii – Peştera Ungurească. Reconstruction of the workshop, in shape of a canopy.

Figure 18. Cheile Turzii – Peştera Ungurească. Trestle for the crucible and ledge for conducting.

Cheile Turzii

Another group of many habitations from the Neolithic period but from other periods too, is located in Cheile Turzii area (Figure 16). Over 100 archaeological points are closed at the mouth of *Valea Miceştilor,* at about 6km south of the salt spring. Most of these habitations have been used temporarily; the area is not suited to lengthy residence or for household activities (animals).

Downwards of the Cheile Turzii, the Hăşdate River falls into the Arieş River (= river with gold). Here much panning for gold has been organized. The road from Cheile Turzii meets the Arieş River to the south, through the ridge area marked by the tumuli (6 km), or through Hăşdate Valley, at 2km (here the gold was washed for the jeweler workshop).

The jeweler workshop at Peştera Ungurească / Peştera Caprelor

This workshop represents one of the most important discoveries. The workshop for gold jewels was set up in the eastern central part of the cave, where the air is expelled.

Figure 19. Cheile Turzii – Peştera Ungurească. Trestle for the crucible and ledge for conducting, other view.

In four to five archaeological expeditions, over 40 golden beads at different stages of work were discovered. There were also golden foils, cuts of golden foils, pieces that decorate clothes or leather (Figure 25). In the workshop copper beads, stone beads or shell beads were also discovered. Some of them were at different stages of work. The workshop was used temporarily, abandoned many times. Firing stages and habitation stages were discovered. In the last stages of the workshop, in one part of the complex, an oven with a hole for a bellows was discovered (Figures 20-21). The oven was used for smelting, purifying or soldering. The main hearths from the first stages had the same functions.

The most important installation of the workshop (Figures 17, 24) was a vault oven with a hole for the bellows that intensifies the fire. The oven was carefully built. Oven dimensions were bigger than the first hearth (with several repairs) found outside the oven. In the first stage the oven has a vault supported by a main pillar (in the shape of an '8'). The vault was made of clay from the cave (or from a shelter located down by the cave).

The oven, located outside the dwelling, has three functional phases. The last one, after the vault fell down, was used as a simple hearth (Lazarovici *et al.*, 2004; Lazarovici and Lazarovici 2007, 271 and the next pages, etc.). The oven mouth was in the interior of the canopy, at about 35cm from the floor when it was built.

In the second phase, in the main area the oven has a 2cm height trestle (Figures 18-19), in the shape of a circular arch, used for different purposes (deposition of the smelting crucibles, cleaning of the gold taken from the river, or for soldering the beads). The trestle, in the shape of a circle arch, served to conduct the golden berries to the oven mouth. The hearth of the oven was downwards to the mouth of the canopy. The air was blown through the left, while the wood and the charcoal (after the oven was warm) were conducted behind the oven and around the prop for the crucible. The most important for the Copper Age is this metallurgic oven with a hole for the bellows, to intensify control and conduct heat (Figures 20-21). In the first stage the hole was used to intensify the fire, to obtain embers (Figures 22-23). In the second one, after the crucible was set up and the embers were arranged, it was used to sustain the embers on order to control the smelting of the crucible contents.

Figure 20. Cheile Turzii – Peştera Ungurească. The hole for bellows with two reparation phases.

Figure 21. Cheile Turzii – Peştera Ungurească. The hole for bellows with two reparations phases, other view.

We have emphasized these details because, up to the present date, the workshop represents a *unique* discovery. The temporary character of the habitation raises some problems. The large amount of animal bones (from those hunted or from those kept domestically) that we have discovered here, prove that the meat cannot have been consumed by a small group of people. They can use two conservation methods, smoking of the meat, using of the salt water or both methods, as used today. Up to the present date we cannot prove either method (our research was especially focused on the workshop).

Figure 22. Cheile Turzii – Peştera Ungurească. Oven, phase I: the wall and vault base.

Figure 23. Cheile Turzii – Peştera Ungurească. Oven, phase I: main pillar.

Most of the pieces discovered (golden, copper, stone, bone or shell jewels) were at different stages of work (Figure 25). Some of them even represented scrap processing, others were lost during the working process; we recuperated them simply because we have sieved all the earth from the area investigated. There are also foils in the different kinds of processing work, cutting fragments and corners from some pieces with traces of leather or textiles. The microscopic investigation is not finished, but we stress that there are many processing traces. In the same workshop two copper tools used for the cutting and processing of the gold were discovered. Some shards from pots with graphite, discovered here, could be used as crucibles (they can support high temperatures).

According to the zoological-archeological analyses of the bones from the cave, in the first place come the small animals, such as sheep / goat at the Zau level (27,9%) and pig (11,4 %). Sacrifice of the animals took place until they were two years old (50%), so salt was necessary during this time.

Figure 24. Cheile Turzii – Peştera Ungurească. Details of the workshop, 1st and 2nd phases.

For the wild animals those that were not mature predominate. They occupied a higher percentage than in other cases (14,7 % in Zau culture; 15,3 % in Petreşti culture; 20,3 % in Bodrogkeresztúr II – Schiebenhenkel / toarte pastilate). Beasts represented 25,8% during the Copper Age. This confirms once more the temporary character of the habitation in the cave, when the workshop was in use.

Figure 25. Cheile Turzii – Peştera Ungurească. Jewels discovered in the levels of the workshop floor.

If sheep and/or goat can be brought in the cave, for which the name *Peştera Caprelor / Goat cave* derives, pigs and other animals cannot climb as far. During Petreşti and Bodrogkeresztúr II – Schiebenhenkel / toarte pastilate cultures the animals prevail (Bindea 2008, Chapter VI).

Gherla – Jichiş

Near the Roman road between Gherla and Buneşti, at the bottom of the hill where the Roman road is located, there is a valley with several salt springs. One, not arranged, forms a small basin with clay and salty iron water. From here, a small salt stream begins, lengthwise, with salt vegetation. Older people who would like to take baths need to make very small preparations here.

At *Tăul Sărat* (Figures 26-27) as well as at *Fântâna cu saramură / Fountain with salt* (Figure 29), small streams start with specific salt vegetation. At 15m distance from here, vegetation is normally different. *Tăul Sărat* is recent. We cannot observe any sort of disturbances over it as in the case of *Fântâna cu saramură*.

Figure 26. Buneşti, Tăul sărat.

Figure 27. Buneşti, Tăul sărat, other view.

Fântâna cu saramură

After the general view, we can see a recent arrangement, deep in the salt strata which reach 1,5m from the soil surface (Figures 30). The fountain is near an older arrangement from which there are many wooden traces of different constructions. An excavation here might be very interesting, raising the possibility of taking dendrology and radiocarbon samples.

Figure 28. Tăul Sărat valley, dried, at about 50m far for the spring.

Figure 29. Fântâna cu saramură Valley.

Figure 30. Buneşti, Fântâna cu saramură, salt vegetation, streams with salt water and landscape.

Several pits dot the coast over the fountain. The area fell because of the salt-water flow. In some areas we can see older human interventions. Attempts had been made to take out the clay to reach the salt rock. At this moment we can observe them only on the margins of the well.

On the margins of the stream, as well as on the salt coast, there is specific salt vegetation that dissimulates the normal vegetation nearby.

In the area of the old arrangements there are over 80cm as deposits, wooden girders and pillars. As the wood is very well preserved, we can see horizontal and vertical structures. Wood brought by rain or settled by people is covered with salt flowers during the warm, dry season.

During the time when the salt fountain is not intensively exploited, from here a small stream flows, at about 3-5 liters / minute, but it depends on the dry or wet season. The archeological area here is very rich, with habitations from different periods (Figure 33). The closest Neolithic site is located on the other part of Someş River, in the main area and south of Gherla town. It belongs to the Zau culture, phase III, perhaps earlier. There were no excavations, but zoomorphic figurines with two heads have been discovered. To its south, four Neolithic sites have also been identified. The earliest site is at Livada (SC IC related to the second migration of the shepherd's) 6km far; Iclod site (belonging to Zau culture, phase IIC-IV) is about 11km distant. The sites identified cover all periods: Neolithic (N), Bronze Age (Bz), Iron Age and Dacian (F, D), Roman period (*castrum*, road, civil dwellings (R), Early Medieval (F), as well as other prehistoric ones, non definite (P) (Figures 33).

We believe that this area was also used to supply human communities from the central Transylvania. Different roads to the Transylvanian Plain start from Gherla. The most important and largest site is Ţaga, located at about 20km distance from the salt source (Figure 32). This salt source is still used today, as farmers from different villages have shown us. Therefore it is possible that different communities, such as Iclod – Petreşti or Zau IV, with an impressive development on the Transylvanian Plain, up to the borders with Bistriţa and Mureş counties, have used these sources.

Beyond Târgu Mureş to the east the salt supplies have been brought from Praid – Sovata; Ariuşd communities have reach Târgu Mureş area, the most western extension.

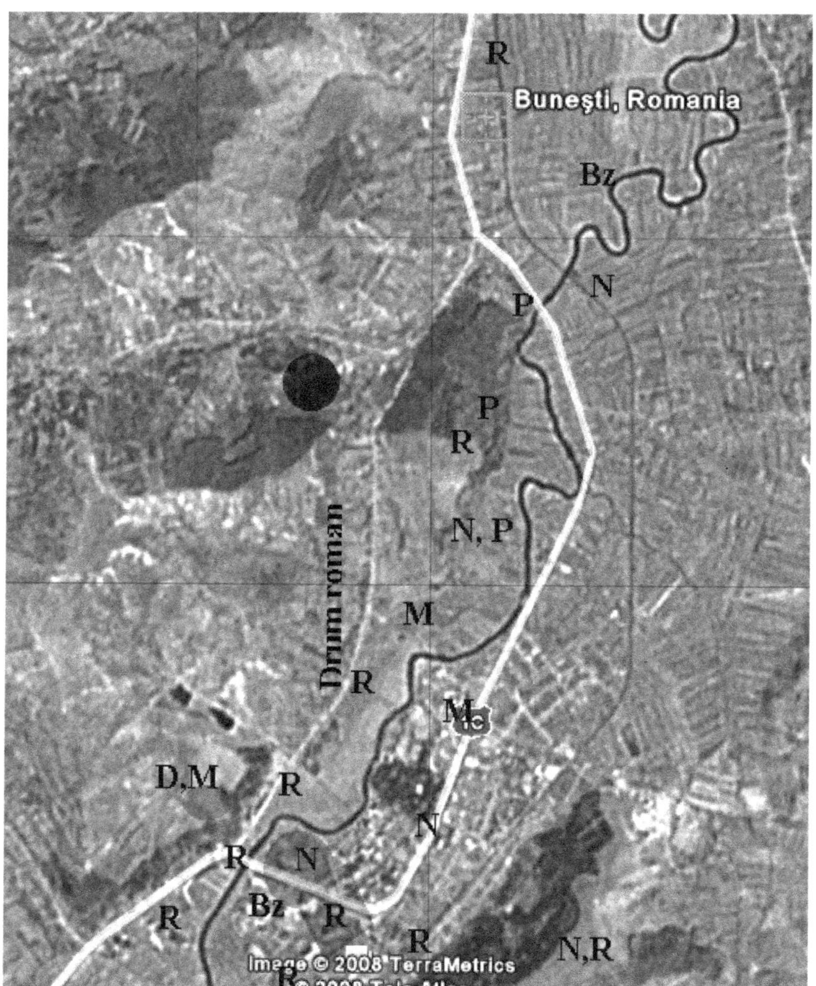

Figure 31. Fântâna cu slatină (red circle) and archaeological points, marked with capitals.

Cojocna – Căian area

In the village Băile (baths) Cojocna, with a low salt concentration (19%), the salt water is used for therapeutic purposes for tourists. As early as the 19th century, M. J. Ackner observed and mentioned traces of salt exploitations of the Roman period. In the area a bronze Roman *fibula* is also mentioned (for the salt source, Rusu 1977, Figure 4; Crişan *et al.*, 1992, 115, no. 6; Wollmann 1996, 243; http://arheologie.ulbsibiu.ro/publicatii/bibliotheca/popa/repertoriu%20c.htm). From the area many valleys with salt and archaeological relics start. We would like especially to emphasize the prehistoric and Neolithic traces (Figure 33).

Even at the spring from the south valley, where the railway goes to Boju (a small low-concentration salt spring), along the valley there are many relics and sporadic habitations related to the 4th and 7th centuries. It is a hilly area, very good for pasturing (Figure 4).

Both areas are at the limit of the salt massive. Valea Florilor continues into Valea Cojocnei, 7km long. At its mouth, upstream of discharge into the Someş River, there is a Dacian site with Dacian - Roman features, with many *villae rusticae* located on all valleys in the direction of the salt sources. Here an unexcavated site has very rich material, belonging to the Iclod – Petreşti synthesis. In the southeastern area there are three tumuli. Towards Căian - Vamă, in a valley that also starts from the salt massive, there is another archeological site with pottery attributed to the same Iclod – Petreşti synthesis (materials in a school collection, as well as in the Cluj museum).

Most of the sites are small, reflecting temporary habitations, specific to such areas. Only the Roman sites are of large dimensions. In each valley there is a *villa rustica*.

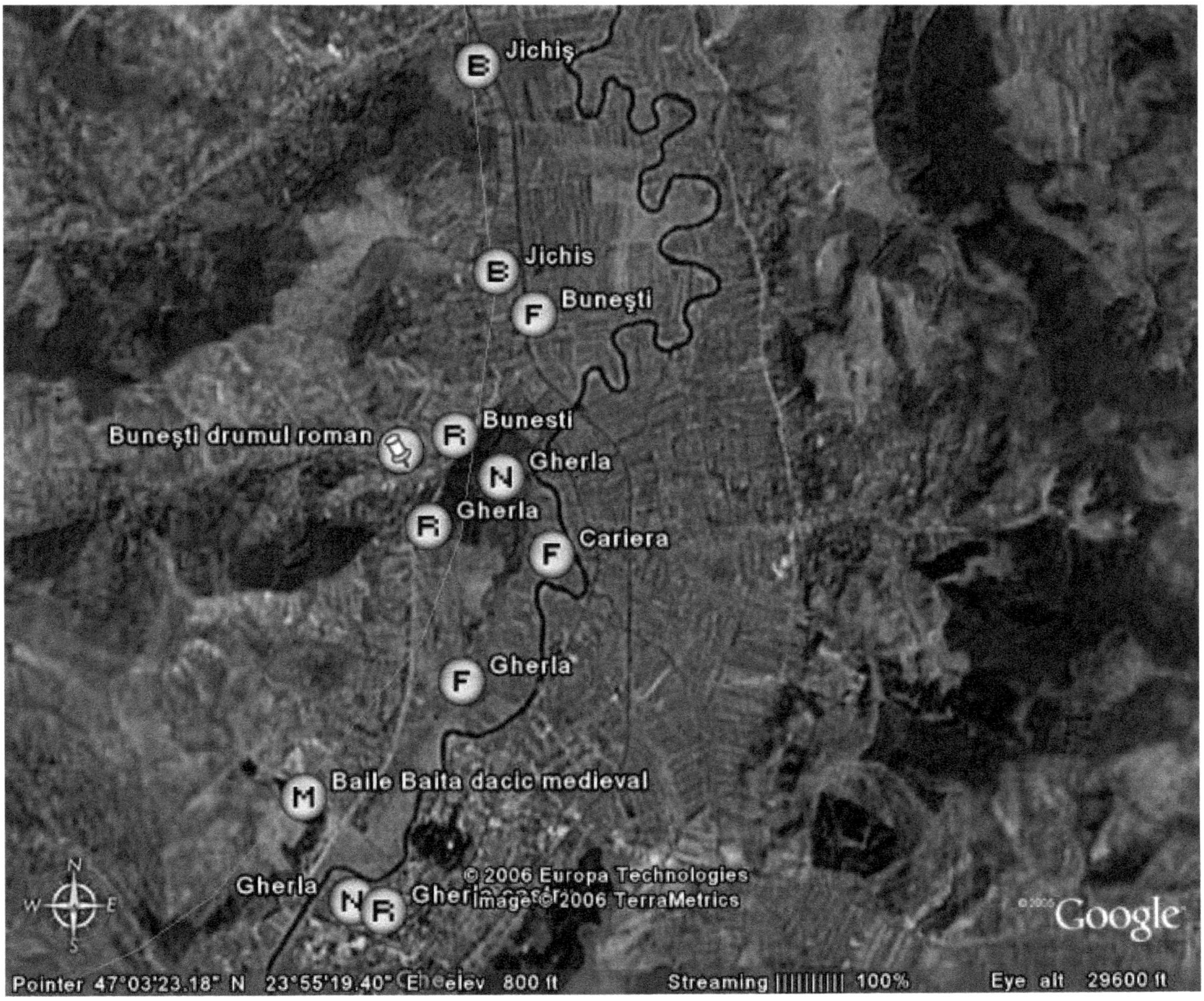

Figure 32. Fântâna cu slatină and archaeological points marked with capitals.

Figure 33. Cojocna Băi massive and archaeological relics from different periods.

Figure 34. Salt traces on Cojocna Băi massive and the valley to Valea Florilor.

Apahida area

The area is very rich in bitterns, in the sense of a salt, a waste product of solar salt operations rich in magnesium sulphate, from Pata (Rusu 1977 no. 40) until after Apahida. All the streams and valleys, from the largest, in Mărăloiu valley, with *'tăul Moerului'* to other smaller valleys brought bitterns to the first part of the Someş meadow under the hills. At Apahida, near the former animal farm, on the road to Căian, there are several lakes in the upper part of the salt mountain (Rusu 1977, Figure 4). On the heights there are no bitterns. There are many traces of several habitations here (Figure 36): 16 for Neolithic times and prehistory, six for the Bronze Age, two Coţofeni habitations, five from the Hallstatt time, many from the Roman periods (ceramics, coins, *fibulae,* lamps, iron objects), three Roman *villae rusticae*, sites related to the Dacian and Latène periods (a Celtic necropolis). From Early Medieval times there are three golden treasures (Figure 35).

There are three tumuli in the railway station area: two on the right side of Valea Mărăloiu, which were destroyed by recent deposits. There was another tumulus on the left side. This one was of a considerable size. Prospected and with an obituary chamber, it belongs to the period of migrations. On the left are the valleys of Someş River. Towards Cluj there are several Roman *villa rustica.*

On the Fânaţe Valley that runs down Apahida, there is a Roman *villa rustica,* four sites from the 4th eds. century, one from the 5th – 6th century, another four from the 8th – 9th centuries and a salt spring. (In the morning it was very salty, after intensive exploitation less so). The main role was played during the period of migrations, when it was a salt supply center for migratory peoples, from Goths to Avars. Three golden princely treasures date from this period (Crişan *et al.*, 1992). The first medieval states (Moravian and Central European) were very interested in salt exploitation and the control of such sources.

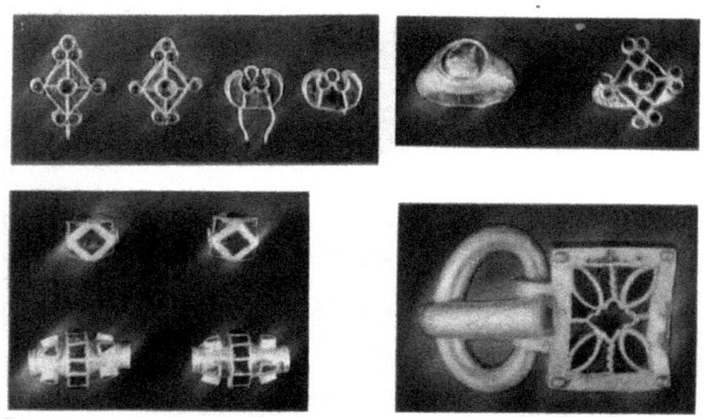

Figure 35. Apahida - Someşeni, jewels from the prince's grave discovered at Apahida 2.

Coasta (commune Bonţida). The brine fountain

The salt massive proceeds to the north on the right side of Someş River. On the interior valleys, in parallel with Someşul Mic, there are traces of bitterns or salty springs. A salty fountain (Figures 37-40) was recently set up between the Sic and Coasta villages. This was on the margin of a small spring or stream that disappears after about 400 m.

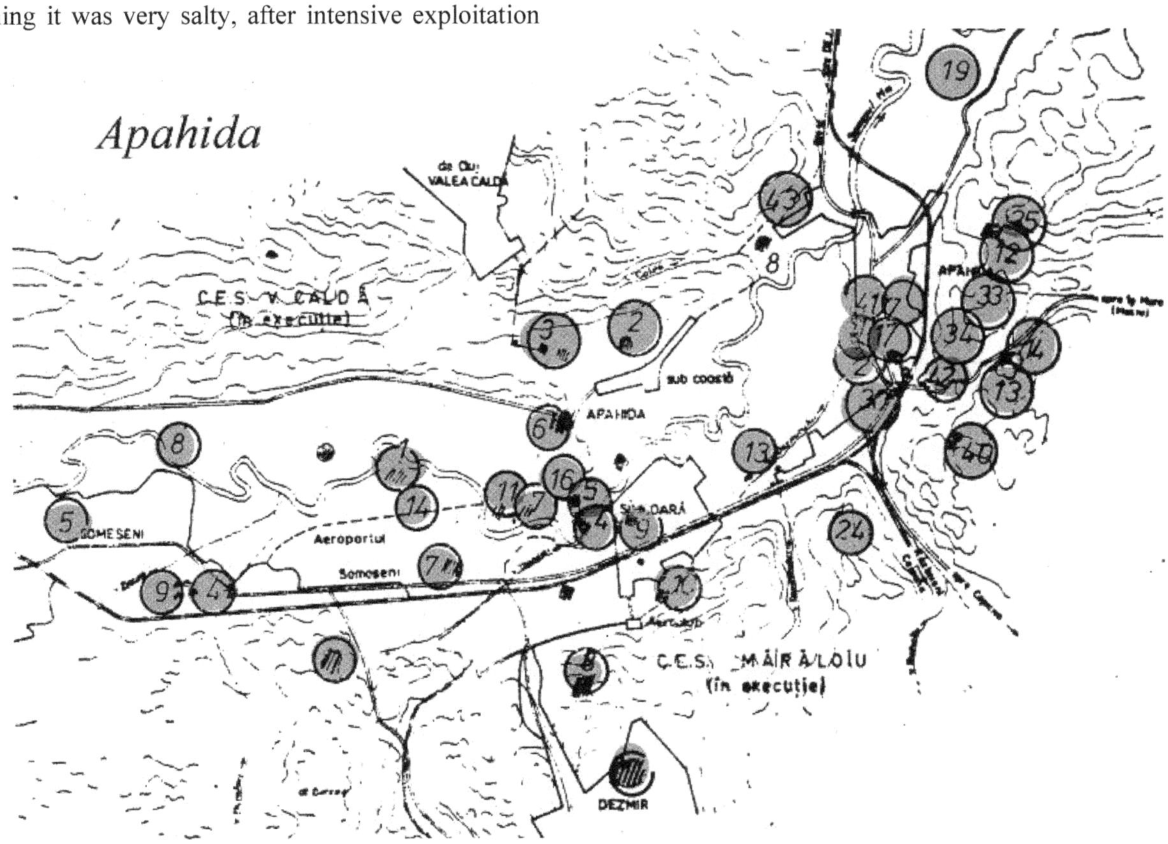

Figure 36. Apahida – Someşeni archaeological area.

Figure 37. Coasta – Sic, Fântâna cu saramură.

Figure 38. Coasta – Sic, Fântâna cu saramură, hook.

Figure 39. Coasta, basin with salt water and mud.

Figure 40. Coasta, basin with salt water and mud, general view.

On its margins there is specific salt vegetation. People from the area use the well occasionally, especially in the autumn. In parallel with it is another dried spring; people using mechanical means have made a basin for salty baths and mud, at about 400m distant from the salty spring (Figures 39). The salt concentration is very high. In the main part of Coasta village, located at 500m distance from the well, many shards related to prehistory and Hallstatt have recently been discovered; earlier several bronze axes and an urn were found here.

On the hilltop that dominates the village there are many salt fountains, belonging to Sic village. Some farmers affirm that there is also a salty lake at *Tăul Coastei.* The valley is marshy. On the margin of salty area lies specific salt vegetation, as well as flowers that never dried.

Sic

Sic Valley is parallel with the Someş Valley which passes through the salty massive (Figure 30). East of the village there is a stream named *Valea Sărată*, near the trout farm. In the village we just carried out a surface investigation; on that day people were in the church. Most of them speak only Hungarian.

It is also a touring area with baths. On the road from Coasta there is a place named *salty fountains / fântânile sărate*. In Sic area (Figure 41) there are 42 points with archaeological relics (Crişan *et al.*, 1992, 342-353; for salt, Rusu 1977, Figure 4) seven Neolithic sites, one from the Copper Age, Tiszapolgár culture, six of Coţofeni culture, etc.). We also have to add Hallstatt sites (one fortification on the hill, another on the valley), one Latène, one Dacian, 11 from the Roman period as well as a *villa rustica* and two Early Medieval sites. St. Ferenczi, who carried out several research visits here, considers that at the salty stream (*Sósrét*) there are salty mine exploitations. On the road to Gherla, near the salty springs there is a Neolithic site, but many others relate to the Roman era.

As can be observed in other places, much information emerges when scientists have investigated the field areas. Archeological determinations are very vague in many cases. Neolithic sites are generally located in areas with fresh water, not in salty ones. Coţofeni habitations are very movable and seem to be related to pasture and also to animal herding and hunting.

The researchers need to undertake larger ethno-archeological studies as well as preparation of the teams beforehand, and the use of some local guides who know the toponimy. Few of them can be found now. In conclusion, in the Transylvanian areas with salty springs, there are many archeological relics, more or less concentrated; to attribute them to the correct cultures will involve an extensive research, in larger teams. They will have to follow and solve very concrete tasks.

Figure 41. Sic, archaeological area.

References:

Biagi, P. and Voytek, B. 2006. Excavations at Peştera Ungurească (Caprelor) (Cheile Turzii, Petreştii de Jos, Transylvania 2003-2004): Preliminary Report on the Chipped Stone Assemblages from the Chalcolithic Toarte Pastilate-Bodrogkeresztúr (Layers). *Analele Banatului* S. N. XVI/1, 177-202.

Bindea, D. and Sângerean, C. 1996. Câteva observaţii asupra materialului faunistic de la Cheile Turzii – Peştera Ungurească. *Acta Musei Napocensis* 33, I, 477-509.

Bindea, D. 2008. *Arheozoologia Transilvaniei în Pre- şi Protoistorie*, Cluj-Napoca, Editura Teognost.

Ciută, M., 2005. *Începuturile neoliticului timpuriu în spaţiul intracarpatic transilvănean.* Bibliotheca Universitatis Apulensis, XII. Alba Iulia, Editura Aeternitas.

Ciută, M., Daisa, B., Breazu, M. and Andrei, Şt. 2000. *Şeuşa, com. Ciugud, jud. Alba. Punct: La cărarea morii. Cronica cercetărilor arheologice din România. Campania 1999*, CIMEC, 101.

Ciută, M., Beldiman, C., Ferencz, I., Mazăre, P., Daisa, B., Breazu, M., Andrei, P., Truţă, C. 2001. Şeuşa –

Gorgane, com. Ciugud, jud. Alba. Punct: La cărarea morii. *Cronica cercetărilor arheologice din România. Campania 2000*, CIMEC, 239–242.

Ciută, M., Antoniu, M., Daisa Ciută, B., Gligor, A. 2005. Şeuşa, comuna Ciugud, jud. Alba. Punct: Gorgan, *Cronica cercetărilor arheologice din România. Campania 2004*, CIMEC.

Crişan, I., H., Bărbulescu M., Chirilă, E., Vasiliev, V. and Winkler, I. (red.) 1992. *Repertoriul arheologic al judeţului Cluj, Bibliotheca Musei Napocensis* V, Cluj.

Ghergari, L., Lazarovici, Gh., Ionescu, C., Tămaş, T. 1999. Studii geoarheologice asupra unor artefacte ceramice din neoliticul timpuriu din România. Staţiunea de la Lunca – Poiana Slatinii, jud. Neamţ. *Angustia* 4, 1-7.

Gherghe, P. 1982. *O nouă aşezare Coţofeni descoperită la Vîrţ, judeţul Gorj. Oltenia. Studii şi Comunicări* 4, Craiova, 38-45.

Lazarovici, C.-M., Lazarovici, Gh. 2007. Arhitectura Neoliticului şi Epocii Cuprului din România. II. Epoca Cuprului. In V. Spinei, V. Mihailescu-Bîrliba (eds.), Editura Trinitas, Iaşi.

Lazarovici, Gh. 1996. *The Process of the Neolithization and the Development of the first Neolithic Civilizations in the Balkans.* In R. Grifoni Cremonesi, J. Guilaine, J. l'Helgouach (eds.), *The Neolithic in the Near East and Europe*, Colloquium XVII, 9, XIII ICPPS, 21-38. Forli, Italia, 8-14, September, 1996. Forli, Abaco.

Lazarovici, Gh. 1998. *About the Neolithization process of the second migration of the Early Neolithic.* In Fl. Draşovean (ed.) *The Late Neolithic of the Middle Danube Region, International Symposium on the Problems of the Neolithic in the Middle Danube Region, June 1997, Timişoara*, seria *Museum Banaticum Temesiense, Bibliotheca Historica et Archaeologica Banatica* XIV, 7-37. Timişoara, Editura Eurobit.

Lazarovici, Gh. 2006. *The Anzabegovo – Gura Baciului Axis and the first stage of the Neolithization Process in Southern – Central Europe and the Balkans.* In N. Tasić, C. Grazdanov (eds.), *Homage to Milutin Garašanin*, 111-158. Belgrade, SASA Special Editions.

Lazarovici, Gh. 2008. *Cultura Zau*, ms. in the scientific volume of Olten (Switzerland) exhibition (in press).

Lazarovici, Gh. 2009. *The Zau Culture.* In Fl. Draşovean, D. L. Ciobotaru, M. Maddison (eds.) *Proceedings of the Conference held at the Museum of Banat on 9th-10th, 2007. Ten years after. The Neolithic of the Balkans, as uncovered by the Last Decades, of Research*, seria *Museum Banaticum Temesiense, Bibliotheca Historica et Archaeologica Banatica* XLIX, Timişoara, 179-217.

Lazarovici, Gh. and Cristea, I. 1979. Contribuţii arheologice la istoria străveche a comunei Uioara de Jos, Ciunga, jud. Alba. *Acta Musei Napocensis* XVI, 431-446.

Lazarovici, Gh. and Bulbuc, A. 1983. Descoperiri arheologice în hotarul comunei Iclod. *Apulum* XXI, 161-166.

Lazarovici, Gh. and Lazarovici, C. - M., 2008. The Workshop for Gold at Cheile Turzii – Peştera Ungurească / Peştera Caprelor. Archaeological Excavation 2003-2004. In E. Popoviciu, M. Ivan, Şt. Ţigan (eds.), *Annals of the Tiberiu Popoviciu Seminar of the Functional Ecuations, Approximation and Convexity*, Cluj-Napoca, Mediamira Science Publisher.

Lazarovici, Gh., Băltean, I., Biagi, P., Spataro, M., Lazarovici, C. - M., Colesniuc, S. and Vrâncean P. 2004. Petreştii de Jos, com. Petreştii de Jos, jud. Cluj. Punct: Cheile Turzii. *Cronica cercetărilor arheologice din România. Campania 2003,* CIMEC.

Lazarovici, Gh., Biagi, P., Spataro, M., Lazarovici, C. - M., Colesniuc, S., Suciu, C., Roman, C., Chitic, O., Angeleski, S. and Tatar, A. 2005. Petreştii de Jos, com. Petreşti, jud. Cluj. Punct: Cheile Turzii. *Cronica cercetărilor arheologice din România. Campania 2004*, CIMEC.

Lazarovici, Gh., Kalmar/Maxim, Z., 1987-1988. Săpături arheologice de salvare şi cercetări etnoarheologice în Munţii Petrindului în anul 1986. *Acta Musei Napocensis* 24-25, 1987-1988, 949-996.

Luca, S. A., Diaconescu, D., Georgescu, A. and Suciu, C. 2006. Cercetările arheologice de la Miercurea Sibiului – Petriş (jud. Sibiu). Campaniile anilor 1997 – 2005. *Bruckenthal Acta Musei* I.1, 9-19.

Luca, S. A., Pinter, Z. K. and Georgescu, A. 2003. *Repertoriul arheologic al judeţului Sibiu.* Bibliotheca Septemcastrensis III. Sibiu, Editura Economică.

Maxim, Al. I. 1971. Un depozit de unelte dacice pentru exploatarea sării. *Acta Musei Napocensis* VIII, 457-463.

Monah, D., 1991. L'exploitation du sel dans les Carpathes Orientales et ses rapports avec la culture Cucuteni-Tripolye. In V. Chirica, D. Monah (eds.) *Le Paléolithique et le Néolithique de la Roumanie en contexte européen*, seria *Bibliotheca Archaeologica Iassiensis* IV, Iaşi, 387-400.

Paul, I., 1981. Die gegenwärtige Forschungsstand zur Petreşti-Kultur, in *PZ*, 56, 1981, Heft 2, S. 197-234.

Paul, I., 1992. *Cultura Petreşti.* Bucureşti, Editura Museion.

Paul, I., 1995. *Vorgeschichtliche Untersuchungen in Siebenburgen,* Alba Iulia, Editura Universităţii "1 Decembrie 1918", seria *Bibliotheca Universitatis Apulensis,* I.

Popa, C. 2009. *Cultura Coţofeni în Transylvania,* Ph.D., "1 Decembrie 1918" University, Alba Iulia. http://arheologie.ulbsibiu.ro/publicatii/bibliotheca/popa/repertoriu %20c.htm

Rusu, M., 1977. Transilvania şi Banatul în secolele VI-IX, *Banatica* 4, Reşiţa, 169-213.

Cavruc, V., Chiricescu, A. (eds.). 2006. *Sarea, Timpul şi Omul. Catalogul expoziţiei.* Sfântu Gheorghe, Angustia.

Weller, O., Dumitroaia, Gh. 2005. The earliest salt production in the world: an early Neolithic exploitation in Poiana Slatinei-Lunca, Romania. *Antiquity,* Vol. 79, No. 306, December 2005 (http://www.antiquity.ac.uk/projgall/weller/index.html #author#author).

Wollmann, V. 1996. *Mineritul metalifer, extragerea sării şi carierele de piatră în Dacia romană / Der Erzbergbau, die Salzgewinnung und die Steinbrücke in römischen dakien.* Seria *Bibliotheca Mvsei Napocensis,* XIII, Editura Muzeul Naţional de Istorie a Transilvaniei, Cluj-Napoca/Deutsches Bergbau-Museum Bochum, Cluj-Napoca 1996, ed. Gh. Lazarovici.

http://arheologie.ulbsibiu.ro/publicatii/bibliotheca/popa/repertoriu%20c.htm

New Archaeological Researches concerning Saltworking in Transylvania. Preliminary Report

Valeriu Cavruc

Muzeul Naţional al Carpaţilor Răsăriteni, Sfântu Gheorghe, Romania

Anthony F. Harding

University of Exeter, United Kingdom

Abstract

In 2005-2008 we carried out researches concerning prehistoric saltworking in the north-eastern part of Transylvania (Bistriţa-Năsăud County), in the sites Băile Figa, Săsarm and Caila.

The Băile Figa site (Beclean district) was discovered by the geologist Dr. Ioan Chintăuan from the Complexul Muzeal Judeţean Bistriţa-Năsăud in 1977. In 2005 he extracted from the Pârâul Sărat stream (salty stream) a wooden trough. Soon after, we visited the site and took two samples from the trough. These produced radiocarbon dates which give a combined date of 1090-970 cal BC (88% probability). The site is placed above a huge deposit of rock salt 1.5-4m from the terrain surface. There were seen three main kinds of evidence on the surface of the site: about 460 timbers visible in the Pârâul Sărat stream; many cavities and mounds of earth resulted from diggings made in different times for salt extraction; small areas with Bronze Age and post-medieval pottery, as well as several isolated artifacts (stone mining tools, wooden pegs, etc.). We also collected in 2006 new samples of timbers for C14 dating and submitted them to the Groningen laboratory. Of these, four fall around 1000 cal BC, one around 250 cal BC and one around 450 cal AD.

In 2007 and 2008 we carried out archaeological excavations in the site and collected 35 samples of timbers for C14 dating (Oxford laboratory). We opened 20 trenches. The most relevant results were obtained in trench I (near the spring of the Pârâul Sărat stream), in trench III (at the northern end of the site), and in trench XV (in the place were I. Chintăuan discovered the trough). In trench I we found the traces from Eneolithic period, Bronze Age, post-Roman and post-medieval periods. The Bronze Age evidence is predominant and most conclusive. It includes some disturbed wooden structures, two wooden troughs, pottery, some wooden small shovels, hammers, a ladder, and stone mining tools. According to the C14 analyses the troughs date to 16th-15th centuries cal BC. There was also uncovered a rectangular construction (3.8 x 2.8m) which seems to be the top of a mine shaft (so far undated). It cut through earlier wattle fences. Trench III revealed a grouping of wattle, pole and plank fences which cover an area of about 19 x 20m. On the east and north-east this grouping is delimited by an arch-shaped wattle fence. In the middle of this grouping we uncovered a room of about 10 x 2.5m which has the shape of three joined circles. In its southern part this room continues by a corridor with parallel walls. This room was roofed by a massive wooden beams and hazel twigs. The artifacts found in this trench include stone mining tools and wooden objects: a socketed axe handle, pegs from troughs, shovels, different hook-shaped objects, a fragment of perforated disc, etc. All these were dated by C14 to the period between 1005 and 915 cal BC. We also found in the trench III and next to it the Early Bronze Age pottery which probably was contemporaneous with the fragment of wood dated by C14 to the beginning of the 3rd millennium BC. Trench XV revealed some vertically placed poles, a trough, a ladder, a stone mining tool and a perforated wooden 'hook' with a small piece of wood inserted in its opening. The Săsarm site covers the valley of the Valea Sărată (salty valley) stream. Along this stream on the length of about 800m various wooden structures and isolated timbers are visible. Some of them definitely come from mining galleries. Most timbers seem to date to the post-medieval period. At the same time, on the left bank of the stream, against the brine well we found a small area with Early Bronze Age and Iron Age pottery, what suggests that people exploited salt in that periods. The Caila site looks similar to Băile Figa. Many wooden structures are visible on the surface, along a length of about 300m, in the brine stream. We discovered a trough in this stream. In the same area a few Bronze Age sherds were found.

Keywords

prehistory, salt, salt mine, tools, wood, Romania

Transylvania is one of the richest salt-bearing provinces of Europe. As numerous written documents show, from the early Middle Ages salt was intensively exploited in this area and widely traded; it was one of the most important sources of money for the authorities and played an important role in economic and political life. In contrast, the evidence of prehistoric and ancient saltworking was limited until recently to a few accidental findings. These included wooden objects used in salt mining found in *Valea Florilor*, *Ocna Dejului*, *Turda*, *Ocna Mureş*, *Băile Figa* and *Valea Regilor* (Wollmann and Ciugudean 2005, 100-101), as well as some votive altars with inscriptions mentioning persons responsible for administration of salt in Roman Dacia found in *Veţel-Micia* (IDR III/3, 119), *Sânpaul* (IDR III/4, 248), *Sărăţel-Domneşti* (Russu 1966, 7-13) and *Boia Bârzii* (Piso 2007). Archaeological researches concerning saltworking have never been performed in

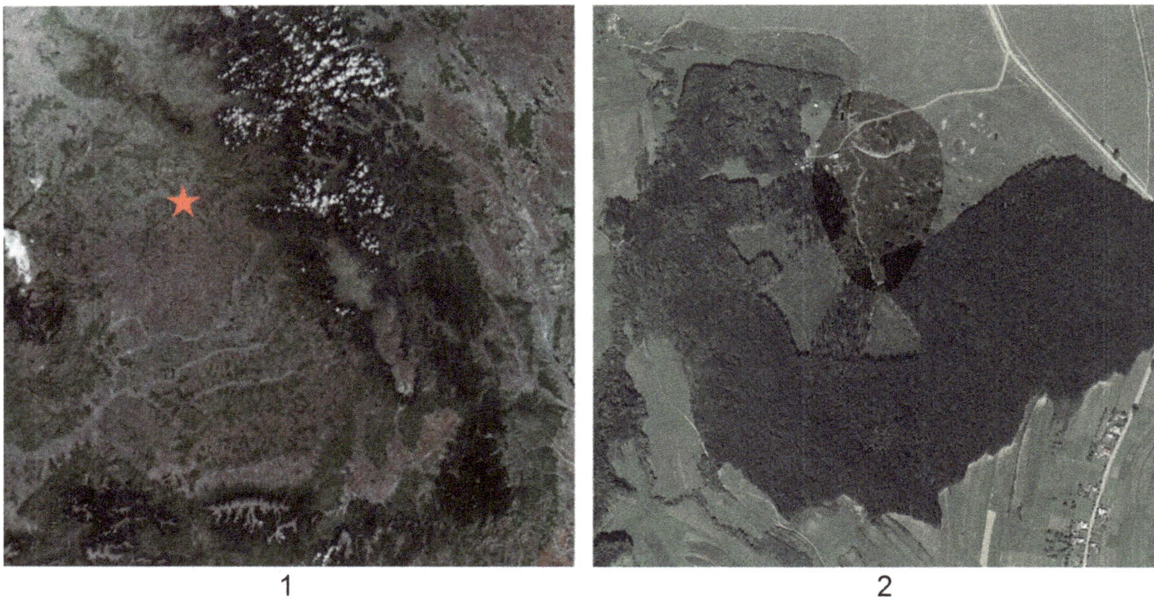

Figure 1. Băile Figa site: the geographical position of the site (1); the aerial photo of the site (2).

Figure 2. Băile Figa site: digital picture of the site (1); visible timbers *in situ* in the stream (2); visible timbers in the site – yellow dots (3).

Transylvania before 2000, when the British-Romanian project regarding prehistoric and ancient saltworking in Carpathian Basin was initiated[1]. In 2005-2008 in the frame of this project we carried out field researches in north-eastern Transylvania – on the territory of the Bistriţa-Năsăud County, in the region around Beclean, which is rich in rock salt deposits relatively close to the surface (i.e. accessible to non-industrial exploitation). We started work on three sites: *Băile Figa*, *Săsarm – Valea Slatină* and *Caila – Sărătura*. The results of these researches have to a great extent overcome the lack of knowledge concerning prehistoric and ancient saltworking in Transylvania. Although analysis is still in progress and many aspects are uncertain, we present here – in very general terms – the first results of this work.

*

The ***Băile Figa*** site (Figure 1-2) is situated in the administrative territory of Beclean town, about 1km south of its south-eastern end, and about 1km north of the north-eastern end of Figa village. The site was discovered by the geologist Dr Ioan Chintăuan in 1977, when he found many traces of wooden structures in the valley of the brine stream named by locals 'Pârâul Sărat' (salty stream), among which he observed the broken end of a trough. In 1988, I. Chintăuan and I. Russu, relying on the similarity between this trough and the one from Valea Florilor (which was believed at that point to date to the 'Dacian La Tène'), dated it to the second period of the Iron Age (Chintăuan and Russu 1988). In May 2005 Chintăuan removed the trough from the salty mud (unfortunately without proper recording and not accompanied by an archaeologist) and transported it to Bistriţa museum (Complexul Muzeal Judeţean Bistriţa-Năsăud). Soon afterwards he published the trough, putting forward the hypothesis that it was used to obtain salt by evaporating brine (Chintăuan 2005).

In summer 2005 we visited the site for the first time and examined it to estimate its archaeological potential.

The geomorphological survey of the site carried out with the support of I. Chintăuan revealed the following. The site overlies a massive rock salt deposit about 1000 x 1000m wide and about 1600m thick. This deposit appears at a depth of 1.5-4m from the surface. The site covers a hemispherical depression about 600m in diameter. The sediment accumulated in this depression, in contact with the salt, transformed itself in time in a consistent layer of salty mud. The mud layer is overlain by a yellowish soil mixed with gravel, about 1m thick. The underground water appears on the surface through several brine springs which, together with the rain water, created the brine stream. The main brine spring of this stream lies at the southern end of the depression. The stream crosses the valley up to its northern end. It joins three tributaries coming from other brine springs on the eastern slope of the depression. At about 350m from its main spring, the stream meets its first tributary, 118m long, that flows towards the north-west-west; about 50m downstream the spring meets its second tributary, about 25m long, that flows to the west; at about 20m north from the second inflow, the stream joins its third tributary, 30m long, that flows towards the south-west-west. About 70m further downstream the Pârâul Sărat veers to the north-west, leaves the depression and brings its waters to the Valea Sărată stream – the affluent of the Meleş River, which in its turn joins the Someşul Mare River. The volume of water in the Pârâul Sărat stream varies significantly, depending of the level of rainfall. During maximum rainfall, the depth of the stream goes to about 0.5m and the width to about 3m. Conversely, during droughts it is almost completely dry and the streambed is marked by a white strip of crystallised salt.

The present morphology of the terrain was strongly affected by human activities, mainly by salt mining, from prehistory up to the present day. Within the site one can thus observe many depressions and mounds that result from salt mining. An ethnographical survey revealed that until the 1970s people from the neighbouring villages extracted salt rock from here and used it mainly as a forage supplement. As the locals say, when they dug for salt they often met pieces of 'black timber', which most likely were the traces of earlier salt mining. There are a few brine wells in the Pârâul Sărat stream; some of them are still in use, others are blocked and abandoned. The locals use brine in order to preserve lard, meat, cheese and vegetables. At the same time, they use the salty mud and brine to cure rheumatic and blood circulation disorders in humans, and for disinfecting wounds on animals. In the summer of 2005, during the survey of the site, with the agreement of the discoverer, we examined the trough (Figure 3) situated at that time in the yard of Bistriţa museum (Complexul Muzeal Judeţean Bistriţa-Năsăud). It was made of a hollowed out tree trunk. Its intact end was closed and the opposite one was broken. The preserved length of the trough is 322cm, the width – 42cm, and the height 37.5cm. The outer part of the bottom of the trough, except its intact end, was straightened by adzing. On its median line, the bottom of the trough is perforated by a row of 17 square-shaped openings. In these holes axially perforated wooden pegs[2] were inserted (Figure 4). The pegs have their upper edges thickened, they are round in section, and are wider than the holes in which the pegs were introduced. Below their tops the pegs are square in section and their thickness similar to that of the holes in which they were introduced. In their lower half the pegs become thinner and round in section. In the perforation of one of the pegs a twisted hemp (?) cord with a knot at one of its ends (Figure 4.2) had been inserted. We examined this cord after I. Chintăuan pulled it out from the peg and therefore we do not

[1] The project was supported by British Academy and the Ministry of Culture and Cults of Romania.

[2] According to earlier publications, the pegs were made of elder and the perforations were produced by removing its core. A recent examination of wood artefacts from Băile Figa by the wood expert Prof. Dr. Ionel Popa, however, maintains that the pegs were made of oak and the perforations bored.

know whether the knot was at the upper or lower end of the peg. We took two samples from the trough for C14 dating at Groningen and in late 2005 we received the dates, which combined fall at 1090-970 cal BC (88% probability).

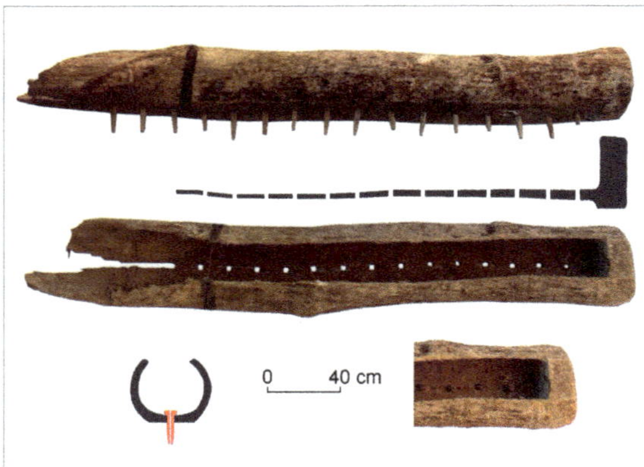

Figure 3. Trough in the Bistrița Museum courtyard.

In 2006, we catalogued and mapped the visible archaeological evidence: the mounds and depressions of presumed human origin and about 460 elements of wooden structures (Figure 2.3). Most of these elements were observed within the Pârâul Sărat streambed and close to it (Figure 2.2). Often the wooden structures went deep under the banks of the stream. Moreover, the traces of several timber structures were visible also outside the streambed, in the recently dug pits. That is why we concluded that in spite of appearances, the wooden structures were not functionally connected to the stream. They are just seen in the stream due to the fact that it removed the soil that covered timbers. We estimated then that the site covers almost the entire valley of about 600m (N-S) x 300m (W-E).

At the northern end of the site, on both sides of the stream, we observed many Bronze Age and 19th century pottery fragments. In the stream we found several isolated artefacts: stone mining tools, a fragment of a stone hammer, three troughs and a few isolated wooden pegs from troughs. Two of the troughs we found a few metres downstream from the main spring of the Pârâul Sărat stream, the third one about 150m north from the spring, close to the point where I. Chintăuan discovered the first trough in 1977. In 2006 we collected further samples of timbers for C14 dating and submitted them to the Groningen laboratory. Of these, four fall around 1000 cal BC, one around 250 cal BC and one around 450 cal AD.

In 2007 and 2008 we carried out archaeological excavations in this site; this was rendered urgent because in the meantime permission had been given to a commercial company to construct a spa complex on the northern part of the site. We opened 20 trenches (mostly by machine, and only five of them extensively investigated). Of these twelve were opened in different parts of the known archaeological site, and eight were positioned at various points around it. These diggings allowed us to delimit more precisely the area in which the wooden structures appear. It covers about 500m (N-S) x 200m (W-E), on both sides of the stream, but predominantly on its right side. The most conclusive evidence was found in trenches I, III, VII and XV, opened along the Pârâul Sărat stream.

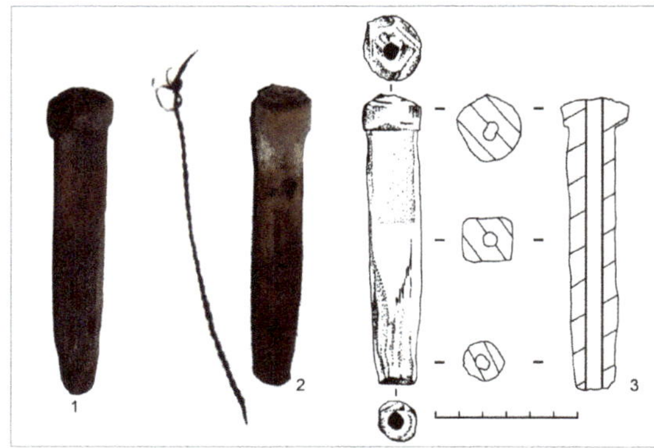

Figure 4. Wooden peg from the trough on the Figure 3 and twisted hemp cord with a knot at one of its ends.

Trench I (S.I) (134 square metres) was opened at the southern end of the site, a few meters downstream from the main spring of the Pârâul Sărat stream. The researches revealed that the main spring of the Pârâul Sărat starts from a pit that seems to have been dug for salt extraction not earlier than the 19th century. The yellow soil excavated from this pit was dumped mainly to the west. At some point, the water from the spring made a channel north-westwards and joined the surface water that runs off the southern slope of the valley. The western periphery of the mound formed from the excavated soil overlay, on the western side of the Pârâul Sărat stream, a layer consisting of 19th century pottery, animal bones and an arrangement made from an oval-shaped boulder discovered in a vertical position, and surrounded by a few small stones.

East of the stream, in the eastern part of S.I, in the infill of the mound a few Eneolithic and Bronze Age potsherds were discovered, without any features which would allow us to date them more precisely. In the same area, in the infill of the mound, as well as in the mud beneath it, we uncovered a rectangular structure (3.8 x 2.8m) built from overlying horizontally placed beams split in half (Figure 5). Two timbers criss-crossed the inside of this structure. These were probably to support the walls against the pressure of the soil from outside. The walls were reinforced by vertical poles stuck into the ground. This construction cut through earlier wattle fences that we discovered in the surrounding clay. Although the investigation of this structure is not yet complete, to judge from its appearance we presume that it was the top of a shaft. As to the dating of this structure, we do not yet have any C14 dates, and the stratigraphical situation offers a wide range of chronological possibilities. On the one hand, it seems that this structure was overlain by the mound made not earlier than the 19th century. On the other hand, this construction cut through earlier structures (wattle fences, probably of the Bronze Age). The complex can thus be dated only within very wide margins – between the Late Bronze Age and the 19th century.

Figure 5. Trench I, rectangular structure built of overlying horizontally placed beams split in half.

Similar constructions, in more or less similar contexts, have been found elsewhere in the Carpathian Basin. Some of them (for example *Valea Regilor – Kiralyvölgy* in Northern Maramureş, today Neresnica, Ukraine, Zakarpatska oblast) were dated to the Late Bronze Age (Wollmann and Ciugudean 2005, 99), others (for example *Loeva* in Sub-Carpathian Ukraine) to the Latest Bronze Age / Early Hallstatt (Krušelnicka 1993, 66-67).

In central part of S.I, along the stream, under the yellow soil, in the salty mud, we uncovered many timbers: poles, planks and parts of wattle fences. Of these only the wattle seems to be *in situ*. So far we do have not dates for all these timbers. One of the posts from the central-southern part of the trench was dated by C14 to the period 330 to 540 cal AD.

Figure 6. Trench I, Wooden troughs.

In the middle of these, in the stream and on its both banks, at a distance of about 0.4m one from another, in the mud, we uncovered two wooden troughs (Figure 6). Both of them were made of hollowed out tree trunks. The trough found on the western bank of the stream was entirely preserved. One of its ends is closed and the other is open (Figure 6). The trough discovered on the eastern bank of the stream has only the closed end preserved so as we could not establish what the opposite end was like (Figure 6; 8).

1 2

Figure 7. Trench I, Wooden ladder.

Although in general these troughs are similar to the one discovered by I. Chintăuan, they show some significant differences. First, at the closed ends of both of the troughs, deep channels were carved around the body (Figure 8), which seems to have been intended for fixing the ties by which the troughs were suspended from some structure(s). Second, both troughs were surrounded by rope made from fibrous plants (ivy?). Third, at the median line of the bottom of each trough there is a row of round holes (Figure 6; 8), in which axially perforated pegs (Figure 9) were inserted (in contrast to the square-shaped holes and square-sectioned pegs on the trough found by I. Chintăuan). In the perforations of some pegs were found traces of twisted cord (Figure 9.1d). In the perforation of one of the pegs of the trough found on the left bank of the stream a wooden needle had been inserted, wrapped in a thin, narrow wooden band (Figure 9.2), which would have provided a tight fit in the hole. The trough found on the eastern bank of the stream has the outside of its bottom rounded (Figure 8), while the trough found on the western bank has the outside of the bottom straightened by adzing (like the trough found by I. Chintăuan).

The trough discovered on the right bank of the stream has the following dimensions: preserved length – 133cm, maximum width – 37.5cm, maximum height – 38.5 cm, the diameter of the holes made on the median part of the bottom varies between 2 and 2.5cm, the distance between the holes is 11-11.5cm.

The trough discovered on the left side of the stream is 172cm long, its maximum width is 34.5cm, the maximum height – 34.5cm; the diameter of the holes, except the central one, is between 2 and 2.5cm, the diameter of the central hole is 3,5cm. The distance between the holes varies from 6 to 9cm.

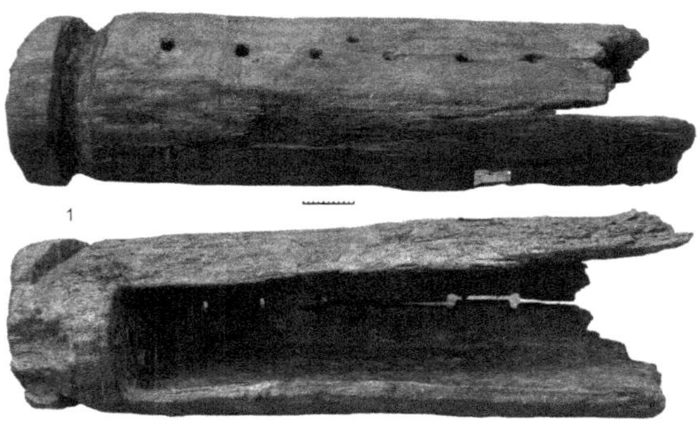

Figure 8. Wooden trough, Trench I.

C14 dates provided by the Oxford laboratory shows that both of the troughs discovered in S.I date to the 16th-15th centuries cal BC. The trough found on the western side of the stream is dated on the basis of body and peg samples, the combined dating of the trough being 1620-1500 cal BC (95% probability). Close to the stream, on its left side, under the intact trough, in the mud layer, we discovered a wooden ladder, about 5.6m long (Figure 7). Interestingly, at Valea Regilor (Kiralyvölgy, Neresnica) a trough and a ladder, both of them very similar to those from Băile Figa, were found together, in the same shaft (Wollmann and Ciugudean 2005, 114, Figure 4). Because both of the troughs were found with almost all their pegs, and bindings were found around their bodies (by which the troughs were probably suspended from some structure), it should be presumed that the troughs were found not far from the place where they were used during the Bronze Age.

Besides these, isolated wooden objects were uncovered: two hammers and two small wooden shovels; stone objects: two mining tools (Figure 10.1) and a fragment of a stone axe. The wooden objects discovered in S.I – the troughs, the ladder, the hammers and the small shovels – have the best and closest analogies in the 'hoards' of mining objects from Valea Regilor (Wollmann and Ciugudean 2005, 114, Figure 4) and Valea Florilor (Maxim 1971, 459, figures 2/4, 10; 461, figure 4). All of these are believed to date to Late Bronze Age. The stone mining tools from Băile Figa show very close analogies in the sites with the evidence for copper mining, for example in *Špania Dolina–Piesky* in Slovakia, where they were discovered in Eneolithic and Bronze Age (Lausitz culture) contexts (Točík and Bublová 1985, 88-99).

As far as the present stage of research allows, we can conclude that the excavations in S.I revealed archaeological evidence from *Eneolithic* (?), *Late Bronze Age*, *Iron Age*, *Post-Roman* period and *19^{th}century*. The Late Bronze Age evidence is predominant in S.I. and it seems to represent traces of salt mining. Future researches and new C14 dates will hopefully allow this information to be expanded and refined.

Trench III (S.III) (384 square metres) was opened in the northern part of the site, on both sides of the Pârâul Sărat stream. Here, in the stream, we observed a grouping of wooden structures: fences made of planks or split poles and numerous poles. Among them were found a few Neolithic potsherds (made from fabric mixed with chaff), numerous Early Bronze Age potsherds (most of them with rusticated surface) and potsherds from the 19th century (wheel-made pottery, often covered by green glaze). Because of the proposed development work this part of the site was extremely vulnerable; this was the main reason why we carried out excavations here in 2007 and 2008. These excavations uncovered a grouping of wooden structures found *in situ* (Figure 9) which covered a surface of about 19 x 21 m. On the east and north this grouping was delimited by an arc-shaped wattle fence supported by poles placed in the ground. The southern and western part of this grouping was destroyed by recent human activities (the bridge in the south and the pool in the west). Small verification trenches excavated west from S.III showed that this grouping extended about 10m west from S.III.

The stratigraphical observations combined with C14 dating showed that the wooden structures uncovered in S.III date to at least three periods. The earliest structures are fences or walls made mainly of massive planks (C14 dating places them between 1050 and 975 cal BC at 68% probability). These are followed by the fences mainly made of wattle propped up by poles stuck in the ground (C14 dating shows their age as 1005-915 cal BC at 95% probability). The latest wooden structures uncovered in S.III are represented by the traces of a well (?) and fragments of large wooden channels; a branch fragment from the well (?) was dated by C14 to the 19th century AD. The earliest timber structures found *in situ* overlie a pit with Early Bronze Age pottery in its infill.

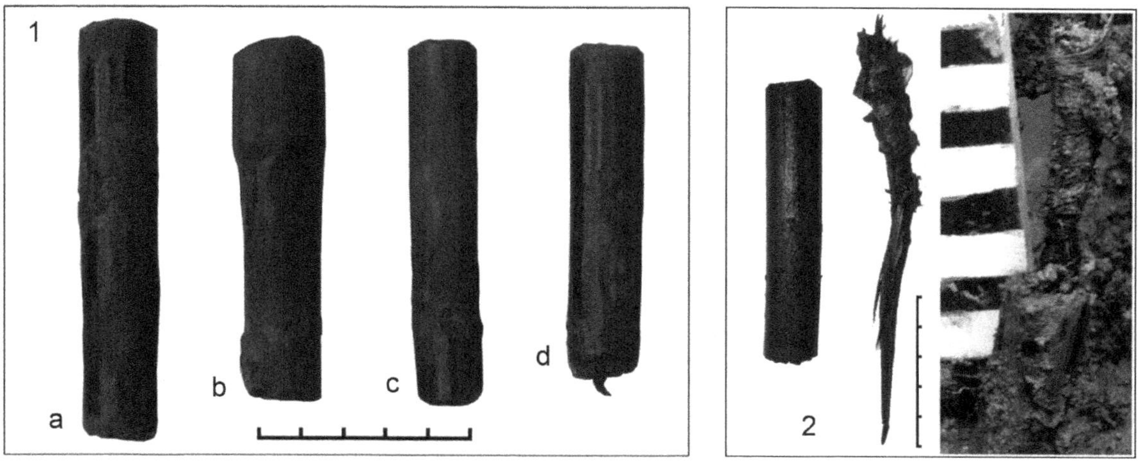

Figure 9. Pegs, Trench I.

Figure 10. Stone mining tools.

Within the grouping uncovered in S.III most of the fences were made of wattle. They are oriented mainly SW-NE. Sometimes massive horizontally placed fork-shaped branches were introduced into these fences, so that their bases were attached to the wattle poles and the other ends to the neighbouring wattle. In the middle of the grouping we uncovered an open space measuring about 10 x 2.5m which in its northern part has the form of three joined opened circles oriented N-S. In its southern part this space continues as a corridor with parallel walls. The space was roofed by massive timber transverse beams perforated at their ends. The tops of the poles by which the wattle was supported were inserted into the openings of these beams. Clusters of hazel twigs were found among these beams. The complex thus seems to have had a roof made of beams and twigs. In many instances traces of leaves were observed between the branches. Among the branches of the northern wall (wattle) of the space we found three hazelnuts (Figure 12.6). All this shows that the complex was made of branches and hazel twigs with leaves and nuts on them. Since the nuts seem ripe, it should be assumed that the room was built somewhere at the end of summer or the beginning of autumn.

In its southern part this space intersects a fence (or wall) made of massive oak planks in vertical position, preserved to a length of about 2.5m, oriented SW-NE (Figure 11.1-2). By its stratigraphical position and C14 dating, this fence (or wall) seems to have been built prior to the room. Most or all of the planks of this structure are made of the same tree trunk and are buried deep in the mud (over 1m deep), while five planks at the northern end are much shorter and less deeply embedded in the mud (0,2-0,4m). Moreover, two of these planks have their lower ends rounded, and on both sides were carved half-round cavities (Figure 11.1-2). It probably indicates that initially, before this fence (or wall) was cut by the wattle construction, there was a niche in this part of the plank fence (or wall).

In the north-western part of S.III a pit with Early Bronze Age pottery was found (see above). It was certainly overlain by the timber structures found *in situ*. This pit has not yet been investigated, but both its stratigraphical position and the pottery found in its infill (similar to the Iernut-Zoltan type pottery, with rusticated surface but without textile imprints) suggest a dating in the Early Bronze Age. This pit may be contemporaneous with the wood sample found out of certain stratigraphical context and dated by C14 to 3020-2925 cal BC (68% probability).

Figure 11. Trench III, wooden structures *in situ*.

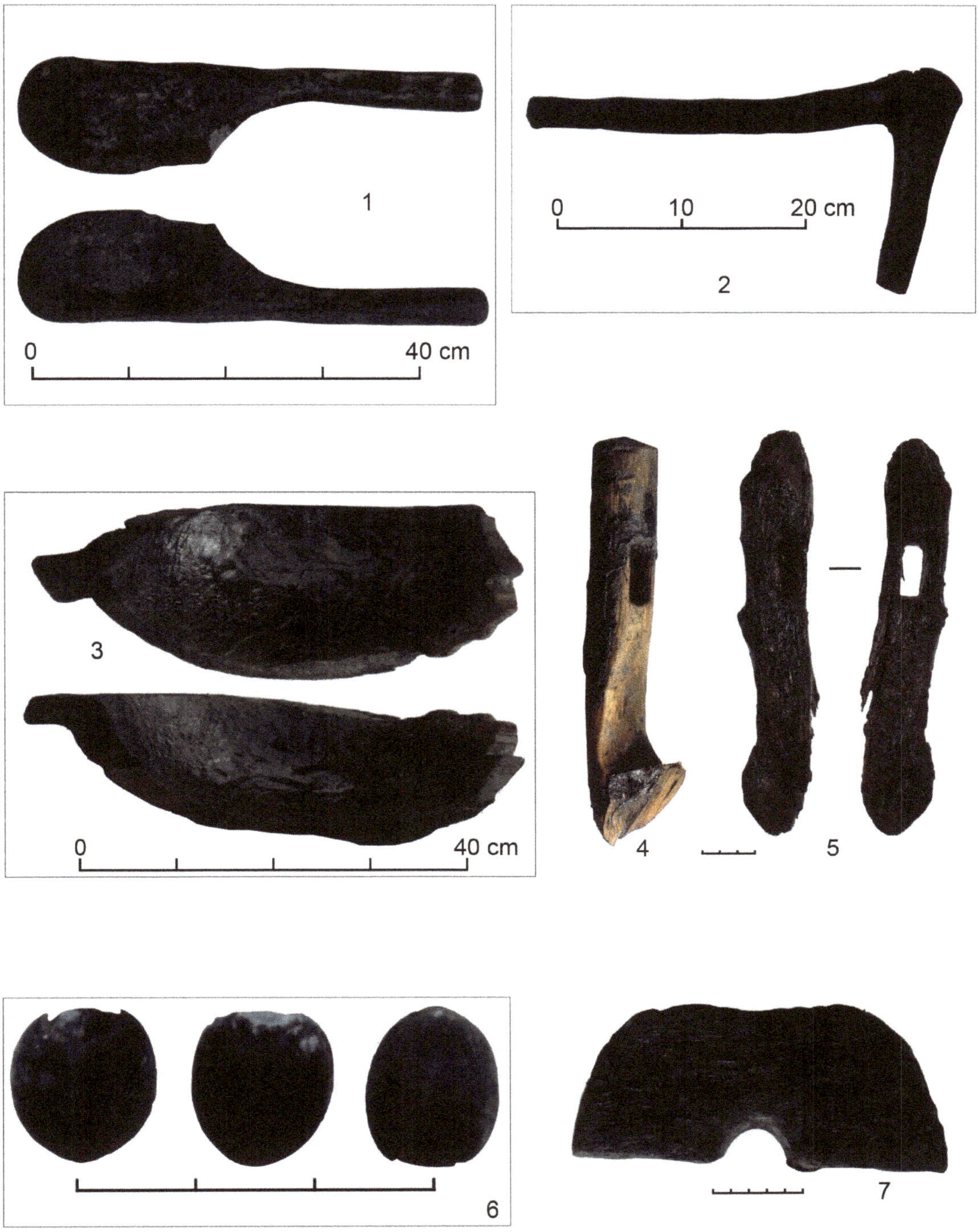

Figure 12. Wooden objects found in the Trench III: left-hand palette (1); socketed axe handle (2); vat (3); hook-shaped objects with rectangular openings on their undersides (4-5); hazelnuts (6); perforated disc (7).

In the western part of S.III a roundish structure about 1.7m in diameter was uncovered, perhaps a well, made of small branches. Its stratigraphical context shows that it was built relatively recently. The pottery discovered around it is characteristic of the 19th century (confirmed by C14 dating).

Several artefacts were discovered among the timbers: a few mining stone tools (Figure 10.2), one of which is painted with red colouring; several wooden objects: a socketed axe handle (Figure 12.2), a left-handed palette (Figure 12.1), two small shovels, two vats (Figure 12.3), several pegs with square section (like those from the trough found by I. Chintăuan), two hook-shaped objects with rectangular openings on their undersides (Figure 12.4-5), a perforated disc fragment (Figure 12.7), a perforated wooden bar, two perforated arc-shaped pieces, small wooden channels and other objects.

Some interpretation of the features uncovered in S.III is possible, but our investigations are not yet concluded, so we prefer to delay a fuller discussion for the time being.

Trench VII (S.VII) (154 square metres) was opened 10m north from T. III, on the eastern side of Pârâul Sărat stream. Two complexes were uncovered within this trench. In the stream bed an arc-shaped wattle structure was uncovered (half of a circular fence?), supported by poles stuck in the ground. On the top of the high right bank of the stream, about 5 metres from the stream, an oval pit was uncovered (about 5 x 3m), the infill of which is made of blackish soil mixed with Bronze Age sherds. In the central part of this pit a cluster of small stones was observed. This complex was left intact in order to be investigated in the next season of excavations.

Trench XV (S. XV) (15 square metres) was opened in the central-southern part of the site, in the Pârâul Sărat riverbed, at about 150m south from S.I, near where I. Chintăuan discovered the original trough in 1977, and where we observed the end of another one. In 2008 we opened a small trench. The excavations pointed out the fact that the trough was in the mud layer, in an oblique position sloping down into the mud, and next to it a little further down, in a horizontal position, there was a wooden ladder, similar to the one discovered in S.I. Nearby there were found a stone mining tool fragment and a wooden object. It looks like a hook made from a branch and at its back end has a rectangular opening in which a small wooden sliver was inserted. Because of the complexity of the situation, T. XV was backfilled in order to preserve the timbers. The excavation in this trench will be continued and the trench extended.

*

In the following section we present two newly discovered sites with evidence of saltworking: *Săsarm – Valea Slatină* and *Caila – Sărătură*.

Săsarm – Valea Slatină (Chiuza commune). The site was discovered in 2005 (V. Cavruc, D. Buzea, L. Vaida, C. Bojica.). It is situated in the western part of Bistriţa-Năsăud county, about 10km north-east of Băile Figa, about 1km north-east of Săsarm village and about 2km west of Chiuza commune. This site includes the valley of the *Valea Sărată* (Salty Valley) stream, at the north-eastern foot of 'Săsarm Hill', which overlies a rock salt deposit. The terrain around the valley is strongly eroded, which seems to be caused by periodic collapses of the galleries dug into the salt massif. According to information provided by local people, salt was mined in this area by non-industrial techniques up until recently.

Along the stream, about 800m in length, numerous traces of wooden structures are visible on the surface, as well as many isolated timbers. At least some certainly come from mine galleries, and many look recent. This is confirmed by the presence at one point of a beam of reinforced concrete. At about the middle of the site, on the left side of the stream, in front of a brine well, between the main course of the stream and its small tributary, in an area of about 40 x 30m, were discovered traces of some Early Bronze Age features as well as several isolated Iron Age sherds. On the surface there are three oval or rectangular areas, each about 2.5 x 3.4m, with an infill made of blackish soil mixed with ash, many small fragments of charcoal and daub, as well as Early Bronze Age pottery. Nearby, north of these complexes, in the steep bank of the stream, we observed a horizontal stripe made of grey soil in the yellow clay, about 4m long and 0.6m thick. Within this stripe Early Bronze Age sherds are visible, as well as a few small fragments of charcoal and unburned timbers. Several isolated Bronze Age potsherds were also found downstream from this place, over an area about 50m in extent. Bronze Age sherds come from coarse pots with rusticated surface. Some fragments come from pots decorated with horizontal cordons; a few sherds are from vessels decorated by impressed twisted cord. Analogies for this pottery come mainly from the Iernut-Zoltan group dated to the final stage of the Early Bronze Age (Cavruc 1997; Ciugudean 2003). In the same part of the site a few potsherds of 'Celtic' type were discovered (4th-3rd centuries BC), and an amphora fragment of the classical period.

Caila – Sărătură (Şintereag commune). The site was discovered in 2007 by the ethnographer Andrea Deak (Chiricescu) from the National Museum of the Eastern Carpathians, during her ethnographical survey concerning the traditional exploitation of salt. The site is situated about 13km NNE of Băile Figa, about 1.5km west of Caila village and about 300m from the track connecting Caila to Blăjenii de Jos. The site lies along a valley called by the locals 'Sărătură' (salty place). This valley is crossed by a brine stream that springs from the foot of the 'Dealul Sărat' (salty hill) hill and brings its waters to the Roşua stream, which is a tributary of the Someş River.

In the summer of 2008 we carried out a survey of this site and catalogued and mapped the visible archaeological evidence on the surface of the terrain. In many respects this site is similar to that of Băile Figa. In the stream along a length of about 300m numerous timbers were observed. In southern part of the site, in its left bank, a fragmentary trough was discovered with square-shaped holes, with inserted perforated pegs, in other words similar to the trough discovered by I. Chintăuan at Băile Figa. A fragment of the base of a trough was found a little further north, lying in the stream. A few isolated Bronze Age sherds were found around the trough. At various points on the stream bed and edges, bindings can be seen, apparently part of wattle fences.

Some preliminary considerations

At the moment, the Băile Figa, Săsarm – Valea Slatină and Caila – Sărătură sites constitute the most direct archaeological evidence for ancient saltworking in Transylvania. Of these, at least for now, the researches we carried out in Băile Figa offered the most conclusive evidence regarding ancient saltworking in eastern and south-eastern Europe. Together with what was previously known, the new discoveries permit a few observations on Late Bronze Age saltworking technology in Transylvania.

Analogies to the Băile Figa findings are known only in the Carpathian Basin – in Valea Florilor, Valea Regilor and Ocna Dejului, all of them in contexts which strongly suggested salt mining. Of all these, the objects from Valea Regilor (Kiralyvölgy) and Valea Florilor show the most relevant analogies to those from Băile Figa.

From the publications of the Valea Regilor findings (Wollmann 1996, 246-247; Rustoiu 2003, 2002, 214, 2005, 354, 364, figure 7; Wollmann and Ciugudean 2005, 99; Cavruc *et al.*, 2006, 43, 49; Lukacs 2006, 13-15, 143), it is known that they were discovered in 1817 during excavations for draining a gallery in an abandoned salt mine. Several objects used in salt mining were found. According to the description and the photograph in the Budapest National Geology Museum Guide published in 1909 and recently republished (Wollmann and Ciugudean 2005, 114, Pl. IV; Cavruc *et al.*, 2006, 49; Lukacs 2006, 14-15, 143), a trough, ladder, hammer, perforated bar, small shovels, jute rope and a block of rock salt with a range of holes performed by water jets were discovered. The trough had been found intact, one of its ends was opened, and the other was closed[1]. According to Lukacs Karoly, *twisted cords* were found in the perforations of some pegs of this trough. Nearby, in 1846 and 1847 timber beams were discovered. E. Preissig related these findings to the bronze hoard from Coştiui and on this base dated them to the Bronze Age (Wollmann and Ciugudean 2005, 100). It should be emphasized that most of findings from Valea Regilor are quite similar, sometimes up to the smallest details, to those from Băile Figa. For example, the wooden ladders at Băile Figa have the same features as the ones from Valea Regilor: the long beams on the ladders from Băile Figa and Valea Regilor are perforated by rectangular holes by which the rungs are fastened. The shape and dimensions of the ends of the rungs inserted into the holes fit them exactly. At the ends of the rungs, where they emerge from the uprights, there are in most cases smaller rectangular openings in which wooden wedges were inserted. In this way the rungs were kept firm and the ladder stable.

The Valea Florilor finds were made in 1938 in an abandoned mine with walls of rounded timbers or wattles (Maxim 1971). Several wooden objects were found in the mine, among which was a trough with both ends preserved, various shovels, etc. Because together with these objects a quern of 'Dacian type' was also found, all the objects were dated to the second part of the Iron Age (Maxim 1971, 457-463; Wollmann 1996, 246; Iaroslavschi 1997, 45-47). Recently V. Wollmann and H. Ciugudean established two very important facts. The first was that they obtained a C14 date for the trough, giving a time bracket of 1420-990 cal BC (95% probability). The second observation was that there were at least three troughs, not one as earlier publications stated. The authors reiterate E. Preissig's hypothesis according to which the troughs were used for swirling the rock salt with water jets (Wollmann and Ciugudean, 2005, 98).

It should be specified that that the above interpretation is just one of the possible hypotheses. Therefore, I. Chintăuan considers the troughs as being used to get solid salt by brine evaporation. There are many possible interpretations and we are exploring them.

We now have 51 radiocarbon dates on timbers from Baile Figa, and more are awaited, as well as on samples from the other sites mentioned here. When we have a complete set of dates we will present a detailed analysis of the picture presented by them. Dendrochronological research conducted by Tomasz Ważny of Cornell University will, it is hoped, add to the possibility of a close dating of these and other sites in the Carpathian region.

Beyond the data that contributes to the elucidation of the technology of salt exploitation, many wooden objects have been found on these sites. The many prehistoric and post-Roman-period timbers from Băile Figa, Săsarm – Valea Sărată and Caila – Sărătură at present represent the richest evidence of its kind in the whole of Eastern and South-eastern Europe.

[1] According to Lukacs Karoly, the pieces discovered at Valea Regilor (*Kiralyvölgy*) were initially brought to Solotvino (Rom. *Slatina*, Hung. *Aknaszlatina*) and for a while they were kept in one of the mines there. At some point the objects were transported to Budapest to the National Museum of History. We do not know how the pieces got from there to the National Geology Museum in Budapest, where according to V. Wollmann and H. Ciugudean they were on show at the beginning of 20th century. These pieces disappeared during the Second World War, but in the summer of 2008 Dr. Carol Kacso, to whom we express our gratitude, informed us that the pieces are at the National Mining Museum in Sopron, information confirmed by the staff of the museum.

References:

Cavruc, V. 1997. The Final Stage of Early Bronze Age in South-Eastern Transylvania. *Thraco-Dacica* XVIII, 1-2, 97-133.

Cavruc, V. Ciugudean H. and Harding A. 2006. Vestigiile arheologice privind exploatarea sării pe teritoriul României în epoca bronzului. In: V. Cavruc and A. Chiricescu (eds.), *Sarea, timpul şi omul,* 41-49. Sf. Gheorghe, Angustia.

Chintăuan, I. 2005. Pan used for salt extraction from brines. *Studii şi cercetări - Geologie-Geografie* 10, 75-78.

Chintăuan, I. and Russu, I. 1988. Consideraţii cu privire la utilizarea sării şi apelor sărate din nord-estul Transilvaniei. *File de istorie* 5 (Bistriţa), 238-277.

Ciugudean, H. 2003. Noi contribuţii privind bronzul timpuriu din Transilvania. *Apulum* 40, 89-122.

IDR, III/3, 1984. Inscripţiile Daciei Romane, vol. III, tom 3, Bucureşti.

IDR, III/4, 1988. Inscripţiile Daciei Romane, vol. III, tom 4, Bucureşti.

Iaroslavschi, E. 1997. *Tehnica la daci.* Biblioteca Musei Napocensis XV, Cluj-Napoca.

Krušelnicka, L. I. 1993. Novi pam'jatki kul'tury Gava-Goligrady. In: *Pam'jatki gal'štatskogo periodu v mežiričči Visly, Dnistra i Prip'jati*, Kyiv, Naukova Dumka, 56-121.

Lukacs, K. 2006. Aknaszlatina. A so, a viz es levego kincses tara, Pecs, Magakiadas.

Maxim, I. Al. 1971. Un depozit de unelte dacice pentru exploatarea sării. *Acta Musei Napocensis* VIII, 457-463.

Piso, I. 2007. Un nouveau *conductor salinarum* en Dacie. *Acta Musei Napocensis* 41-42/2004-2005 (2007), 180-182.

Russu, I. I. 1966. Sclavul Atticus. Contribuţii la istoria economică a Daciei. *Studii şi cercetări de istorie veche* VII, 7-13.

Rustoiu, A. 2003. Sarea Maramureşului şi aşezările dacice de pe Tisa Superioară. *Marmaţia* – Arheologie-Numismatică – 7/1, Baia Mare, 201-217.

Rustoiu, A. 2005. The Salt of Maramureş and the Dacian Settlements alongside of Upper Tisa. *Mousaios* X, 353-367.

Točík, A. and Bublová, H. 1985. Prispevok k vyskumu zaniknutej tazby medi na Slovensku. *Studijné Zvesti* 21, 47-135.

Wollmann, W. 1996. *Mineritul metalifer, extragerea sării şi carierele de piatră în Dacia romană/Der Erzbergbau, die Salzgewinnung und die Steinbrüche im Römischen Dakien*, Cluj-Napoca/Klausenburg.

Wollmann, W. and Ciugudean, H. 2005. Noi cercetări privind mineritul antic în Transilvania. *Apulum* XLII, 95-116.

The Beginning of the Salt Exploitation in Spain: Thinking about the Salt Exploitation in the Iberian Peninsula during Prehistoric Times

Jesús Jiménez Guijarro
Comunidad de Madrid, Spain

Abstract
We present in this paper a synthesis of the salt use during the Prehistory in the Iberian Peninsula. We put our attention in natural use and exploitation of salt among prehistoric groups, especially around the concepts of salt gathering and salt making. We think about the impossibility to speak for salt exploitation before the Chalcolithic and Early Bronze Age times. We speak around the hypothesis of direct relation between the called 'Revolution of secondary products', Bell-bakers culture and salt exploitation during Prehistory.

Keywords
neolithic, salt gathering, salt exploitation, mining, Europe, Iberian Peninsula

Introduction
All the herbivorous need salt for their life. The man needs salt too, we know, but, for understand the salt use and the salt making process, their differences and similarities, we need to answer some questions. They were the developments to manufacture and to use the salt during Prehistory always the same? Can we compare salt use in the East and South West of Europe?

In our paper, we try to think about the real evidences of salt exploitation in the Iberian Peninsula, especially during prehistoric times. We make a review of evidences and we attend the dates, cultural ascription and the real context for some archaeological sites, and we'll revisit the existence or not of clear remains for salt exploitation in this region of Southwest Europe.

Therefore the investigation of this matter is based only in circumstantial and very poor evidences. A lot of archaeological sites that we suppose directly relate with salt exploitation have only a scare and very poor archaeological remains and salt making evidences too. In a lot of cases, these places are badly acquaintances or we know only that they are founded in a geological substrate rich in salt. Because of it, we assume a noticeable role in the production and in use of salt resources for these places.

The extensive archaeological literature, written in English and French, presents some differences in the employment of certain linguistic terms. Thus, we have some problems with the translation of terms like '*exploitation*'. In English, this word are related with the use of something, but in French either in Spanish, the word '*exploitation*' or '*explotación*' are related with different senses of mining –or agricultural- work or in more large sense, is related with the obtaining of something working with them. The sense of these similar words really outlined some fine and very important difference. Thus, in Spanish language, we use the term '*uso*' but we should understand that does not imply the idea of salt use as industry or manufacture. This does not imply the idea of salt making or production, but simply its use.

The same concepts were in the mind of some classical writers. Thus, Pliny did some distinction between '*natiuus*' (XXXIX, 77) and '*facticius*' (XXXIX, 81). The first ones correspond with a natural production – may be gatherering- and the second ones correspond with a man making or industry. We think that '*natiuus*' concept is related with our 'salt gathering' and the '*facticius*' are related with our 'salt production or salt making'. This one is related in Spanish or French languages neither with the concept of salt 'exploitation'.

We have in mind the idea of natural use of salt like a natural use of some foods. Thus, is possible to understand the first salt use like another type of gathering. Since the first prehistoric times – from Paleolithic we suppose- until the first cultures that use metals – Copper and Bronze Ages- was the time for salt gathering. We think that salt was not 'domesticate' at the same time that vegetables – like wheat, rye or barley – or that some animals – like goat, sheep, pig or cow. On the contrary, salt was domesticated when the control over the first domesticated vegetables and animals was perfectly founded. We know that the process of domestication and control over food implements was a so difficult and certainly large process. Thus the salt control was related with tame and especially with hunting. When the human group goes on with hunting practices they need a low level of salt implementation on their diet. The salt, like we said, was taken from the preys and its meat. And the preys eat his salt needs from springs and natural salt sources too.

But, when do the human groups need to complete their diet of salt? We think in tree different cultural and economical ways:
1-When they have specialization in stockbreeding, which implies a great importance of herding in theirs economies. We are thinking about certainly herdsman specialized economies.
2-When the herds are enclosed in sheepfold or in a cowfold.
3- Finally, when the human group needs to make an intensification of production and, specially, a production increasing of milk or its derivative

products. This one can be related with human consume or better with young animal needs for life.

We explain in our paper our disagreement about making salt during the first Neolithic times in Spain and our belief in the close relation between salt making and herdsmen economies.

Nevertheless, we have certain confidence about the development of salt gathering, but not complex salt exploitation or salt making, during Neolithic times in Iberian Peninsula. We have certain indirect evidences that denounce the existence of a natural salt gathering as complement inside of the earlier agricultural economies. We can emphasize, for example, the recurrent association between some archaeological places and the salt-mines.

A lot of evidences presented as clear remains as salt mining during the Early and Middle Neolithic in Spain are only circumstantial probes of the advantage of these tribes over the rich surrounding ecological niches. Thus, the high profile sites occupied by Neolithic farmers, and especially shepherds, are related, normally, with salt areas. But we are convinced that this association was not related with salt exploitation, but with its natural use or salt gathering.

These 'neolithic' sites are normally broad occupation areas where Chalcolithic and Bronze Age evidences are the most common archaeological remains. This matter turn critical our investigation, because we need to know which evidences are directly linked with the salt exploitation and which not.

Summary, we have a new archaeological problem and we need to do a new approach point of view. Certainly, we need more and new archaeological sites and we need to do new field research focuses to find unmistakable evidences of salt exploitation.

We are grateful with our colleagues from Romania and Bulgary for their unstimated aid for understanding the thin differences between the concept of exploitation, use and gathering of salt in different languages. They made us to think long time after this colloquium about this matter. The travel to some archaeological sites of Romania, and the possibility to make a direct visual analysis of some pots remains (specially some firedogs) from salt 'mines' or sources gave us a more accurate idea from the great difference that exist between the eastern and south-western Neolithic. These differences were greater around the salt use, the salt gathering and the exploitation and making of salt. Maybe, salt making in Spain since Bell Baker Chalcolithic and Early Bronze Age could be explained as foreign development and this maybe explained by some kind of cultural diffusion.

Spatial and cultural context

Iberian Peninsula has more than 8000km of seacoast. Some sites of the eastern coast have remains of sea salt workings and some of these sites are still in use. But Iberian Peninsula has not only these kind of seacoast salt exploitations (Figure 1). On the contrary, Spain has a lot of locates with remains of inner salt resources like salt springs, salt sources and the great salt mountain, and certainly rare site, of Cardona (Weller 2002; Fíguls *et al.*, 2007; Weller and Fíguls 2008). The exploitation of inner salt today is really scarce. Thus, recent works suggest that the most productive workings are mineral salt mines (64'5 %), sea-salt mines (32'6%), and only 2'9% corresponds with salt springs of inner lands. These evidences contrast with the number of sites in use for extract these resources and give us an idea about the salt springs profit value. To take out the 32'6% of Spaniard salt we know 22 evaporation seacoast. On the contrary, they are 27 inner salt springs to take out the 2'9% of Spaniard salt production.

But, can we suppose the same salt profit behaviour value during Prehistory? Surely not. Thus, we need to think, first, in two different concepts: salt use (we said salt gathering) and salt exploitation and we need to think about the ancient use of salt in Prehistory too (Jiménez-Guijarro 2007)

It's necessary to think about the salt distribution across the territory There was availability of saline resources?. Was easy the access to these resources? Aside from these questions we need to search for the arrangement of salt around the world and we need to think about salt resources facilities too.

Salt and domestic herding

The quantity of salt that needs a head of domestic herd is directly related to the need of the human group to increase or not the production (Figure 2). Thus, the lack of sodium, chlorine and potassium in domestic herd incite reduction of the appetite, loss of weight and the decrease of the milk production –all of this related with a notable production decreasing-. Besides, iodine low level causes problems with the gland of the thyroid and reduces the fertility of the animals.

Also are some differences inside of animal species respect to their mineral needs. Thus, in the cattle, the contribute of minerals –around 15- with the purpose to guarantee an adequate nutrition and to assure an efficient production of milk surpasses the needs of, for example, the ovine cattle.

Proper animal nutrition means giving the animals the proper amount of all nutrients necessary for optimum production. This involves knowledge of the nutrients themselves, factors that affect the requirements of animals, and the feeds used to deliver those nutrients.

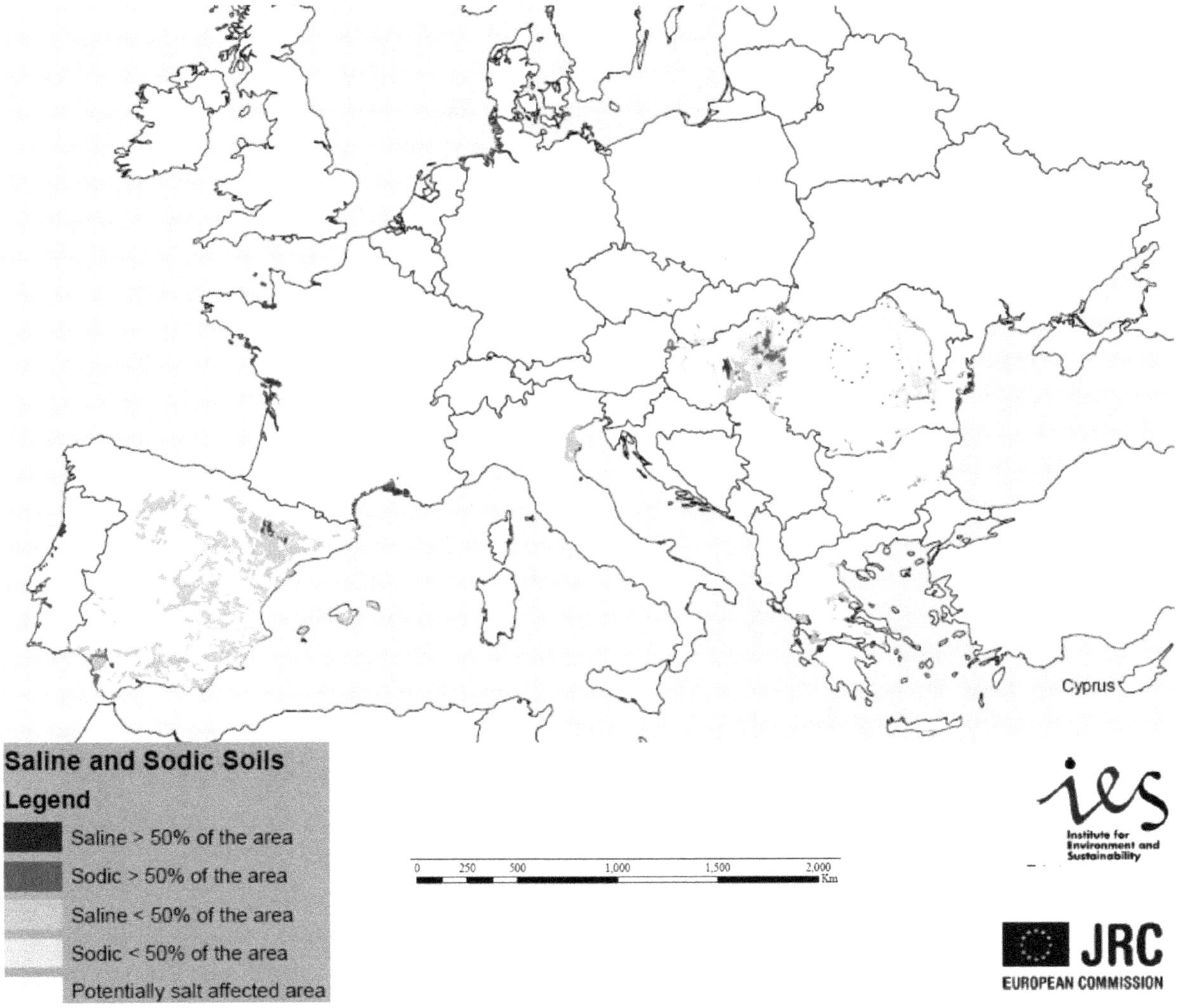

Figure 1. Principal saline and sodic soils of Europe.

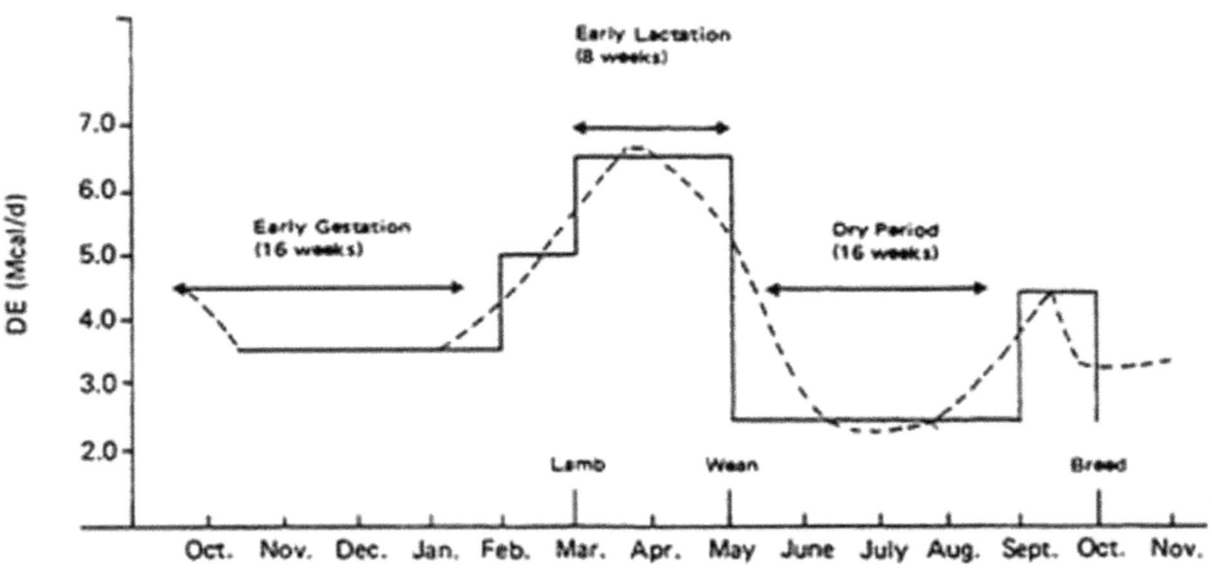

Figure 2. Approximate daily digestible energy (DE) requirements of 65- to 70-kg breeding ewes at various production stages.

For the ewe flock, proper nutrition involves giving animals all the good quality forage they want, and supplementing that with nutrients that may be deficient. So the basics of animal nutrition are good forage management, such as proper fertilization, a mixture of grasses and legumes, maintaining forage at a nutritious stage of growth, and providing forage in adequate quantities. We can think about the following matter: agriculture exploitation for herds?

Attending to the criteria established currently and referring us exclusively to the minerals related with salt, we can establish the following board of daily needs for the cattle (Figure 3).

In the case of the sheep, has been notified besides that the absence of minerals as the Sodium, drift not only in problems of growth, but can give rise to demonstrations of depraved appetite and even, *in extremis*, to demonstrations of cannibalism, above all toward the young ones.

Mineral	Maintenance (1)	Growth/grow fat (2)	Lactation (3)
Sodium (gr.)	4-8	6-8	5
chlorine (gr.)	5-9	9-12	12
Iodine (gr.)	0,4-0,6	0,5-0,8	1

1= individuals from 350 to 400kg of weight
2= Individuals from 300 to 450kg of weight
3= To produce 10 litter of milk

Figure 3. Board of daily needs for the cattle.

In Figure 4 are shown the various minerals and vitamins of concern, levels found in good forage, and the requirements for these nutrients by various classes of sheep. The requirements are based upon the Nutrient Requirements of Sheep, Sixth Edition (1985).

Macrominerals
There are many minerals that are required in the diet of sheep. Macrominerals are required in larger amounts, with that requirement expressed as a % of the diet or as grams per head per day. In table 1, above, they are shown on the first 6 rows of the table. Some of these are already in sufficient quantity in forages, so supplementation is not needed. Others are never in adequate amounts, so must always be in a supplement. Finally, there are those that are marginal, meaning amount in the forage and amount needed are close to each other, thus supplementation is sometimes needed, and sometimes not.
Adequate Potassium
Deficient Sodium (when combined with Chlorine, makes salt)
Marginal Calcium, Magnesium, Phosphorous, Sulfur

Intake of Mineral
Sheep do not eat the same amount of mineral throughout the year. They have a craving for salt, and consume a complete mineral to get salt. Intake is higher when consuming lush fresh forage, such as in the early spring. During the dry summer months intake is lower, this is also the case when sheep are eating hay. If a water source is nearby intake is higher than when water is a great distance away. In addition to nearby water, intake is higher if mineral feeders are located in shady areas or along paths frequently traveled by sheep.

Traditional stockbreeders knows that when feeding high rates of cereal grains, for over a month, the sheep will become deficient in calcium and sodium. The solution is to add 1.5% stocklime and 0.5% salt (unless the water is salty) to the grain.

Sodium (Na) and chlorine (Cl) serve many functions in the body. The maintain osmotic pressure, regulate the acid-base balance, and control water metabolism in tissues. Sodium occurs primarily in extracellular fluids and bones. Chlorine is found within cells, in the body fluids, in gastric secretions such as hydrogen chloride, and in the form of salt (Underwood 1981).

Animals that are deprived of adequate salt may try to satisfy their craving by chewing wood, licking dirt, or eating toxic amounts of poisonous plants. Inadequate salt may result in inappetence, growth retardation, inefficiency of feed use, and increased water consumption (Hagsten, Perry and Outhouse 1975; Underwood 1981).

		Mature Ewe		Young Lamb
Nutrient	Good Forage	Early Pregnancy	Nursing Twins	Fast Gain
Calcium, %	.45	.25	.4	.55
Phosphorous, %	.40	.2	.3	.25
Potassium, %	2.0	.5	.8	.6
Magnesium, %	.25	.12	.18	.12
Sulfur, %	.25	.15	.25	.15
Sodium, %	.0005	.10	.15	.10
Iron, PPM	100	40	40	40
Copper, PPM	8	10	10	10
Manganese, PPM	70	40	40	40
Zinc, PPM	30	30	30	30

Selenium, PPM	.15	.3	.3	.3
Vit A, IU/lb DM	50,000	1000	1200	500
Vit D, IU/lb DM	500	100	100	100
Vit E, IU/lb DM	10	7	7	7

Figure 4. Class of Sheep and Their Requirements (in diet Dry Matter).

Several feeding and metabolism studies have been conducted to determine the sodium and/or salt requirement of sheep. McClymont *et al.* (1957) reported that the addition of 1.2 to 2.6g of sodium per day (as sodium chloride) to the diet of very thin wethers fed a low-sodium grain diet increased growth rate. They concluded that the sodium requirement was greater than 0.9g/d (0.06 percent of the diet). From balance data, Devlin and Roberts (1963) estimated the sodium requirements for maintenance of wether lambs to be 1.01g/d (0.18 percent of the diet). Hagsten, Perry and Outhouse (1975) concluded that the dietary salt requirement for growing lambs ranged between 0.33 and 0.43 percent of the air-dry ration (90 percent dry matter). They further stated that since most sheep rations contain approximately 0.2 percent salt, a supplemental level of 0.2 percent is adequate. Based on the maintenance of a normal Na+:K+ ratio in the parotid saliva, Morris and Peterson (1975) concluded that a dietary sodium level of 0.09 percent met the requirements of lactating ewes. Apparently no feeding trials have been conducted in which the requirement for chlorine can be assessed independently of sodium; thus, the chlorine requirement is unknown.

When adding salt to mixed feeds, it is customary to add 0.5 percent to the complete diet or 1.0 percent to the concentrate portion.

The Range operators today commonly provide 220 to 340g of salt per ewe per month as a salt lick. Drylot tests show lambs consume approximately 5 to 10g of salt daily (Denton 1969). Mature ewes in confinement consume 15 to 30g of salt daily when it is offered free choice (Jordan and Hanke 1982). Salt may safely be used to limit free-choice supplement intake if adequate water is available. Such mixtures are usually 10 to 50 percent salt depending on the desired amount of ration to be consumed.

On the basis of research conducted by Meyer and Weir (1954) and Meyer *et al.* (1955), the maximum tolerable level of dietary salt for sheep was set at 9.0 percent (National Research Council 1980). Jackson, Kromann and Ray (1971), however, reported a linear decrease in weight and energy gains of growing-finishing lambs as salt content increased from 1.8 to 7.6 percent of the diet.

But we must not forget the real importance of not artificial access to the macrominerals and salt surpluses during Prehistory. Thus, the better hunting grounds are commonly the places preferred by a lot of herbivorous and they are those places associated with water sources that have some salt concentration. We find these places plenty of wild herbivorous during spring and autumn months. Here, when a great part of water is evaporated, these animals consume rich plants in salt (halophyles) and lap clay to intake their needs of salt.

All these keep us to make the most important difference inside the salt needed to explain its real power in Pre-History. We need to make some kind of differences between salt use or 'salt gathering' and salt exploitation or 'salt making'. These differences are clearly collected in some classical texts, like the 'Natural History' of Pliny. He made the first contrast among natural and artificial salt:

> 'The salt born or it's made (...) by two causes: condensation and drying up (XXXIX, 73) (...) Moreover, sheep, cows and mules excite their appetite eating salt, and they gives more milk and the cheese are better' (XXXIX, 88).

For ancient writers the relation among salt and civilize way of life was clear, and the relation with animal production intensification was so clear too.

But salt has some conservative properties too. These ones are specially related with the fish and meat preservation. We can suspect a close relation between fisher economies and the use of salt to preserve and store this kind of food surpluses. Possibly, the daily observation of these salt properties led people (H/G tribes) to made experimentation with the advantages of salt use for increase food-storage. We think that may be exist a close relation among salt use and prehistoric sailor cultures because the conservation of food was necessary to carry out large trips through sea but also for do trips through the coastal line.

Chronology and contextualization

But, can we speak about salt exploitation (better salt making) from Neolithic times or this concept was related with Early Bronze and Iron Age in Iberian Peninsula?

In Iberian Peninsula, we'll see, herds increasing were related with the beginning of Bronze Age. At the same time it was a great population increasing. Moreover, true archaeological evidences for salt making are related only with Bronze Age in this area like Santioste, Molino Sanchón II (Delibes *et al.*, 2007a), Espartinas (Valiente *et al.*, 2002; Valiente Cánovas and Ayarzagüena Sanz 2005), El Perical (Malpica *et al.*, 2008).

We found some economical and cultural implications related with the salt production. Thus, salt making is proportional with population increasing as we saw. It's no doubt about the direct relation between salt making and herding like Pliny said and like some archaeological evidences show.

The access to the technical information and the know-how of salt production, are building through the patient and long time experimentation. Because this, is not possible to do a direct association between salt making and the early Neolithic in Spain.

Therefore, salt production and salt trade was related with the possibility or impossibility to give the control over the natural salt sources by social groups or tribes. The tribe needs to have control over the territory and over the raw material and food-surpluses too. Only through the direct access and the strong control over salt sources the human group would be able to dominate and have control over salt production. This step, that implies the continued and recurrent access to the resources, should be the main base for the start of salt commerce and exchange.

We think that we need to put the salt exploitation process in relation with the general mining cycle (Jiménez-Guijarro 2007).

This one involves the existence of a mineral surplus (salt) and the necessity of this raw material by the own group and for other neighboring groups (Figure 5). A trade chain is set up among these groups like a part of a more complex social and economical model. Salt making implied some type of specialized working. In Iberian Peninsula the whole complex mining process was ever related with Sherrat's idea of secondary products revolution. This one was bounded up with the demographic increase happened inside of the complex Chalcolithic and Earlier Bronze Age societies.

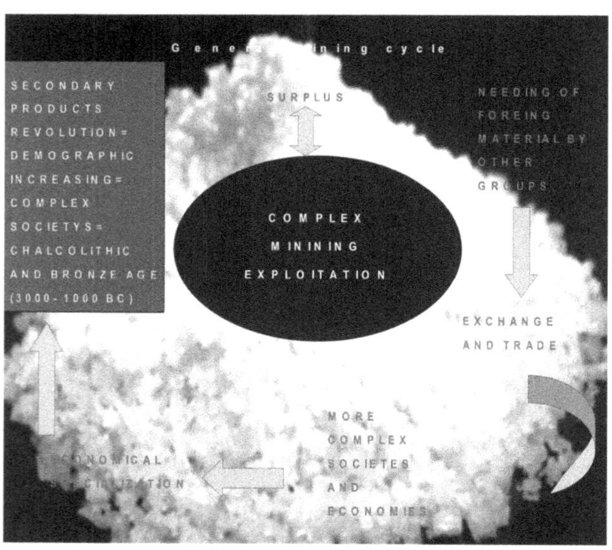

Figure 5. General mining cycle.

But, what kind of archaeological remains point out the salt culture on Iberian prehistoric times?

Following A. Harding, we can signal six direct evidences for salt making:

1.- Brines (from surface salt sources or coming from pits)
2.- Great fireplaces surfaces
3.- Big ceramic pots (normally plain) for cock the brine under fire and make salt precipitation.
4.- Small, and normally, thin pots of crude or highly porous clay used like moulds for make the salt cakes.
5.- A lot of clay firedogs used like supports for the vessels in the salt making process and mostly concentrated around fireplaces areas.
6.- A lot of broken vessels, named *briquetage* or rubble pots, located near the fireplaces too.

We can note some indirect evidences for salt making like a great concentration of archaeological sites found around the salt sources (Delibes *et al.*, 2007b, 53). Sometimes this is so hard to see, but is a very interesting for discover prehistoric settlement pattern (Alexianu *et al.*, 2007; Weller *et al.* in this volume).

Iberian Peninsula has the most important concentration of saline and sodic soils in Europe (Figure 1). Practically all of these soils are located in the inner lands and they are related with salt sources and salt springs. At the same time Spain has one of the highest indices of sunstroke from Europe. All of this would be able to be put in relation to the importance of salt making through the natural evaporation. Since 2008 we are working in the study of this question in the area of the Salty River Valley in Guadalajara (Malpica *et al.*, 2008; Malpica *et al.* in this volume). We begin to have some certainly interesting results about the natural exploitation (from evaporation and precipitation methods) of the salt in the margins of streams, gaps and estuaries. Also we have new evidences about the construction over limestone of some facilities for salt evaporation (Figure 6 to 8).

The exploitation system is well known. Salt water, obtained from pits or from salty rivers and streams, is conducted to some extensive and little deep surfaces called ponds. Once the water is contained here, we left to act the sun that induces water evaporation and the consequent precipitation of the salt. This production system has been in use since the Modern Age in well known sites like Imón, La Olmeda (both in Guadalajara) or Añana, in Navarra, North Spain. The system of inner-salt-production has clear parallels with the exploitation and production of coastal and marine salt mines. Attending to the appointments of the classical writers we can imagine the existence of similar systems during the Ancient History (especially during Roman Empire times) and possibly during the Middle Age.

Our hypothesis presumes the existence of similar systems during later prehistoric times (maybe Chalcolithic times) attending to the recent data obtained as part of the Research Project developed in the Salty River Valley, in Guadalajara. Here, we have been able to document a complex kegs system related with rock carving channels over limestone. The disposition of the structures and the inclination of the limestone surface would permit to carry out a process of water evaporation and salt precipitation. The porosity and facility for warming-up of the limestone through the sunlight would facilitate and accelerate the process. This archaeological assembly is associated with pottery remains dated since Prehistory (Bell Baker Chalcolithic and Early Bronze Age) to Roman Empire.

Figure 6. Detail of limestone carvings to make salt by sun evaporation system (El Perical, Guadalajara, Spain).

Figure 7. Zenital view of salt exploitation area from El Perical (Guadalajara, Spain).

Now well, in this case we have no evidences of Neolithic archaeological remains recover around the site. Besides, the purpose of the limestone construction is clear and, if we confirm its use for salt-making by evaporation, we should conclude that exists some intention by manufacturing salt in certain quantity. We could be demonstrating the strong control over salt resources and territory too. Both purposes remain far from the interests that we have documented for earlier prehistoric neolithic tribes. Also, this method of salt-making, in which the action of fire seems played no role, collides with salt production models for the Bronze Age from Santioste (Delibes, Viñé and Salvador 1998), Molino Sanchón II (Delibes *et al.*, 2007a) and Espartinas (Valiente *et al.*, 2002) where salt-making by fire is well demonstrate.

Therefore, did were complementary these two salt production systems? Following N. Morère and her interpretation of Pliny, we must to say that yes (Morère 2008), at least in Ancient Greek and Roman times.

Figure 8. General view of limestone rokc with carvings to make salt, located near of a salt source. (El Perical, Guadalajara, Spain).

Figure 9. Salty clay inside of a salt source in summer times (Salt River Valley, Guadalajara, Spain).

A question that should keep in mind is the possibility that the salt was obtained, in Early Prehistory, directly from clay. We have had the opportunity to verify how precipitates the salt upon being sunlight evaporation in the margins of rivers, streams and salty lagoons (Figure 9). Today, a lot of sheepherds go to the salty sources and lick the salty clay near these areas removing the muddy clay with their legs.

We have no one direct or indirect evidences of salt making in Neolithic times across the whole Iberian Peninsula. Its true that we know several neolithic sites located near of salt springs but we have no evidence of salt making similar to the archaeological remains recover in other salt springs located in Eastern Europe.

However, we know a scare sites related directly with salt sources (Figure 10): Santioste- Molino Sanchón II - Villafafila (in Zamora- Northwest-Central Spain); El Perical and Salty River Valley (in Guadalajara, Northeast-Central Spain); Espartinas (in Madrid. Central Spain); La Marismilla (in Sevilla, Southwest Spain); Cardona (in Catalonia, Northeast Spain).

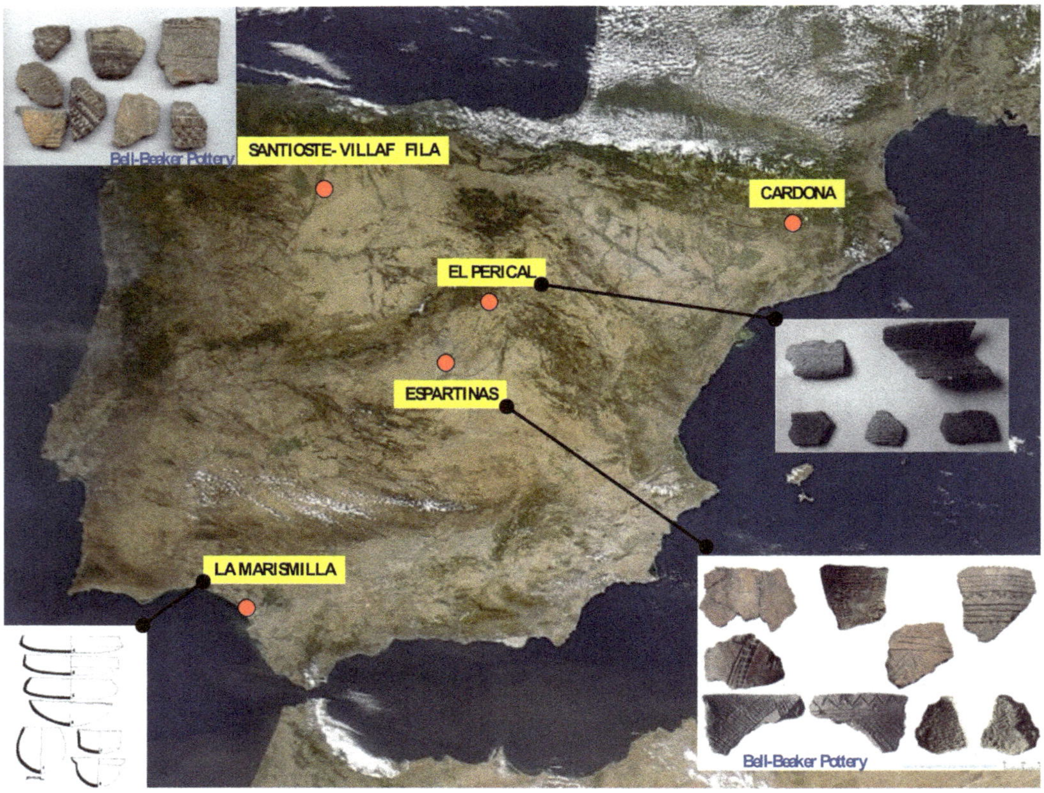

Figure 10. Chalcolithic and Bronze Age salt making sites with clear archaeological evidences from Prehistory in the Iberian Peninsula.

All of these sites are related with great salt areas but no one of these has evidences of salt making during Neolithic times. Only a very scare number of sites have clear evidences of neolithic settlement too. In Villafafila we know only two fragments of clear neolithic pottery located in surface explorations (Rodríguez Rodríguez, Larrén Izquierdo and García Pozas 1990). Now well, these archaeological remains are not related with the archaeological deposit of Santioste or Molino Sanchón.

Certainly around some of these sites, like Espartinas or Cardona we know large concentration of Neolithic sites, but really we cannot to put them in relation exclusively with salt exploitation.

Only Weller and Fíguls spoke about some indirect evidences to put in relation the salt mountain of Cardona with the Middle Neolithic culture called 'Pit graves Culture', but really these evidences are still very scare (Monah 2007, 127) and, in our opinion, haven't clear and direct relation with salt exploitation.
Moreover, all of these sites are well dated in Chalcolithic (like La Marismilla) and Bronze Age (Espartinas, Santioste, El Perical). By other side, in the area of Salty River Valley, in Guadalajara, that is one of the most important inland salt areas of Spain, we have no evidences of Neolithic sites related with the inland salt sources and little salt lakes (Malpica *et al.*, 2008; e.g. in this volume). Our full surface explorations allow us to know a discreet Chalcolithic and Bronze Age sites. All of these are located near of the salt springs, normally over little mounds or small hills around salty areas controlling the better valleys and the natural paths. Instead, we have no evidences of briquetage or pots related with salt making.

Was this area no importance during Neolithic times?. Was this area uninhabited from Paleolithic to Chalcolithic times? We have some doubts about it. In the summer of 2008 we discovered some scattered flint remains, specially made up for small flint sheets and geometrics that may be belongs to later Paleolithic times –our called Epipaleo-Mesolithic horizon (Jiménez-Guijarro 2008 and 2010)- or, may be, early neolithic times. Nevertheless no one of these remains is clearly related with habitation sites or with salt sources neither. We interpreted these archaeological remains like some kind of flint waste or lost tools related with tribes hunting medium and large mammals nearest of the salt sources and little salt lakes.

We know four hypothesises for salt gathering and salt making in Spain during prehistoric times.
The first, wrote by Cassen, Labriffe and Ménanteau (2004) and Escacena, Rodríguez, and Ladrón De Guevara (1996), spoke about the Pits Culture like a salt specialized groups. They interpreted the near coast open sites with a lot of pits like some kind of prehistoric 'salt factories'. Lamentably, are a lot of similar sites, nearest or so far from coastal line, with the same kind of pits, with some channels dig in the ground and with a similar material culture around Spain, Portugal and France that are not related with salt-making.

Cassen and his colleagues explain besides the careening cups and pans and plain plates as bowls for brine heating and they said that this new pots shapes are related with salt production. Moreover, these researchers saw a clear control action over the salt production and trade since V milennia BC in western France and Southern Spain.

Our hyphothesis, suggest by the contrary, that the presence of plain plates discovered inside of dolmenic and tumular funerary households furnishing could be corresponds with some kind of 'funerary feast' remains. They can be related with support vessels (may be essences burners) related with a death men complex wares but not with salt-makers like they said (Cassen, Labriffe and Ménanteau, 2004). Thus, the open walls, careening and plain bowls were part of the new cultural evidence and the main part of a new culinary concept: the roasted meat for eat, and for the banquet and funerary feast too.

The second hypothesis, defended by Weller (2002 and 2004) and Fíguls, Mata-Perelló and Sanz (2003) showed the direct relation of Middle Neolithic culture called 'Pit graves Culture' with the control and exploitation of salt. The green stone pearls made with variscite had a great importance like exchange and prestige items and we know that one of the most important deposit of variscite are located in the north-east Spain, near the town called Gava (Barcelona). They speaks about some close relation between the trade of these elaborated green beads, some first class bright brown flint, big stone axes and salt. Some of these materials, especially green beads, bright brown flint and big stone axes are households located inside of pit graves and normally related with some plain pots. Weller and Fíguls explained this wealth of households in relation with the trade of salt and the powership of some leaders. They saw a clear control action over salt production and salt trade since the IV milenia BC.

The third hypothesis suggest, following Delibes, that the Bell-Beaker called Ciempozuelos are clearly related with the production, exploitation and trade of salt around second milennia BC. For Delibes there is a clear existence of control over salt production since III milenia BC. This control was made by some kind of 'princes' or leaders (Delibes 1977, Delibes *et al.*, 2007b, 60).

We are completely in agreement with Delibes and we have the certain conviction about the close relation between Bell Bakers and the first salt making. Thus, Santioste, Molino Sanchón II, El Perical and Espartinas presents inside of their archaeological remains more or less wide pots decorated with Bell Baker techniques. Moreover, these three sites show evidences of Ciempozuelos style, the incised version of bell baker pottery in South-west of Europe dated in Early Bronze Age.

The Spanish incise bell baker horizon (called Ciempozuelos) was normally related with a sheep herd culture. Moreover, only in Santioste and Espartinas (Valiente Cánovas and Ayarzagüena Sanz, 2005) we have today some clear evidences of firedogs and pots supports used to make salt by fired system. No evidences of Neolithic potteries are present here again.

Interpretation and conclusions

Based on the archaeological remains and evidences, we need to answer some questions about the use of salt in Spain:

1- Salt was not used during Neolithic times in the Iberian Peninsula.
2- Salt making had a punctual and circumstantial use in Neolithic context being more common 'gathering' methods for extraction and use of salt.
3- Salt making during Neolithic times was made by evaporation and not by ignition.
4- We have no evidences of salt making in Neolithic contexts but it could be discover in future.

Some researchers explain their fully belief about the importance of salt trade inside of neolithic cultures (Weller *et al.*, 2007). We consider that they give overmuch importance to salt. Thus they see the salt like one of the principal power in the economical structure, especially when it is inside of the trade and exchange circuit. But, what we know about economical structure of 'pit graves culture'? Were they an essential herdsman culture? Had they inside of their economical structure more importance herder way of life than agriculture? In other words, were the 'pits graves' economy based on shepherd society? And, which was the approximate number of their sheep and goat or cows flocks?

We need to think about our first presumptions around the relation between salt and animal domestication. We said before that prehistoric groups need to increase their salt amount for the animals when they wants to increase their production –specially their milk production- or when they have the herds in sheepfold or in a cowfold. But, we have no evidence of this kind of use among the earlier neolithic cultures in Iberian Peninsula. In our point of view, it's clear that these tribes did not need salt making for increase the salt intakes for their herds.

Finally, we can conclude our work with six relevant questions:

- Existence of two different uses of salt with great chronological, technical and cultural differences during prehistoric times:

  Natural use of salt, that we named as 'salt gathering' (until IV- III milenia BC)

  Exploitation (and trade), that we named as 'salt making' (since III milenia BC)

- Salt making must be included inside of the complex mining cycle. Salt production seems to be a clear exponent of the herding intensification and specialization and shows an exponential growth of human and animal population in prehistoric times. Salt

making and salt trade implies some kind of production intensification among prehistoric groups too.
- We haven't still direct evidences for the relation between salt making and Neolithic in whole the Iberian Peninsula.
- Salt making coincides with the growth and consolidation of villages and farmer lifestyle. This one is relate with economies based on farming and on shepherd and cowsfold.
- The production, control and first trade of salt are contemporary with Bell-Beaker's leadership. This one happens around the transition from third to second milennia BC and with the beginning of Bronze Age.
- Salt remains was included, since fourth milennia BC, inside the control of trade chain and the general exotic prestige goods distribution like a green stone pearls, Ivory, amber, obsidian and copper. Salt become a new economical and commercial element and a new power resource too, out of the domestic circle of culture.

Acknowledgements
This work was made inside of the Spanish Research Project MEC I+D- HUM 2007-66118 'Salt exploitation and organization of territory since Ancient times to the beginnings of feudal society in the Area of Sistema Central (Spain)', under the leadership of Dr. A. Malpica.

References:

Alexianu, M. T., Weller, O. and Brigand, R. 2007. Approche etnoarchéologique de l'exploitation des sources salées de Moldavie: Les enquêtes récentes. In N. Morère Molinero (ed.), *Las salinas y la sal de interior en la historia: economía, medio ambiente y sociedad. Inland saltworks and salt history: economy, environment and society*, 407-434. Madrid, Dykinson.

Cassen, S., Labriffe, P. A., and Ménanteau, L. 2004. Sels de mer, sels de terre. Indices et preuves de fabrication du sel sur les rivages de l'Europe occidentale, du V au III Millénaire. *Cuadernos de Arqueología Universidad de Navarra* 12, 9-49.

Delibes, G. 1977. Sal y jefaturas. Una reflexión sobre el yacimiento del Bronce Antiguo de Santioste en Villafáfila (Zamora). *Brigecio, Revista de estudios de Benavente y sus tierras* 3, 33-46.

Delibes, G., Fernández, J., Rodríguez, E. and del Val, J. 2007 a. Molino Sanchón II: Un salín de época campaniforme en las lagunas de Villafáfila (Zamora). In N. Morère Molinero (ed.), *Las salinas y la sal de interior en la historia: economía, medio ambiente y sociedad. Inland saltworks and salt history: economy, environment and society*, 47-70. Madrid, Dykinson.

Delibes, G., García, R., Larrén, H. and Rodríguez, E. 2007 b. Cuarenta siglos de explotación de sal en las lagunas de Villafáfila (Zamora): de la edad del Bronce al Medioevo. In A. Fíguls, O. Weller (eds.), *1a Trobada internacional d'arqueologia envers l'explotació de la sal a la prehistòria i protohistòria. 1st International archaeology meeting about prehistoric and protohistoric salt exploitation, Cardona, desembre 2003*, 47-70. Barcelona, IREC.

Delibes, G., Viñé, A. and Salvador, M. 1998. Santioste, una factoría salinera en el inicio de la Edad del Bronce en Otero de Sariegos (Zamora). In G. Delibes (coord.) *Minerales y metales en la Prehistoria Reciente. Algunos testimonios de su explotación y labereo en la Península Ibérica*, Col. Studia Archaeologica 88, 155-198. Valladolid, Universidad de Valladolid.

Denton, D. A. 1969. Salt appetite. *Nutrition Abstracts Review* 39, 1043.

Devlin, T. J. and Roberts, W. K. 1963. Dietary Maintenance Requirement of Sodium for Wether Lambs. *Journal of Animal Science* 22, 648-653.

Escacena, J. L., Rodríguez, M. and Ladrón De Guevara, I. 1996. *Guadalquivir Salobre. Elaboración Prehistórica de Sal Marina en las Antiguas Bocas del Río.* Sevilla, Confederación Hidrográfica del Guadalquivir.

Fíguls, A., Mata-Perelló, J. M. and Sanz, J. 2003. Neolithic exploitation of halite at the 'vall salina' of Cardona (Catalonia, Spain). In A. Fíguls, O. Weller (eds.), *1a Trobada internacional d'arqueologia envers l'explotació de la sal a la prehistòria i protohistòria. 1st International archaeology meeting about prehistoric and protohistoric salt exploitation, Cardona, desembre 2003*, 199-218. Barcelona, IREC.

Hagsten, I., Perry, T. W. and Outhouse, J. B. 1975. Salt requirements of lambs. *Journal of Animal Science* 40, 329.

Jackson, H. M., Kromann, R. P. and Ray, E. E. 1971. Energy retention in lambs as influenced by various levels of sodium and potassium in the rations. *Journal of Animal Science* 33, 872.

Jiménez-Guijarro, J. 2007. ¿Aprovechamiento o explotación?: Reflexiones acerca de la minería y uso de la sal durante la Prehistoria. In N. Morère Molinero (ed.), *Las salinas y la sal de interior en la historia: economía, medio ambiente y sociedad. Inland saltworks and salt history: economy, environment and society*, 185-216. Madrid, Dykinson.

Jiménez-Guijarro, J. 2008. *El Proceso de neolitización del interior de la Península Ibérica.* Ph. Thesis. Universidad Complutense, Madrid.

Jiménez-Guijarro, J. 2010. *Cazadores y campesinos. El proceso de neolitización del interior de la Península Ibérica.* Bibliotheca Archaeologica Hispana, 31. Real Academia de la Historia. Madrid.

Jordan, R. M., and Hanke, H. E. 1982. Effect of mineral additions to trace mineralized salt on daily intake of salt and minerals. *Proceedings of 54th Sheep and Lamb Feeders Day.* Minnesota University, 182.

Malpica, A., Morère, N., Fábregas, A. and Jiménez-Guijarro, J. 2008. Organización del territorio y explotación de la sal en el área del Río Salado (Sigüenza, Guadalajara, España): Antigüedad y Edad Media. Resultados de la I Campaña 2008. In *Actas del XI Encuentro de Historiadores del Valle del Henares*, Guadalajara, 49-62.

Mark, L. Wahlberg and Scott, P. 2006. Minerals and vitamins for shepp. http://www.ext.vt.edu/news/periodicals/livestock/aps-06_10/aps-373.html. Last update, October, 2006.

McClymont, C. L., Wynne, K. N., Briggs, P. K., and Franklin, M. C. 1957. Sodium chloride supplementation of high-grain diets for fattening Merino sheep. *Australian Journal of Agriculture Research* 8, 83.

Meyer, J. H., and Weir, W. C. 1954. The tolerance of sheep to high intakes of sodium chloride. *Journal of Animal Science* 13, 443.

Meyer, J. H., Weir, W. C., Ittner, N. R., and Smith, J. D. 1955. The influence of high sodium chloride intakes by fattening sheep and cattle. *Journal of Animal Science* 14, 412.

Monah, D. 2007. Le sel dans la Préhistoire de la Roumanie. In N. Morère Molinero (ed.), *Las salinas y la sal de interior en la historia: economía, medio ambiente y sociedad. Inland saltworks and salt history: economy, environment and society*, 121-163. Madrid, Dykinson.

Morère Molinero, N. 2008. Une nouvelle approche de Pline sur le sel et l'eau salée. In O. Weller, A. Dufraisse and P. Pétrequin (eds.) *Sel, eau et forêt: hier et aujourd'hui*, 365-380. Cahiers de la MSH Ledoux 12 (coll. Homme et environnement, 1). Besançon, Presses Universitaires de Franche-Comté.

Morris, J.G. and Peterson, R.G. 1975. Sodium requirements of lactating ewes. *Journal of Nutrition* 105, Issue 5, 595-598.

National Research Council. 1980. *Mineral Tolerance of Domestic Animals*. Washington D.C., National Academy of Sciences.

Rodríguez Rodríguez, E., Larrén Izquierdo, H. and García Pozas, R. 1990. Carta Arqueológica de Villafáfila. *Anuario del Instituto de Estudios Zamoranos Florián de Ocampo* 1, 33-76.

Underwood, E. J. 1981. *The Mineral Nutrition of Livestock*. Slough, Commonwealth Agricultural Bureaux.

Valiente Cánovas, S. and Ayarzagüena Sanz, M. 2005. Cerámicas a mano utilizadas en la producción de la sal en las Salinas de Espartinas (Ciempozuelos, Madrid). In O. Puche and M. Ayarzagüena (eds.), *Minería y Metalurgia históricas en el Sudoeste europeo,* Actas del *II Simposio sobre minería y Metalurgia históricas en el Sudoeste europeo (Madrid 24-27 de junio de 2004*), 61-70. Madrid.

Valiente, S., Ayarzagüena, M., Moncó, C. and Carvajal, D. 2002. Excavación arqueológica en las Salinas de Espartinas (Ciempozuelos, Madrid) y prospecciones en su entorno. *Archaia* 2, 33-45.

Weller, O. 2002. The earliest rock salt exploitation in Europe: a salt mountain in the Spanish Neolithic. *Antiquity* 76 (292), 317-318.

Weller, O. 2004. Los orígenes de la producción de sal: evidencias, funciones y valor en el Neolítico europeo. *Pyrenae* 35 (1), 93-116.

Weller, O., Brigand, R., Nuninger, L., Dumitroaia, Gh. and Monah, D. 2007. Analyses et modélisation spatiale autour des sources salées de Moldavie précarpatique durant la Préhistoire. In N. Morère Molinero (ed.), *Las salinas y la sal de interior en la historia: economía, medio ambiente y sociedad. Inland saltworks and salt history: economy, environment and society*, 165-184. Madrid, Dykinson.

Weller, O. and Fíguls, A. 2008. Prèmier extraction de sel minier: Place et rôle du sel de Cardona dans les échanges communautaires du néolithique moyen catalan. In M.S. Hernández Pérez, J. A. Soler Díaz, J. A. López Padilla (eds.), *Alicante 27 al 10 noviembre 2006, IV Congreso del Neolítico Peninsular*, Vol. 1, Marq, 353-360.

Part III. Ancient Texts and Salt

Salt in the Antiquity: a Quantification Essay

Bernard Moinier
Consultant Sel et Santé, Paris, France

Abstract
Ancient economy is to a large extent depending upon estimating, where possible, which is the market size and how supply meets its various requirements. The following pages constitute an attempt to quantify salt demand in combining ancient population figures with salt major uses, provided the former are available. Irrespective of despondent gaps and the scarcity of statistics, it makes sense to delineate a framework within which literature, supported by archaeological and epigraphic studies, could be more precisely evaluated with this respect. Although certain developments deserve an amount of criticism all of them are aimed at contributing towards a wider discussion on the salt consumption dynamics, at least in the late Roman Republic and the Early Empire.

Keywords
population, salt production, salt uses, bread, fish salting, animal feeding, hide curing, diet, statistics, measure units

We find reasons to be optimistic in reading the monograph of Pliny the Elder and the developments concerning salt in Antiquity left by Aristotle and Strabo while we initiate a reflection about this substance. '*Thanks to Strabo and Pliny we are reasonably well informed about how salt was produced in the Roman Empire*' (Adshead 1992). However, we rapidly discover that occurrences are heterogeneous and epigraphy and archaeology do not add much to the fragmentary character of information in the literature. The spatial-temporal value of salt fails to be perceived due to the lack of statistics. Thus, it is indispensable to study what could have been, in a given period and region, the volume of the offer (salt marshes, rock salt deposits, salt lakes and springs) taking into account both the identified and non-inventoried sites, and that of the demand. We should take into consideration the diversity of the applications of salt, be they food (in the kitchen or at the table, for food preservation, salt meat and fish sauces, animal feeding), or non-food (hide and leather curing, colour fixing, metallurgy, cosmetics, crenotherapy, pharmacy).

The fact that we want to have an order of magnitude concerning the corresponding quantities of salt is not a caprice. For example, at the end of Augustus' reign (who died in 14 AD), Italy would have represented a market of about 50,000 tonnes and it is very probable that the supplying could not have been ensured only by the salt marshes of Ostia... We would also want to examine to what extent, and with which reservations, it would be possible to make up for the lack of statistics to orient more positively some of the commentaries inspiriting the study of salt in Antiquity.

A dead-lock and uncertain tracks
Certain experts have thought of mathematical models. Their scientific robustness does not mask for long their little adequacy in the technological, climatologic and sociologic context in which salt was produced by the Etruscans, the Ionians or the Turdetans (where, when, how).

- *Mathematical modelling*

Of course, some proposal does deserve attention if it offers methods to the archaeologists allowing the calculating of salt production in the most ancient times. This is the case of a study (Akridge 2008) which, regarding the salt production techniques through evaporation, already practiced in the preindustrial societies, is applied to determine the limits of carrying them out starting from brine. Even though they have variants, they are founded either on solar evaporation, or on ignigenous evaporation. Their choice depends essentially upon the geographic and geological factors. Thus, it appears unusual to come with the idea that the quantity expected by a population in a certain place could drive the choice of the salt production technique. Man produces salt in a place where he can have access to the resource, or where it is easy to get it operated. Nevertheless, for the Antiquity and the Prehistory, the descriptions of the technique in places where the exploitation is attested are rare, as the remnants of salt works. Moreover, we know nothing about the effectiveness and the quantities obtained. As for the quality, even less is available...

Akridge tries to show that there are mathematical formulas that allow us to calculate how much crystallised salt we can get from brine. He uses simulations to get them exemplified. In the case of La Placita salt pans (Mexico), he says that '*it takes five days of solar evaporation to concentrate the brine sufficiently to begin collecting salt. About 25-30 kilograms are collected from each era (18 in all, 3m sideways). An average of 7 tons of salt can be produced at each salt making site during the dry season*'. This type of observation is available for a certain site at a certain moment in time. We do not see how we could transfer it to the Ostia salt marshes when the Etruscans exploited them or at the beginning of the Empire. Besides, the formulas cannot be applied as the parameters that they integrate are not valid for the Antiquity, mostly S – the initial concentration of salt in brine. The heating of brine in the case of an ignigenous fabrication (applied external heat to a vessel) depends on the temperature of the environment and also on the heat applied to the vessel in order for the brine to boil. Could the discovery at Shavington (Cheshire) of a lead pan cut into eight pieces be enough to understand the technique used in Britain during the Roman age when, as it was not identified in situ, we do not take into account the fact that it was the heating vessel? How can we then calculate the evaporation rate of brine? What about the effectiveness of heating? And what could we take from the theoretical explanation of the authors in order to apply it for the salt spring that the Autariatae and the Ardiaians were fighting for (Strabo *Géo.* VII, 5, 11), or concerning another at Tragasae?

Of course, these calculation methods do refer to the title of the article, but they do not correspond to the objectives announced in the introduction and reiterated in the conclusion. Further to the analysis of an enterprise during which they were going to regret the fact that the formulas proposed were to be outside their preoccupations, the archaeologists or the historians can only question its benefits in order to estimate the efficiency and the total salt production in the sites exploited in Prehistory or in Antiquity. In return, the demographic researches should get into attention as the population provides a practical calculation basis, mostly concerning the diet. The salt requested for human consumption remains rather steady and constant as quantity; the inter-individual variations are not significant in the context of a global approach. If we have information concerning the daily or annual ratio, are we not allowed to calculate the salt needs for a population? Quantifying salt demand provides its lessons within its limits.

- *Ration*

In *De Agricultura* (DA), Cato the Elder (234-149) mostly gives the following indications for capacity or weight to be converted in IS (International System) measures. See Figure 1. We prefer to keep the term *modius* and not to translate it (bushel) in order to avoid any metrological confusion. Taking into account the salt granularity, the density varies from 0.6 for coarse salt to 1.2 for fine salt (vacuum salt). In order to take into account the predominance of common salt produced during the 2nd century BC, then during the Empire, we are going to consider as minimal basis 7kg for a *modius* of salt. See Figure 2.

In order to avoid the difficulties brought along by the conversion of capacity units in mass units, we chose to take into consideration the metrology works in France after the adoption of the metric system (Tarbé 1813; Marion 1989). Before that moment salt was measured in volume, not in weight. We see that a French *setier* means 208 litres and 163.2kg. We take into account 1.2 litres for 1kg. Thus, 7.5 litres of salt mean 7.3kg. It appears that taking as indicative value 7kg/*modius* avoids any risk of overestimation when we calculate the salt needs starting from demographical statistics. This approach seems to eliminate the reserves formulated by the historians that tried out this quantification exercise, even though they did not know what conversion basis they should use…

- 1 sextarius = 0,54 litre	436g in salt equivalent
- 1 semodius = 4,37 litres	3,5kg in salt equivalent
- 1 modius = 8,75 litres	7kg in salt equivalent
- 1 libra = 327g	327g

Figure 1. Conversion in IS measures to get equivalents in salt.

Even though the information is still limited, Cato the Elder exposes that giving a modius of salt per year, meaning around 7kg/year to a slave that works in farming of small areas is enough. '*Salis unicuique in anno modium satis est*' (*De Agricultura*, LXVII). In the same category as oil and olives, this ration is aimed at supplementing the diet of slaves, who worked hard and suffered from the heating sun. In this case, the rations were somewhere around 18g/day. Having opted for a density of 0.97, Robert Etienne gets to 22.8g/day. He underlines that this ration means more than the double of the contemporary rations. Anyway, he does not wonder, whether the salt provisions and salt intake may present certain differences. The same remark is valid for Jacques André who gets to 18.5g/d: whole 7-8g/d are enough, in his work upon the diet in Rome (1981). In comparison to dietary salt intake, what represent the food use and the real ingestion of the salt?

Information expressed in measure units that we may find in *De Agricultura* allows us not only to know the ration allocated to slaves that worked in farming of small areas during the 2nd century BC, but also to seize its complexity in relation to the alimentary salt provisions. Besides, salt used in the kitchen in order to cook the meals or at table having recourse to salt cellars, there is also the salt used to preserve olives or for pork salting, mostly for hams, about which we know (Polybius, Varron) that they were intensely exchanged by the Celts, including in the Cisalpine Gaul.

Designation	Measure DA	IS Measure	Observations
Salt ration given to slaves	1 modius/year	7kg/ year	
Ham curing (per unit)	½ modius/ year	3.5kg/ham	4.5kg presently
Salted olives (per modius)	½ pound	163g	
Olives in brine (*muria dura*)			See brine
Concentrated brine - Sea water - Salt	 1 quadrantal 1 ½ pound	 709g 490g	 27g x 26.26 litres

Figure 2. Statistics taken from *De Agricultura*.

Otherwise, thanks to *Historia Augusta* (Claudius' life, XIV, 1-13), we may try to calculate the daily salt ration on the basis of the annual pay attributed to the tribune of a legion by emperor Claudius (268-270). We are generally talking about 20 *modii* of salt per year to be divided into around fifteen persons (who lived in his household). This endowment thus means 140kg/year (7kg for a *modius*), which gives us a daily average of 25g. Nevertheless, we should take into account the fact that the tribune also had horses (barley and hay are mentioned) and that he had guests at table. Thus, we may believe that the actual salt quantity used by the tribune and his family was not that big. There are three developments which, in *Historia Augusta*, give us information upon the ordinary diet. We tried to make such a comparison in Appendix A. It is very difficult as the authors did not pay much attention to the lists. The influence of Vegetius (that Chastagnol talks about in his *General Introduction* to this work) is not enough to make order. Of course, Vegetius recommends a list of details upon the state of the troops and of the expenses necessary to keep them. Nevertheless, he does not give more details on the subject. '*We have to make sure to have enough wood and fodder during the winter, enough water during the summer and, all year long, wheat, wine, vinegar and salt* '. With a *sextarius* of salt shown in Aurelius' life (IX, 6), meaning 436g, the administration ensures a 22g ration for 20 days or for twenty people...

In the case of alimentary rations for a legion, it seems that such a quantity is not excessive, as the soldier eliminates a lot of sodium through perspiration. The reconstitution of the salt capital requires around 5-6g of supplementary salt. This remark of Perea Yebenes (2006) is valid for anybody who works in farming of small areas during the summer heat. We should also remember that the salt given by Cato the Elder to his slaves is directly or indirectly related to their diet, as it enters the *pulmentarium familiae*. It goes with the bread or the pancake, it helps preserving the fallen or over-ripped olives and for the traditional pork salting. Nevertheless, we are not sure that by 'slave' we should understand a single individual. Couldn't we assume that there was also a female companion (*contubernium*), or maybe one or two children living with the slave? Following this hypothesis, the salt ration that was attributed only to the slave could get down to 12g/day. Covering around 86ha, Cato's domain is valorised by a farmer and 28 single slaves (Lévy 1981).

We find at Aristophanes (*Acharnians*, 810-815) a very interesting indication, like that left by Cato. Diceopolis buys a sow from Megarian for a choenice of salt, meaning 875g (1 medimnus of salt = 42kg and there are 48 choenices in a medimnus). This quantity is enough for one person during a trimester. In this case, the ration is around 9.6g/day. Without forgetting the delicate problematics of salt provisions, shouldn't we also wonder about the actual salt intake?

- *Salt intake*

This notion allows making the balance sheet of the actual sodium intake, coming from different sources (Na 17 mmol = NaCl 1g). By convenience, it is generally expressed in salt equivalents (g/d). Presently, it would be useful to make the distinction among the following:

The naturally present sodium in certain raw aliments expressed in salt 1-2g/d
The salt which remains from cooking or use at table 2-3g/d
The salt which remains from the industrial or artisanal preparations 3-4g/d
The sodium present in water and sodium salts besides the chloride c. 1g/d

The salt intakes are between 7 and 10g/d for most of the population. The big consumers are defined by salt intakes >12g/d. We should mention that the kidneys and the heart regulate them within a range of 4 to 18g/d, while the minimum requested by the organism is 3-4g/d. The sodium proportions excreted in the urines in 24h means an average of 8g/d. From these statistical elements we should retain the difference which exists between the dietary use of salt and its intake. The phrase 'salt which remains' means not taking into account a certain amount of salt discarded before the ingestion. This could represent currently more than 50%.

Could such a balance sheet be valid for the times of Cato the Elder or Pliny the Elder that their information represents examples that can be generalized? In a study upon the diet of the slaves, Robert Etienne does not take that risk. Jacques André does not, either. We could nevertheless take it into consideration just in order to make a comparison:

Monthly ration	Equivalent in litres	Equivalent in kg	Daily ration
Soldier (150 BC) = 2/3 medimnus	34.56	25.90	0.9kg/d
Citizen (73 BC) = 5 modii	43.77	32.83	1.1kg/d
Slave = 5 modii	idem	idem	1.1kg/d

Figure 3. Wheat ration.

The naturally present sodium in certain raw aliments expressed in salt (2g/d?)
Discretionary salt (use of the salt cellar, garum and salt-based spices) (2-3g/d?)
Salt that remains from meat and fish salting (3g/d?)
Bread, cheeses, water and other sources of salt (3-4g/d?)

Of course, there may be great differences according to the eras and territories, and there is a tendency to use less salt in cities, developing with the evolution of the lifestyle in the regions incorporated in the confederation later to be confounded with Italy at the beginning of the social war (91-89 BC). Thus, the population of Rome seemed to have more than 180,000 inhabitants en 270, then 375,000 in 130, and a million during Augustus' reign (Garnsey 1988), meaning 1,250,000 inhabitants when we calculate it according to the wheat distributions (Oates 1934), level confirmed by Carcopino (1939) and by many others. We are proposing, with the following reservations, a calculation basis of 12g/d for the last three centuries and 10g/d for the Empire until the Antonine dynasty. We are talking here about a minimalist approach which does not take into consideration the salt for human consumption that is actually absorbed by the organism.

What should be the estimations concerning the uses of salt?

The objective being that of getting to acceptable orders of magnitude, we should found our approach on the whole of salt uses. Besides the human diet, the sectors known as needing much salt are animal feeding, leather tanning, dyeing, second fusion metallurgy, cosmetics, crenotherapy and pharmacy. As salt is not consumed as such, but only to give taste to meals, to contribute to the success of recipes (bakery, cheese industry) and for meat and fish salting, it seems right to examine, first of all, the most well known element within the alimentary ration, meaning the wheat and bread consumption. Besides, what could be better for someone who is very hungry than bread and salt? Nothing could, according to Horace (*Sat.* II, 2).

- *Bread and salt*

Cato the Elder says the following on the matter: for the slaves working in the fields, the monthly ration consists, during the winter, of 4 *modii* and during the summer 4 ½ *modii*; for the farmer, the farmer's wife, the foreman, the shepherd, for any season 3 *modii*. We do not see many differences among the ratios coming from other sources. Polybius (VI, 39.12-15) states 3.9 *modii* as minimum for soldiers. The average is 5 *modii* during the first century AD, which is confirmed by Seneca (*L. Lucilius* IX, 80, 7). Contemporary testimonies valid for Egypt attest a ration of about an *artabe*, meaning circa 5 *modii*. The table below presents the information retained by André in his book concerning the diet in Rome (1981, 192).
Could we take into account a ration of 5 *modii*/month for all categories, and if so, for which period? During Aurelian's reign, the distribution of annona bread seems to have been reduced to 3 *modii* (*Historia Augusta – Aurelian* 35, 1), together with pork and salt (Zosimus *Hist. Nouv.* I, 61 – Note 90).

According to the author of *Epitome de Caesaribus* (who writes at the beginning of the fifth century AD), during Augustus' reign, Egypt provides every year for Rome 20,000,000 *modii* of wheat. We know that during this period Egypt's part is, according to Flavius Josephus (*The Jewish War* II, 386), of about a third of the provisions within the City. Taking into account the provisions coming from Africa, Sicily (> 10 M *modii*), and Sardinia (>7 M *modii*), we total 60,000,000 *modii* (Virlouvet 1995). If we assume that the annual consumption is of about 60 *modii per capita*, we calculate a theoretical number of 1,000,000 inhabitants. We know that, some time ago, Oates (1934), evaluated it at 1,250,000 inhabitants. On this basis, he comes up with a typology that we present in the table below (Figure 4).

We took it for inspiration when elaborating the following table (Figure 5) in order to get to a more precise measure of the salt within bread consumption by the population of Rome upon modulation according to sex and age.

	Inhabitants	*Modii* of wheat
Men	491,800	29.5 millions
Women	408,200	19.6 millions
Children	350,000	12.6 millions
Total	1,250,000	61.7 millions

Figure 4. Population of Rome and wheat consumption.

	Inhabitants	*Modii* of salt	Equivalents in tonnes
Men	491,800	442,620	2,310
Women	408,200	285,740	1,350
Children	350,000	175,000	840
Total	1,250,000	903,360	4,500

Figure 5. Population of Rome and salt consumption.

Thus, we see that the salt contents related to bread represent, in Rome, in Augustus' time, around 4,500 tonnes/year. It does not seem to be valid for Italy, indeed. Taking 20 pounds as average weight of the modius (327g x 20 = 6.54kg), Goujard (1975), who comments upon Cato, calculates that 4 *modii* of wheat are enough for 106 pounds of bread, and 4.5 *modii,* for 120 pounds. The monthly ration of bread is thus somewhere around 33.6 and 39.2kg (a little more than the minimum for a soldier, meaning 33.2kg). If we assume an annual bread consumption of 420kg, which means 4kg of salt for its processing, we may calculate, theoretically, what the bakery represents out of the uses of salt in Italy. If we take into consideration a minimum of salt 4kg, the needs for this sector get to around 24,000 tonnes at the beginning of the first century for 6 million inhabitants. The grains and the salt are also objects of frauds as certain operators put various substances or even water in it. It appears that the controls had as purpose checking the capacity and not the mass. Thus, the epigraphic documentation may be of little help if trying to analyse the weight of the salt.

- *Fish salting*

We call fish salting products of the sea that have been put in salt or in brine, and then left, for a certain period of time, under its action... The global proportion fish/salt is 3/2 (Sternberg 2000). For other archaeologists, it is not simple to evaluate the quantity of salt necessary for salting, but we may situate it somewhere between a third and half of the fish quantity. For example, 250 tonnes of salt are necessary for the salting vats of Lixus, if we take into account the lowest salt proportion (1/4) (Hesnard 1998). At Tirytake, salting about 12,800 quintals of fish per year requires (Etienne and Mayet 2002) about 300 tonnes for a more important capacity.

Fish salting facilities are an important market for salt because of the nutritional (through the preservation ensured by the salt) and speculative character (taking into account the high value of specialties such as *garum*) of this sector within the ancient economy. In this sense, we should make the difference between the following:

1. The domestic salting, for vegetables, olives and meat (mostly pork). They are part of the everyday life in the country. We cannot forget the indications left by Cato the elder: half of modius to salt a ham, meaning about 4kg, half a pound per *modius* to salt the olives, meaning > 160 g. We also have Columelle's indications (*Res Rust*. X). This repartition does not mean that vegetable salting never gets out of the rural economy. During the first century AD, the *negotiatores salsari leguminari* bring such vegetables to the Vindonissa camp (Windisch, Argovie).
2. The commercial salting, where fish and meat compete. For reasons related, among others, to archaeology (aliments processed in brine or fish-based sauces were transported in amphorae), fish seems to have won over meat (no packaging for hams, but for loin?). Another reason seems to be that of the numerous persons involved in such activities as fishing, salt industry and the production of amphorae. In many Roman provinces, the production of salt fish started developing during the first century, as in the case of the counters that the Greek cities established on the Pont shores. The most ancient garum factories are those of Corinth and Delos (Amouretti 1997). And the reputation of *tarichos* started to be built here a long time ago.

The comprehensive studies and the excavation reports are numerous from now on. Nevertheless, it is impossible to collate the statistics for this sector. We have to hope that, through classifying, upon regions and centuries, the establishments inventoried and documented by the archaeologists, we shall determine the needs in salt immediately available in the salt marshes near the salting and salt fish factories. For this purpose, we have to make efforts in order to establish a corpus of definitions. Over the course of the readings, we have the impression that there is a lot of ambiguity in concerning this matter. This is the case of the *muria* (brine), that we should read, in the context of the commercial exchanges, as salted fish and not as salt in solution, corresponding to a salt saturated water (salt marshes) or to a more or less concentrated brine (salt springs) which could not be transported in amphorae down in the hold. When analyzing such hypotheses, it is important to avoid the anti-economic and irrational features and to favour what is easy and cheap to produce (Pons Pujol 2006).

Information concerning the treatment capacity of fish and other alimentary resources in the salting factories allows us to reflect upon the scale change which affected salt marshes along several centuries (end of the Republic, Early Empire). We should nevertheless keep certain reserves...

a) Only a few factories and complexes have been totally excavated and there is an uncertain chronology concerning the capacity used in the inventoried sites.
b) The salt/fish proportion is still a subject of debates among the researchers.
c) It is difficult to establish a common denominator in the regions where the agriculture or the salt culture is practiced given the mutations observed in terms of markets.

For the first century AD, we mention the following classification for the sites where the salt is collected after solar evaporation:

- 200 t/year and less,
- up to 2000 t/year,
- up to 4000 t/year,
- more than 4000 t/year.

If we use the same nomenclature for the Italian peninsula, we observe an incontestable revolution in comparison to the third century BC (appendix B).

- *Animal feeding*

The ancient agronomists emphasize how important salt is for animals. In relation to the role played in their health state, Aristotle (*Hist. Anim*. VIII, 10) says they are more at ease when they can drink as much as they want and when they get salt regularly. The distribution of salt to animals prevents the risks of mineral deficiency (loss of appetite, slow growth, reduced fertility, mediocre lactation, increased morbidity). As in the case of the human being, we are talking about the hydro-mineral balance. For the flocks, he recommends a *medimnus,* meaning around 42kg of salt, for a hundred animals. He says that, during the summer, it should be renewed every five days. The calculation of the daily ration gives 84g/d the level of

the ration seems too high for such a small animal, as it is close to that for a ruminant.
Only this occurrence referring to the animal feeding offers a statistical anchorage. Nevertheless, the salt needs are very well-known nowadays and we could even suggest that they have not changed that much along the centuries. In return, we know nothing about the importance of the flocks. Some studies upon the transhumance mention high levels of magnitude for certain regions (100,000 sheep in Crau, several hundred thousand in Central Italy), which could be an indication of the sheep in motion (summering on mountain pastures). However we cannot make the same suppositions as in the case of the population.

Concerning the livestock farming in Greece, Chandezon (2003) argues concerning the transhumance. The positions depend upon its definition. He says that the size of the flocks and the distances are relevant determinants (but he does not give any statistic). Two of the inscriptions he analyzed refer to a very modest scale in comparison to the estimations for other regions, as between 70,000 and 100,000 sheep seem to have belonged to the sovereign of Knossos (Faure and Gaignerot 1980). Thus, Zenon of Kaunos, representing Apollonius, *dioiketes* of Ptolemy II, has 1863 sheep and 122 goats. Eubolos of Elateia is authorized to pasture 1000 sheep and goats, as well as 220 cows and horses… We may, count them only by snippets, indeed.

- *Hide curing*

Concerning hide curing, the salt used represents 35-45% of non-treated hides. The example of the tents necessary for the Army allows an approach of the salt needs for this sector. *Historia Augusta* (Claude XIV, 3) suggests an endowment of *pellium tentoriarum decurias triginta*.

In his commentaries regarding the setting of a military camp, the author of *De munitionibus castrorum* talks about the way the cohorts set their tents. He says that a tent is enough for eight people. A century has 80 soldiers: this means 10 tents. However, if two centuries camp one in front of the other, the soldiers don't set more than eight tents per century (Pseudo-Hygin). Since Vespasian's reign, the number of soldiers for the first cohort doubled. During Marcus Aurelius, the number of legions reaches 30. On a theoretical basis of 6,000 men per legion, we have 180,000 men in total. Thus, they need these 225,000 tents. In order to cure the 70 hides necessary for a tent, we need 210kg of salt as we need 3kg of salt for the curing of one goat hide. Thus, the salt needs are > 4700 tonnes. We may admit that not all tents were made of cured hides; some of them were made of cloth. But we know nothing abut how much they lasted. The replacement ratio depended upon the circumstances.

The auxiliary troupes count around 200,000 men at the end of the first century AD (Cosme 2007). According to the observations made at Vindolanda, a tent is enough for 10 soldiers. Thus, theoretically, they need 20,000 tents. We may admit that only 15,000 are made of hides. On the same bases as earlier, they needed around 4500. Even though these calculations do not totally reflect the reality, they orient the reflexion towards the real dimensions of salt market. Only to satisfy the needs of the army, hide and leather curing is a very important market. But, only by knowing a market size, we cannot know the general annual needs. We repeat it just in order to make ourselves clear.

Which demographic bases?

Contrary to given ideas, ancient cities did have censuses; this way, we at least know the numbers of the elements available are too fragmentary to be used in determining a population level for a given year. As for the Roman Republic, the *census* (every five years) allows, in theory, to dispose of a series. Between 508 BC and 14 AD, there had been 37 censuses. It is not very reliable as the counting methods change upon the evolution of the territory. In addition, *Roman Italy* is a false phrase for 89 BC (Nicolet 1988). It seems acceptable to work upon the estimations made or retained by demographists (Corvisier and Suder 2000). If the documentary frailty of the historical demography makes it arguable, its deliberate omission means lacking an instrument used to better understand the evolution of the salt economy in Antiquity.

We should note that the current state of the ancient historical demography (Corvisier 2001) makes the object of a balance sheet tentative in *Annales de démographie historique*. The persons interested in the subject may use it as reference point.

- *Italy*

For the first estimations concerning Italy, we suggest the use of the following range: 12-24g/d for the salt at the end of the Republic and 10-20g/d for the first two centuries of the Empire. The conclusions are presented in the table below (Figure 6).

Given the quantities of salt, we pass from an average of 40,000 to 34,000 tonnes, which is not that surprising as, with the population growth, we also note a progress of the importations of products which involve salt: salt meats, leather, make-up, etc. The agriculture crisis that started in the third century BC grows stringer and transforms into social crisis. The wars having as purpose the construction of the Empire allow masking the misery of the victims of the crisis under the destruction of the defeated and the contribution of the new provinces. They attract as locusts, the *negotiatores* and publicans. In order to fight against their excesses, the Emperor institutes a monopole over the mines and the army gets its own factories in the occupation zones.

The production sites which benefited from the demographic boom disappear bit by bit in Italy up to the Late Empire. The sea salt production probably decreased as, in certain regions affected by the ruin of the small farmers and the abandonment of the lands;

salt marshes are no longer reconditioned. From here the pestilences (malaria) evoked in the literature.
In the meantime, fish salting boosted salt production in Occidental provinces (Baetica, Lusitania, Mauretania Tingitana) and in Africa Proconsularis.

Republic	Italy: total population (in M/inhabitants)	Salt needs (1000 t) Weak hypothesis	Salt needs (1000 t) Strong hypothesis
225 BC	4,5	20	40
70-69 BC	6,0	27	54
28 AC	7,5	33	66
Empire			
14 AD	6,0	22	44
47 PC	7,0	25	50
96-192	8,7	31	62

Figure 6. Relative estimation of salt needs in Italy.

The question now is to know whether the demand may be satisfied by the offer if it depends mainly upon the sites exploited according to solar evaporation methods (natural salt deposits exploited through simple crop, salt marshes or water circulation which allows the collection), or according to mining techniques (dry mining, brine boiling from salt springs). An approach by region allows emphasizing the sites to be evaluated by taking into account the information available concerning this subject. It is the object of a table (Figure 7) where there is an estimation of the capacities according to the production types and to the number of estimated sites.

Production	*Modii* (7kg)	Tonnes	Number (E)	Total (tonnes)
Minimal	800	5.6	3,600	20,160
Notable	2,400	16.8	12	201
Important	9 600	67.2	2	134
Exceptional	38,500	269.5	1	269

Figure 7. An estimation of the capacities susceptible of producing salt in the third century BC.

We obtain a theoretical capacity of 20,700 tonnes which corresponds, taking into consideration the variations, a range of 16-24,000 tonnes. We should note that the estimation of the needs of the Italian confederation gives us 18,700 tonnes upon the basis of 1 *modius* x 2.7 million inhabitants. Appendix B completes it, corresponding to the end of the third century BC, and the table (Figure 8) corresponding to the first century AD. The volumes indicated are just an approximation. They do not segregate sea salt and rock salt and are expressed in tonnes.

We should note that the marine salt harvests can face important variations (in a proportion of 1 to 2), and not necessarily symmetrical. For the first century we have a range of 50-70,000 tonnes which only makes easier the reflection upon the production and the satisfaction of the demand. We should also elaborate a series of tables corresponding to different periods within the history of Rome. We may note that it is an uneasy task. If the researchers have demographic reference points concerning Rome and Italy, they do not generally agree, the most optimistic situating the total population somewhere between 15 and 20 millions, including women, children, slaves and other persons with no rights. All is in the method! It seems useful to integrate the demography in an ensemble of human sciences and to get used to approximate concerning life expectancy, fertility and birth control, mortality and the impact that military expeditions (conscription) and civil troubles (displacements) could have had upon the family as basis cellule. These elements are determining concerning the socio-economic conditions of the city progress, and especially the adequacy of the means of support to alimentary needs.

	Weak hypothesis	Strong hypothesis	Observations
Tuscany	3,000	5,000	
Sardinia	5,000	8,000	
Latium	19,000	25,000	Ostia...
Campania	2,000	3,000	
Bruttium	4,000	6,000	Taranto...
Sicily	3,000	5,000	
Apulia	8,000	12,000	Salapia...
Veneto	5,000	9,000	Spina...
Total	49,000	73,000	

Figure 8. The salt production: plausible hypotheses.

- *Empire*

There are as many estimates concerning the population of the Empire. According to Durand (1977), at the end of Augustus' reign (who died in 14 AD), the Empire counts 56.8 million inhabitants, maybe a little more... See more details in the table (Figure 9). The estimates for Gaul vary considerably. The most common figure is that of 6 millions. At the beginning of the first century, the population of the Empire counted at least 55 million inhabitants.

Europe	31.6	26.4
Asia	14.0	19.5
Africa	11.2	11.5
Empire	56.8	57.4

Figure 9. Population within the Empire at the beginning of the first century.

Thus, the gross salt needs would be somewhere between 397,600 and 401,800 tonnes, calculated on the basis of 7kg per person and per year. After adjusting it by taking into consideration the number of women and children (6 and 5kg *per capita*), we obtain a 340,400 – 351,800 tonnes span. This gives us an average of 373,000 tonnes.

At the end of the Antonine dynasty (192 AD), the population of the Empire makes the object of even more disputed estimates. At the death of Emperor Hadrian (138 AD), the population of the Empire (3,300,000 km^2) would have been between 60 and 70 million inhabitants (Jerphagnon 2002). According to Homo (1947), we may even count up to 80 million inhabitants. This opinion was later strongly contested, so that 60 millions seems more realistic. See Figure 10.

	Homo[1]	Corvisier[2]
Occidental part		
Italy	7	8.7
Spain	8	6.6
Gaul (Brit. Germ. incl.)	12	11.4
Danube countries	3	4.1
Islands (Si. Sa. Co.)	7	1.1
Africa	8	4.2
Oriental part		
Balkan countries	5	2.8
Asiatic East	22	11.5
Egypt	8	6.0
Cyrenaica	{	0.6
Total Roman Empire	80	57.0

Figure 10. Population of the Empire at the end of the Antonine dynasty (million inhabitants).

The second series (according to the authors that Corvisier reminded of) is based upon minimalist evaluations. For example, the Oriental kingdoms reach to 11.5 million inhabitants, while the Asian provinces, including Syria and Cyprus, have a rather large population, around 19.5 million inhabitants. According to recent works upon the Hellenistic economies, the Seleucid kingdom would have had between 20 and 25 million inhabitants at its peak, from which 5-6 in Mesopotamia (Archibald 2001).

The power the attraction exercised by big cities, and we are not talking only about Rome, should be taking into account as it involves other ways of living, and especially a change in the dietary habits. At the end of the first century AD, Rome counted more than a million inhabitants. Alexandria had between 500,000 and 750,000 inhabitants, comparing to 300,000 or 400,000, a hundred years before (Ballet 1999). Strabo (*Géo.* XVI, 2, 5) estimates that the population of Antioch on Orontes is less important than those of Alexandria and of the Tigre Seleucia.

How does this heterogeneous population evolve starting from the second century? We see a contraction to the West by using other evaluations (authors quoted by Harl KW. *Population estimates*, 1998) that we reproduce below (Figure 11).

Starting from these estimates, the gross salt needs (7kg *per capita*) would be around 420,000 and 400,000 tonnes during the second century AD. After adjusting it by taking into consideration the number of women and children, and also the new alimentary behaviours (6kg *per capita*), we arrive to a figure between 360,000 and 340,000 tonnes. This gives us an average of 350,000 tonnes. During the second part of the fourth century, we see that the average figure drops to 300,000 tonnes/year (328,000 at the beginning, 272,000 at the end).

	Antoninus (138-161)	Constantine (306-337)	Theodosius (379-395)
Occidental part	36.1	23.6	22.0
Italy	8.7	4.0	4.5
Spain	6.6	5.0	4.0
Gaul (Brit. Germ. incl.)	11.4	7.5	9.5
Danube countries	4.1	3.0	{
Islands (Si. Sa. Co.)	1.1	1.6	1.5
Africa	4.2	2.5	2.5
Oriental part	20.9	31.0	23.4
Balkan countries	2.8	5.5	5.0
Asiatic East	11.5	18.5	12.0
Egypt	6.0	6.5	6.0
Cyrenaica	0.6	0.5	0.4
Total Roman Empire	57.0	54.6	45.4

Figure 11. Evolution of the population of the Empire (138-395) (million inhabitants).

Such a globalization, of around 400,000 tonnes (second century) or 300,000 tonnes (end of the fourth century), presents only a relative interest. The estimates made in terms of the provinces or of the regions within the Empire could correspond to individual quantities. Thus, we should more likely wonder whether the sites are in a position to satisfy the local needs (Baetica, Narbonensis, Mauretania, Egypt, etc.) taking into account the population estimated at a certain moment and the salt-open markets, and not only to sea salt... The experts should study it with cautiousness and determination. It is worth mentioning a tentative concerning the salt consumption in Celtiberia (Talavera Costa 2007). Knowing that salt did not have a great value on the market, the motivation is more related to numerous sites near which there were fish salting factories, hide curing factories and to the future implantation of such sites that to an increase in the capacity of the privileged sites. Besides the researches regarding the production of salt (sites and techniques), we should also have studies, variable in terms of the provinces within the Empire and of the periods, concerning the usages of salt (food conservation or

[1] Homo 1947, 398.
[2] Corvisier and Suder 2000, 87. We are talking about evaluations according to McEvedy and Jones 1978, 20-33.

transformation, animal feeding, cosmetics, artisanal applications) and we should analyze the evolution within the economic and political context on the basis of the works available.

Finley's disciples will certainly find in this incomplete essay the proof of the pointlessness of such an illusory, even dangerous arithmetic, when the reader trusts the results without assessing the validity of the enterprise (Finley 1982). Is all quantification intention due to fail only because there are no statistical yearbooks for the Antiquity? We know that the statistics are false, mostly when they enter into details. However they are still necessary. Without pretending to have reconstituted certain series, searching for something that the Ancients did not seem to be interested in, it seems judicious to refine the common reflection upon the effective importance of salt despite the limitations of the law of numbers, of the '*ignominious reality of the corrupted texts*' and of the lack of sufficient elements. The ceramic fragments do not offer more information either.

References:

*** 1994. *Histoire Auguste* (trad. Chastagnol A). Paris, Robert Laffont.

Adshead, S. A. M. 1992. Antiquity, *Salt and Civilization.* London, Macmillan.

Akridge, G. 2008. Methods for calculating brine evaporation rates during salt production. *Journal of Archaeological Science* 35-6, 1453-1462.

Amouretti, M. C. 1997. Villes et campagnes grecques. In *Histoire de l'alimentation,* Paris, Fayard.

André, J. 1981. *L'alimentation et la cuisine à Rome.* Paris, Les Belles Lettres.

Archibald, Z. H. *et al.* 2001. *Hellenistic Economies.* London, Routledge.

Aristote. 1964-1968-1969. *Histoire des animaux* (trad. Louis P). Paris, Les Belles Lettres.

Ballet, P. 1999. *La vie quotidienne à Alexandrie.* Paris, Hachette (Pluriel).

Carcopino, J. 1939. *La vie quotidienne à Rome à l'apogée de l'Empire.* Paris, Hachette.

Caton l'Ancien, 1975. *De Agricultura* (trad. Goujard R.). Paris, Les Belles Lettres.

Chandezon, Chr. 2003. *L'élevage en Grèce (fin Ve-fin Ier siècle AC) – L'apport des sources épigraphiques.* Paris, De Boccard.

Corvisier, J. N. and Suder, W. 2000. *La population de l'Antiquité classique.* Paris, Presses Universitaires de France.

Corvisier, J. N. 2001. Etat présent de la démographie historique Antique. *Annales de démographie historique*, 101-140.

Cosme, P. 2007. *L'armée romaine.* Paris, Armand Colin.

Durand, J. D. 1977. *Historical estimates of the world population: an evaluation,* Population Studies Centre Philadelphia.

Etienne, R. and Mayet, F. 2002. *Salaisons et sauces de poisson hispaniques.* Paris, de Boccard.

Etienne, R. 1981. Les rations alimentaires des esclaves de la *familia rustica* d'après Caton. *Index,* 66-77.

Faure, P. and Gaignerot, M. J. 1980. *Guide grec antique.* Paris, Hachette.

Finley, M. I. 1982. Le document et l'histoire économique de l'Antiquité. *Annales* 37, 697-713.

Flavius Josèphe. 2009. *Guerre des Juifs* II. Texte établi et traduit par A. Pelletier. 3e tirage, 386. Paris, Les Belles Lettres.

Garnsey, P. 1988. *Famine and food supply,* (1996. Famine et approvisionnement dans le monde gréco-romain). Cambridge, Cambridge University Press.

Hesnard, A. 1998. Le sel des plages, *Mélanges de l'Ecole Française de Rome – Antiquité.* 176.

Homo, L. 1947. *L'âge d'or de l'Empire romain.* Paris, Fayard.

Horace. 1823. *Oeuvre completes* (trad. Batteux Ch.). Paris, Dalibon.

Jerphagnon, L. 2002. *Histoire de la Rome antique.* Paris, Hachette (Pluriel).

Lévy, J.-Ph. 1981. *L'économie antique.* Paris, Presses Universitaires de France.

Marion, M. 1989. *Dictionnaire des institutions de la France*, Paris, Picard.

McEvedy, C. and Jones, R. 1978. *Atlas of World Population History.* New York, Penguin Books.

Nicolet, Cl. 1988. *Rendre à César – Economie et société dans la Rome antique.* Paris, Gallimard.

Oates, W. J. 1934. The population of Rome. In *Classical Philology*, 29-2, 101-116.

Perea Yebenes, S. 2006. El uso de la sal en el ejercito romano y su abastecimiento en epoca altimperial. In A. Morillo Cerdan, *Arqueologia militar romana en Hispania*, 345-359.

Polybe. 1970. *Histoire* (trad. Roussel D). Paris, Gallimard.

Pons Pujol, L. 2006. L'importation de l'huile de Bétique en Tingitane et l'exportation des salaisons de Tingitane (I-IIIe siècle PC). *Cahiers Glotz*, 61-77.

Pseudo-Hygin 1979. *De munitionibus castrorum*. Paris, Les Belles Lettres.

Sénèque 1993. *Lettres à Lucilius* (traduction Henri Noblot, revue par Paul Vayne, Robert Laffont). Paris, Bouquins.

Sternberg, M. 2000. Données sur les produits fabriqués dans une officine de Neapolis (Nabeul, Tunisie). *Mélanges de l'Ecole Française de Rome – Antiquité,* 135-153.

Strabon. 1966. *Géographie* Livre III. Paris, Les Belles Lettres.

Talavera Costa, J. 2007. La sal de la comarca de la Sigüenza arévaca: ¿Riqueza natural – riqueza social? In N. Morère Molinero (ed.), *Las salinas y la sal de interior en la historia: economía, medio ambiente y sociedad. Inland saltworks and salt history: economy, environment and society*, 241-261. Madrid, Dykinson.

Tarbé, S. A. 1813. *Manuel pratique des poids et mesures*, Paris, Merlin.

Varron. 1985. *Economie rurale* – Livre II (trad. Guiraud Ch.). Paris, Les Belles Lettres.

Virlouvet, C. 1995. La consommation de céréales dans la Rome du Haut-Empire: les difficultés d'une approche quantitative. *Histoire et mesure,* 261-275.

Appendix A – Rations according to *Historia Augusta* (*Probus* IV, 6; *Aurélien,* IV, 6; *Claude,* XIV (Ed. Chastagnol, 1994)

Food products	***Per diem* (*Probus*, IV, 6)**	***In incerto* (*Aurélien*, IX, 6) For?**	***Per annum* (*Claude*, XIV) For 15 persons**
Beef	… librae	40 librae	na
Pork	6 librae	30 librae	na
Goat meat	10 librae	na	na
Chicken	1 every two days	2	na
Bacon	qs	nd	2 000 modii
Salt	**qs**	**1 sextarius**	**20 modii**
Salt fish	nd	1 sextarius	na
Green vegetables	qs	qs	na
Dry vegetables	qs	qs	na
Wheat	na	na	3 000 modii
Barley	na	na	6 000 modii
Bread	na	16 loafs bread	na
Table wine	10 sextarii	40 sextarii	3 500 sextarii
High quality oil	na	na	150 sextarii
Ordinary oil	na	1 sextarius	600 sextarii

qs: quantum satis
na: not available
1 *modius*: 8.75 litres – salt equivalent around 7 kg
1 *sextarius*: 0.54 litres – salt equivalent around 436 g
1 *libra*: 327 g

Annexe B – The regions of Italy at the end of the third century BC: salt needs (weighted estimate) and capacities (strong hypothesis)

	Population (1000 inhabitants)	Territory (1000 km^2)	**Salt: need estimates** (tonnes)	**Salt: optimal capacity to satisfy them** (tonnes)
Rome	923	35.6	6 500	8 000
Latium	431	10.6	3 000	3 000
Samnium	390	10.3	2 700	-
Apulia	284	17.1	2 000	3 000
Abruzzi	172	7.4	1 200	-
Etruria	274	19.1	1 900	3 000
Umbria	111	7.2	800	1 000
Lucania	167	10.4	1 200	2 000
Total	2 752	117,7	**19 300**	**20 000**

The estimates are calculated on the basis of a modius *per capita* and *per annum* (7kg/year).
Data related to the population and the territory are taken from Corvisier and Suder 2000, 73.

Hypotheses, Considerations – and unknown Factors – regarding the Demand for Salt in Ancient Greece

Cristina Carusi

Università di Parma, Italy

Abstract

The ration of salt which Cato the Elder advised should be given to slaves in **De agricultura** (1 **modius** or 8.7l **per capita** per year) can be considered – with some precautions – a good indicator of individual consumption of salt in the ancient world. By applying this figure to the estimated population sizes of various Greek **poleis** of the 5th and 4th centuries BC, it is possible to get an idea of a particular community's demand for salt for food and domestic purposes. Unfortunately, it is currently impossible to put figures on other uses of salt, in particular animal rearing, in order to reach a total for overall demand. Nonetheless, the estimates which are possible suggest a few general considerations. For **poleis** where there were not industrial activities demanding large quantities of salt, it seems that salt demand was fairly limited and so local resources were generally able to guarantee the supply for the community. In contrast, the figures found for the production capacities of a number of fish-salting factories from along the Mediterranean coastline suggest that their demand for salt would have been decidedly higher. In this latter case it is necessary to ask how these various sites would have been supplied, bearing in mind both local resources and the possibility of importation.

Keywords

salt consumption, Cato the Elder, Greek poleis, demography, alimentation, animal rearing, ancient saltworks, fish-salting industries

The main difficulty in assessing an ancient community's salt requirements is the lack of contemporary quantitative data regarding the human consumption of salt and its use in key activities such as animal rearing and fish salting.

In the few attempts that have been made to date, the common view is to use modern data to calculate human consumption: for example, Giovannini's calculation for salt demand in central and southern Italy in the Republican era (Giovannini 1985) and Mangas and Hernando's similar calculation for a pre-Roman community on the Iberian peninsula (Mangas and Hernando 1990-91) both choose to use the current annual estimate of 2.5kg *per capita* as their starting point.

However, to reach a figure for overall demand requires a calculation of the additional quantities of salt needed for animal rearing and food conservation, for which the guidelines are much less clear. Giovannini states that, in addition to the 2.5kg consumed by each individual, we should add 2.5kg for ovine consumption and 10kg for a cow per three inhabitants; doubling the figures to include food preservation produces an annual demand per capita of 30kg, or 70,000 tons for the entire population of central and southern Italy in the Republican era (around 3,500,000 people). Mangas and Hernando also state that an ancient community would have needed 30kg per capita per annum, comprising 2.5kg for human consumption, 2kg for a sheep, 20kg for a cow and 5.5kg for food preservation; a Celtiberian community of 440 people, 5 horses, 300 sheep, 100 goats, 50 cows and 300 pigs would consume over 8 tons of salt in a year.

It is clear that this is a task which involves numerous variables, some of which are very difficult or impossible to reconstruct using available information – not least of which being demographic data, which on its own is a source of great debate. Bearing in mind the aforementioned difficulties, I would like to discuss in this article some of the factors which come into play when estimating the demand for salt by a Greek *polis*, indicating some factors which in my view form the limits of such an investigation and suggesting a few considerations on this topic.

Taking human consumption first, it is useful to remember that we do possess quantitative data handed down from antiquity. I refer to the portion of salt which Cato the Elder advised should be given to slaves in the section of *De agricultura* which is dedicated to slave nourishment: besides a precisely measured ration of flour or bread and wine, a *familia*'s meal should consist of discarded olives (or, if they are out of season, *hallec* or vinegar), a *sextarius* of oil per head each month and one *modius* of salt per head for the year (*de agr.* 67).

As one *modius* is about 8.7 litres, this ration may seem excessive in relation to current figures, which estimate the amount of salt eaten by one individual in the space of a day – including both naturally occurring salt and added salt in prepared food – to be around 8-9g. Although the weights of different types of salt vary depending on their chemical and physical makeup, using an average density of 1.2kg/litre for evaporated classic fine salt (Colas 1985, 20-21) gives an annual figure for modern consumption of around 2.5 litres, equivalent to less than a third of Cato's suggested ration.

We must bear in mind, however, that the 8.7 litres allocated to each individual are not intended solely for table consumption: salt was not only used as a condiment, but also for cooking or as a preservative in various foodstuffs such as bread, vegetables and cheese – processes which entailed the loss of a certain percentage of the substance – as well as for other domestic tasks such as the treatment of animal skins for clothing and the administering of medications.

We must also consider that the variety and richness of modern-day European alimentation are in complete contrast with the diet of antiquity in which cereals were by far the dominant component, providing around 70-75% of calorific intake (Foxhall and Forbes 1982, 72-75). In this context, salt took on a crucial function in making a particularly monotonous and insipid diet more appetizing (Gallo 2001, 463-464). Indeed, even when cereals continued to play a central role in the modern era, daily consumption of salt is estimated to have been double that of today (Braudel 1979, 178).

At any rate, it is probable that the amount of salt allocated to slaves in *De agricultura* was a little higher than that generally consumed by a free man: indeed, cereals and bread played an even greater role in the meals of a *familia*, constituting around 90% of calorific intake (Étienne 1981). The slaves on an agricultural estate, furthermore, were principally manual labourers and as such used up the body's salt reserves more quickly, necessitating a diet with a higher proportion of sodium chloride.

Cato's ration should be evaluated within this context, in which the slaves' diet – including their quota of salt – was calculated in order to reap maximum returns in terms of labour at minimal cost: this profit-orientated mentality pervades the entire work (Martin 1971, 85-88). It is not a surprise, then, that even an estate-owner who was generally fairly tight-fisted in his relations with his slaves would have been generous regarding windfall olives and salt, whose price was decidedly low (Carusi 2008, 162-165).

On this basis, I do not feel that there are valid reasons to question the reliability of Cato's figure; on the contrary, I believe that this data, treated with the precautions seen above, constitutes the best currently available indicator of the relative importance of salt for individual consumption in the ancient world. Estimates which are based on current ingestion of salt fail to consider the radical differences between the modern and ancient diet, and risk applying figures to the ancient world which are based on entirely foreign parameters. In contrast, we should recognise in *De agricultura* a valuable and authoritative eyewitness of ancient reality.

I feel that it is possible, based on this data, to suggest estimates of salt consumption in ancient communities – estimates which, following the above interpretation of Cato's ration, will refer not to salt ingested on its own, but to salt used also for other alimentary purposes (flavouring, cooking and preserving) and for other domestic purposes (clothing, medication, etc.).

In this discussion I will limit myself to using the figure provided in Cato's text to ask how much salt would have been needed in those Greek *poleis* of the 5th and 4th centuries BC for which we possess estimates of the number of inhabitants. These estimates, which are generally felt to be reliable, have been taken from ancient sources on military contingents. I have chosen to concentrate on individual *poleis* as the fact that they are communities with a limited geographical and demographic size makes calculations more manageable and less open to significant variation.

With the precautions explained above, I will argue that Cato's figure can be used to provide a reliable indication of the amount of salt consumed by an individual within the sphere of a Greek *polis* of the 5th or 4th century BC. This conviction is based not only on the fact that *De agricultura* was inspired by Greek sources regarding similar topics (e.g. Xenophon, Democritus and Theophrastus) (Goujard 1975, xl-xli), but above all on the considerable homogeneity of the Mediterranean diet, from the dominance of cereals (Foxhall and Forbes 1982, 51-65) to the central role of salt in the preservation of food (Horden and Purcell 2000, 195-197). It is clear, then, that if the application of modern figures to the ancient world is confusing, the same cannot be said for using a second-century Roman text to evaluate how much salt would have been needed by a Greek *polis* of the 5th or 4th century BC.

Using population estimates for Attica in the 5th and 4th centuries – respectively over 300,000 and between 200,000 and 250,000 (Hansen 1988, 12) – and combining these with Cato's ration, it is possible to find that salt consumption for domestic use in Athens was around 2600m^3 in the Periclean age and between 1750m^3 and 2170m^3 in the 4th century.

The same calculation can be repeated for other *poleis* for which we possess similar demographic data (Hansen 2006, 93-96): for Argos, whose population reached *c.* 70,000 inhabitants at the start of the 4th century, domestic consumption of salt can be put at *c.* 600m^3; for Megara in the first half of the 5th century, as for Ambracia in the opening years of the Peloponnesian War, *c.* 30,000 inhabitants would have used *c.* 260m^3; on the eve of the same war Corcira's population of *c.* 55,000 produces an estimate of *c.* 480m^3; whilst Eretria at the start of the Hellenistic age comprised *c.* 15,500 people, with a salt consumption of *c.* 135m^3 (Figure 1).

However, these estimates refer only to salt used in food and domestically: we have seen that total salt consumption by a community is driven in addition by the use of salt in animal rearing and other activities, the most important being the production of salted fish and fish sauces.

One of the unknown factors which most hinders a calculation of total demand is the difficulty of assessing animal consumption, due not so much to a lack of information on feeding – which, unlike the human diet, can be extrapolated from modern data (*c.* 10g/*die* for a sheep = 31g per annum according to Kaufmann 1978, 459-460) – but more from the impossibility of calculating the number of animals in a given area.

The prevailing idea in the most recent studies on the practice of animal rearing is to split Classical and Hellenistic Greece into three areas, broadly corresponding to three different climate types. In the Aegean zone, comprising Attica, the Cyclades and southern Ionia, animals – for the most part sheep and goats – had to occupy a relatively small space, and were strongly integrated within agricultural activities. In central Greece (Thessaly, the Peloponnese, Aeolis and Crete), meanwhile, environmental conditions allowed a more diffuse form of farming, which was mainly concentrated in the hilly regions on the edge of the civic territory. It was only in Macedonia and north-west Greece, however, that a true form of extensive animal rearing was able to develop, helped by political systems which, unlike the *poleis*, covered a greater area and permitted the practice of transhumance (Chandezon 2003, 402-404).

In this context the size of flocks varied noticeably from case to case, from the citizens of Keos who owned a few sheep each (Ael. *de nat. anim.* 16, 32), through the Athenian Panaitos, from whom 84 sheep and 67 goats with their offspring were taken in 414 (*IG* I³ 426, ll. 72-74), to Euboulos of Elatea who at the end of the third century was granted the rights of pasture by the city of Orchomenus for 220 cows or horses and 1000 sheep or goats (*IG* VII 3171, ll. 37-40).

In the case of Attica, extensive animal rearing was not practised and it is known that the people were able to evacuate their livestock to Eubea and nearby islands on the eve of the Peloponnesian War (Th. 2, 14): it is therefore probable that the number of animals was no higher than a few tens of thousands, with a salt consumption of a few hundred metres cubed per year – a figure much lower than that for estimated human salt consumption.

Even if there are not precise figures available that permit more accurate calculations, it is acceptable to posit that in communities with limited space for animal rearing, animal consumption of salt would not have greatly influenced overall consumption.

More worryingly, though, it is not possible to suggest any estimates for regions which practised extensive animal rearing, where the consumption of salt was significantly higher not only through its use in animal feed but also through the making of animal-based products.

I would suggest that future research efforts should concentrate on the animal-rearing sector, using more approachable local situations – and even modern data – as a base to form a reliable indicator of the extent of salt consumption by animals.

With the exception of the production of salted fish and fish sauces, which we will consider later, the above reasoning can be applied to other activities which used salt (treatment of animal skins, some metallurgical processes, the production of perfumes and ointments, etc.) – activities for which we do not have exact details regarding the quantity of salt required, even if we can assume that they had a minor impact on the total consumption of salt in each individual community.

Confronted with this series of unknown factors, it is only possible to present some general considerations: in particular, we can posit that in communities in which industrial activities were absent and thus large quantities of salt were not needed, the use of salt in food and domestic work would have constituted almost all of the total demand.

From this point of view Athens presents a particularly useful example, as there is no documented evidence from Attica of extensive animal rearing, a salting industry of any great size, or the use of salt in the metalwork of Laurium.

	Population estimate	Estimate of annual salt consumption for alimentary and domestic purposes
Athens (Periclean age)	over 300,000	2600m³
Athens (4th century BC)	200,000-250,000	1750-2170m³
Argos (start of the 4th century BC)	70,000	600m³
Megara (1st half of the 5th century BC)	30,000	260m³
Ambracia (eve of the Peloponnesian War)	30,000	260m³
Corcira (eve of the Peloponnesian War)	55,000	480³
Eretria (start of the Hellenistic age)	15,500	135m³

Figure 1. Estimate of annual salt consumption for alimentary and domestic purposes in some Greek *poleis* of the 5th and 4th century BC.

Nevertheless, we should bear in mind that within the spectrum of *poleis* in the Classical Age, Athens represented an exception as far as its population's size was concerned, so that its demand for salt, as we have seen, was far greater than any other contemporary community. It has been calculated, in fact, that out of the approximately 1000 *poleis* present in Greece in the 4th century BC, only 10% had a territory larger than 500km^2 and more than 27,000 dwellings (Hansen 2006, 24-34). This means that only a hundred or so cities would have had a demand for salt – for food or domestic use – greater than 235m^3, which leads to the conclusion that the majority of Greek *poleis* would have had a relatively low demand for the substance.

Unfortunately, it is not possible to combine these estimates with figures for the productivity of ancient saltworks in order to create a direct link between the demand of a particular community and the supply of salt provided by local resources. Nonetheless, for purely indicative purposes we can recall some figures regarding modern saltworks, whose methods of harvesting and production may have been not so different to their ancient precursors.

At the end of the 18th century AD, the German naturalist Peter Simon Pallas stated that in certain years two natural salt lakes in the district of Pérécop, in the Crimea, the Staroe Osero ('Ancient Lake') and the Krasnoe Osero ('Red Lake') could produce between them 200,000 to 800,000 *pouds* of salt, equivalent to around 3276 tons (*c.* 2730m^3) and 13,100 tons (*c.* 10,900m^3) respectively (Baladié 1994, 155). Production at the saltworks at Tragasai, meanwhile, reached 100,000 *stai* (3600m^3) in 1817, and was estimated to be 1,558,307kg (*c.* 1300m^3) in 1894 (Cook 1973, 222-224), whilst the small Cretan island of Kaudos harvested around 77q of salt (*c.* 6.4m^3) in the 1920s without a proper system of production (Guarducci 1930, 477-479). Thermisi, in the south of Argolis and on the territory of the ancient city of Hermione, was producing 20,000 tons of salt per year by 1849 (*c.* 16,000m^3) (Jameson, Runnels and Andel 1994, 311).

Although these few examples can only give a partial view of the true productive potential of ancient saltworks, the impression remains that the extent of their activity was anything but negligible, and should in no way be undervalued.

In this work it is enough to state that environmental and climate conditions favoured the formation of marine salt in the Mediterranean basin to a great extent: it is not a surprise, then, that ancient sources suggest that salt was widely known throughout the Greek world and that the sea was its principal source (Carusi 2008, 31-32). Furthermore, in the Archaic and Classical eras the majority of Greek settlements could be found near to the coastline – even in the rare instances when a town did not have direct access to the sea, the sea was only ever a short distance away. This situation is reflected in the famous metaphor uttered by Socrates in *Phaedo* (109b), in which the Greeks, from Colchis to Hercules' Column, were like frogs or ants around a pond.

It is probable, then, that the majority of Greek cities could rely on local resources to fulfil their internal needs – needs which, as we have seen, would not have been especially high in relation to production on the coast. Furthermore, in many cases the demand for salt for food, domestic tasks and animal consumption could have been fulfilled through the use of salt simply harvested from the coastline, without the need for dedicated production areas. An ethnographic parallel is the case of the Cretans who, until the early decades of 20th century AD, used to collect and sell the salt which formed naturally in the rocky crevices of the coastline, thereby going against the state-run monopoly (Davaras 1980, 2-4). Even the direct use of seawater can replace salt in many cases (Morère 2008, 371-373): Pliny states by way of example that those in coastal regions frequently made bread by kneading the dough with seawater in order to save salt (*H. N.* 18, 68).

The situation would have been decidedly different in the case of industries requiring a large quantity of salt. The most obvious example is the fish-salting factories, especially those where production reached considerable levels and was geared towards exportation. Thanks to reliable elements from ancient sources and modern and archaeological data, it is possible to suggest estimates of the demand for salt in some of these production centres.

According to a recipe in the *Geoponica* (20, 46, 3), to make *garum* in the style of the Bithynians required two *sextarii Italici* of salt (around 1.09 litres) for each *modius* of fish, with a salt: fish ratio equivalent to around 1:8 and the mixture fermenting in the sun for two to three months. However, we should note that the preparation of fish sauces similar to *garum*, such as the *nuoc-man* of Indochina or the *gharos* of Constantinopole, uses a salt: fish ratio closer to 1:2 or, more frequently, 1:4 (Grimal and Monod 1952, 32-33, 37).

As we do not possess indications of how much salt was needed in the salting of slices of fish, it is necessary to use estimates derived from the production of the mentioned fish sauces, bearing in mind that the different products would have required different preparation techniques.

The figures above can be combined with estimates of the capacity of salting vats which have come to light in various locations around the Mediterranean, with the warning that only a small percentage of the discovered factories – which in turn only represent a limited part of the factories in action in ancient times along the Mediterranean coast – have been excavated and written on in detail. As a result, the available estimates can only provide a limited snapshot of the true production capability of individual factory sites.

One of the best-know sites, Lixus in Morocco, contained ten different factories, with an overall capacity of around 1013m^3. A production cycle of three months, assuming that all of the salting vats were in use at the time, would have required between 126m^3 and 506m^3 of salt. It seems that other production centres, such as Sexi and Baelo Claudia, both in Spain, had a more limited capacity, producing 500m^3 and 269m^3 respectively, and requiring *c.* 350m^3 and *c.* 130m^3 of salt respectively (Étienne and Mayet 2002, 95-96).

The industrial complexes of Neapolis (Nabeul) in Tunisia (Slim *et al.*, 2007, 39-40) and Sabratha in Libya (Wilson 2007, 175-177) display an even lower capacity (*c.* 183m^3 and *c.* 100m^3) and consequently a lower demand for salt (*c.* 90m^3 and *c.* 50m^3 respectively).

Regarding the area around the Black Sea (Højte 2005, 142-148, 150-153), finally, the salting vats which have come to light at Tyritake suggest an overall productive capacity of 457m^3, with a potential maximum demand of *c.* 220m^3 of salt, whilst those at Chersonesos suggest a capacity of 2000m^3 and a potential maximum demand of *c.* 1000m^3 (Figure 2).

	Estimate of the capacity of excavated salting vats	Estimate of the demand for salt for a cycle production of three months (salt:fish ratio equivalent to 1:2)
Lixus (Morocco)	1013m^3	506m^3
Sexi (Spain)	500m^3	350m^3
Baelo Claudia (Spain)	269m^3	130m^3
Neapolis (Tunisia)	183m^3	90m^3
Sabratha (Lybia)	100m^3	50m^3
Tyritake (Crimea)	457m^3	220m^3
Chersonesos (Crimea)	2000m^3	1000m^3

Figure 2. Estimate of the demand for salt in some fish-salting centres around the ancient Mediterranean world.

Even if these estimates refer to installations from the Imperial Age, when the production of salted fish and the construction of suitable factories seem to have undergone a widespread and dramatic period of growth all around the Mediterranean, we cannot ignore the fact that even in the 5th century BC the areas of Pontus and Gadeira had already reached a productive capacity which was far from negligible. This productivity supported a flourishing export industry, whose volume of goods was probably not much lower than those calculated earlier for more recent installations (Carusi 2008, 182-184).

Comparing these figures with those found earlier for food and domestic use in individual *poleis* creates the impression that the demand for salt in a medium-large salting centre would have been relatively consistent, and that on its own it could have equalled or exceeded the demand for salt for food or domestic use by a small or medium *polis*.

In this case it is sensible to ask whether local resources would have been enough to supply such a demand, and, if so, how they would have done so. After all, despite the lack of information on production levels of ancient salting factories, it is plausible to assume that the running of a reasonably large complex would have demanded an intensive exploitation of salt resources.

It is not a coincidence, in any case, that the main centres of production – above all on the northern coast of the Black Sea and the southern coast of the Iberian peninsula – developed in areas which favoured not only fishing but also the farming of salt. We should bear in mind that salt is a relatively heavy foodstuff, with a low cost in relation to its volume, and so the cost of transporting it would have been particularly heavy (Carusi 2008, 168-169).

This does not mean that we should always assume a direct link between a large salting factory on a particular site and a large supply of salt nearby: it is always essential to question how the single centre would have been supplied, balancing the twin possibilities of local resources on the one hand with importation from neighbouring areas on the other (bearing in mind distance and the cost of transport).

In my opinion this is a particularly interesting line of enquiry, in which individual, local cases can be used to lead towards more general points for reflection on the production and circulation of salt within the ancient world. An initial and significant example of this form of investigation is found in Lagóstena Barrios' recent study on the production of salt along the Betican coast during the Roman era (Lagóstena Barrios 2007).

By overlaying a map of the towns which housed key salting factories in ancient times with that of saltworks known to have existed along the coastline from the Middle Ages onwards, the author has found that some of the principal centres for salted foodstuffs, such as Baelo Claudia and Malaka, would not have had sufficient salt resources on their own territory to supply their activities. As a result, they probably made use of imported salt from other sites with a better supply. It has been suggested, based on this observation, that the supply of salt to the many factories which developed along the Malacitan coastline in the 1st century AD depended on frequent shipments of salt from the Atlantic coast of Betica. This creates a scenario which is very different from the commonly held notion of a close link between salt harvesting site and factory.

In my opinion, this example clearly shows the possible objectives and implications of an investigation which aims to estimate an ancient community's demand for salt. In order to quantify individual consumption, total food consumption by humans and animals, and demand from other sectors in a way which is not a sterile exercise in itself – sterilised further through the number of unknown factors in play – it is not only necessary to bear set criteria and limits in mind in creating estimates, but also to place these estimates within a wider framework: using them as an impetus to lead to more generally interesting questions on the role of salt in the ancient world.

References:

Baladié, R. 1994. Le sel dans l'antiquité sur la côte nord de la Mer Noire. A propos d'un paysage des Histoires d'Hérodote et à la lumiere des voyageurs de l'époque moderne. *Il Mar Nero* 1, 145-166.

Braudel, F. 1979. *Civilisation matérielle, économie et capitalisme, XVe – XVIIIe siècle. I. Les structures du quotidien: le possible et l'impossible.* Paris, Armand Colin.

Carusi, C. 2008. *Il sale nel mondo greco (VI a.C. – III d.C.). Luoghi di produzione, circolazione commerciale, regimi di sfruttamento nel contesto del Mediterraneo antico.* Bari, Edipuglia.

Chandezon, Ch. 2003. *L'élevage en Grèce (fin Ve – fin Ier S. a.C.). L'apport des sources épigraphiques.* Bordeaux, Ausonius.

Colas, A. 1985. *Le sel.* Paris, Presses Universitaires de France.

Cook, J. M. 1973. *The Troad.* Oxford, The Clarendon Press.

Davaras, K. 1980. Kritikes Epigraphes III. *Archaiologiki Ephemeris*, 1-42.

Étienne, R. 1981. Les rations alimentaires des esclaves de la «familia rustica» d'après Caton. *Index* 10, 66-77.

Étienne, R. and Mayet, F. 2002. *Salaisons et sauces de poisson hispaniques.* Paris, De Boccard.

Foxhall, L. and Forbes, H. A. 1982. Sitometreia: the Role of Grain as a Staple Food in Classical Antiquity. *Chiron* 12, 41-90.

Gallo, L. 2001. Appunti per una storia del sale nel mondo greco. In S. Bianchetti, E. Galvagno, A. Magnelli, G. Marasco, G. Marotta and I. Mastrorosa (eds.), *Poikilma. Studi in onore di M.R. Cataudella in occasione del 60° compleanno*, 459-471. La Spezia, Agorà.

Giovannini, A. 1985. Le sel et la fortune de Rome. *Athenaeum* 63, 373-386.

Goujard, R. 1975. *Caton. De l'agriculture.* Paris, Les Belles Lettres.

Grimal, P. and Monod, Th. 1952. Sur la véritable nature du garum. *Revue des Études Anciennes* 54, 27-38.

Guarducci, M. 1930. Ordinamenti dati da Gortina a Kaudos in una iscrizione inedita di Gortina. *Rivista di Filologia e Istruzione Classica* 8, 471-482.

Hansen, M. H. 1988. A Note on the Growing Tendency to Understimate the Population of Classical Attica. In M. H. Hansen (ed.), *Three Studies in Athenian Demography*, 7-13. Copenhagen, The Royal Danish Academy of Sciences and Letters.

Hansen, M. H. 2006. *The Shotgun Method. The Demography of the Ancient Greek City-State Culture.* Columbia, University of Missouri Press.

Højte, J. M. 2005. The Archaeological Evidence for Fish Processing in the Black Sea Region. In T. Bekker-Nielsen (ed.), *Ancient Fishing and Fish Processing in the Black Sea Region*, 133-160. Aarhus, Aarhus University Press.

Horden, P. and Purcell, N. 2000. *The Corrupting Sea: a Study of Mediterranean History.* Oxford, Blackwell.

Jameson, M. H., Runnels, C. N. and Andel, T. H. 1994. *A Greek Countryside. The Southern Argolid from Prehistory to the Present Day.* Stanford, Stanford University Press.

Kaufmann, D. W. 1978. *Sodium Chloride.* Washington, D.C.

Lagóstena Barrios, L. 2007. Explotación de la sal en la costa meridional hispánica en la Antigüedad romana. Aportación al estudio de la cuestión. In N. Morère Molinero (ed.), *Las salinas y la sal de interior en la historia: economía, medio ambiente y sociedad. Inland saltworks and salt history: economy, environment and society*, 301-323. Madrid, Dykinson.

Mangas, J. and Hernando, M. R. 1990-91. La sal y las relaciones intercomunitarias en la Península Ibérica durante la Antigüedad. *Memorias de Historia Antigua* 11-12, 219-231.

Martin, R. 1971. *Recherches sur les agronomes latins.* Paris, Les Belles Lettres.

Morère, N. 2008. Une nouvelle approche de Pline sur le sel et l'eau salée. In O. Weller, A. Dufraisse and P. Pétrequin (eds.) *Sel, eau et forêt: hier et aujourd'hui*, 365-380. Cahiers de la MSH Ledoux 12 (coll. Homme et environnement, 1). Besançon, Presses Universitaires de Franche-Comté.

Slim, L., Bonifay, M., Piton, J. and Sternberg, M. 2007. An Example of Fish Salteries in Africa Proconsularis: the *Officinae* of Neapolis (Nabeul, Tunisia). In L. Lagóstena, D. Bernal and A. Arévalo (eds.), *Cetariae 2005. Salsas y Salazones de Pescado en Occidente durante la Antigüedad. Actas del Congreso Internacional (Cádiz 2005)*, 21-44. Oxford, BAR Publishing.

Wilson, A. 2007. Fish-salting Workshops in Sabratha. In L. Lagóstena, D. Bernal and A. Arévalo (eds.), *Cetariae 2005. Salsas y Salazones de Pescado en Occidente durante la Antigüedad. Actas del Congreso Internacional (Cádiz 2005)*, Oxford 2007, BAR Publishing.

Historical Development of the 'salinae' in Ancient Rome: from Technical Aspects to Political and Socio-Economic Interpretations

Nuria Morère Molinero
Universidad Rey Juan Carlos, Madrid, Spain

Abstract
This paper analyzes the earliest appearances of the term 'salinae' in Latin literature. A chronological and historical approach is used to ascertain the knowledge of these productive facilities in Ancient Rome and to argue that the interpretation of the term '**salinae**' occasionally went beyond the mere framework of a production site, and occasionally became – as demonstrated by ancient toponymy – economic centers fed by road networks whose wealth revolved around salt.

Keywords
'salinae', salt works, latin writers, Roman history, Hispania, Galia, Britannia

Salt and salt works in Ancient Rome are being documented increasingly by means of texts and written documentation thanks to the classic works by Giovannini (1985) and Traiana (1992), in addition to recent work and new interpretations by Alexianu (2008), Carusi (2006; 2007; 2008; 2009), Delibes *et al.* (2007), García Vargas and Martínez Maganto (2006), Lagóstena (2007a; 2007b, 2007c), Napoli (2007), Morère (1994, 2002, 2008). Our understanding of the techniques of sodium chloride production is constantly improving; based on epigraphy, studies have been carried out on the economics, politics, and society in relation with it. However, in order to derive the most benefit from the thematic focus of this monograph, it is our aim here to analyze, through specific examples, the meaning of the term '*salinae*' for peninsular and provincial Ancient Rome in terms of its interpretation and its political and economic value. In other words, we shall analyze the 'knowledge and practice' of '*salinae*' with an emphasis on chronological factors. The documentation is vast, epigraphic, and literary; however, due to space limitations, in this paper we will only analyze certain literary documents, leaving other aspects of this analysis for in press publications (Morère 2010, 1465-1473). This is, therefore, not an exhaustive analysis, rather a proposal for possible interpretations of this term and its historical contexts.

Prior to focusing on Antiquity itself, to provide an introduction as well as a historical *ex cursus*, we would like to examine the term '*salinae*', '*salina*' (salt works) in other periods where there is a longer history of research and other documentary information in order to gain an understanding of the political and socio-economic significance of this term. Salt works are first documented in the oldest Christian texts written in Latin in the Iberian Peninsula. They are linked to the first feudal lords, bishops, and kings, but in a historical context of the entire process of the concentration of property – that is, of the salt works – based on a monopoly of salt. Previous studies, for example, accept that in the Iberian Peninsula small pieces of property existed which slowly disappeared over time as feudalism developed in parallel with the conversion of salt into a form of royal tax payment. This development followed the conquest of the various territories by the Christian monarchs (Castille in the 12th century, Ibiza in the 13th, and so on). Regulation of incomes arising from salt will occur gradually over time before that of the actual production sites (salt works). The first edicts date from the 12th century, and, in the 16th century, finally culminate in the complete control over salt production by Philip II of Spain (1564). There is also evidence of an entire series of geographical and complementary terms, but these refer to the obtaining of brine and/or salt production, not only in relation with coastal sea salt, but also to inland salt. For example, the terms 'wells' and 'areas' appear in connection with donations in 10th century Añana (Plata Montero 2008, 62), 'brine wells', and 'salsa' in the region of Cabezón de la Sal in the 9th and 10th centuries in relation with exploitation or a system for exploitation (Pérez Bustamante 1977-1978, 149-150). In the inland, '*pausatas*' appears in 917 and '*salinas*' by the year 962 (Rodríguez Rodríguez 2007, 502), '*salinae*' in Sigüenza-Atienza in the 12th century (Minguella y Arnedo 1910, 348 ff.), and '*salina*' or '*salinae*' in Ibiza as of the 13th century. These dates advance in time in step with the conquest's advance, although prior mentions are made (Cirer Pons 1999, 11). 'Wells' and 'salt works' are identified in the abovementioned Royal Decree of Philip II (1564), thereby giving the term 'well' the meaning of production site, as well as the possibility of its also referring to a brine extraction procedure, for example, in Spanish Mexico ('tapextle' well – Reyes Garza 2007, 858). Finally, it is in the 17th century when the Covarrubias dictionary of 1611 consolidates the definition of '*salinae*', identifying '*ciertos lugares donde el agua de los poços tendida en unas eras se convierte en sal*' - certain places where the water from wells that is spread out in areas is converted into salt- (1611: see 'salinas' in Hernando García-Cervigón and Alonso Sutil 2007, 766), as compared to prior hesitation towards a definition by Marineo Sículo (2004) (15th - 16th centuries). But it has been a very long process in which salt works and wells have signified and shown, within a context of their monopolization and control by the Crown, stages or developments in the process of appropriation. Nor should we disregard sociological interpretations or those having to do with labor organization (Beltrán Costa 2007, 899 ff.).

In the end, the salt works will take shape as planned constructions that are characteristic of a reasoned architecture that reaches its peak in Spain's Bourbon period, in a milieu of enlightened exploitation. A more reasoned production will be sought, as well as a more supervised and controlled exploitation, which will be reflected in an entire series of technical improvements and in the perfection and improvement of the warehouses, the buildings where the Royal control of salt by the Bourbons will be carried out. The warehouses, enormous in their dimensions and perfectly constructed, will be a form of prestige for a centralized government and for absolute authority, and an element of archeology and architecture, serving as testimony for the occupation of a territory and of its wealth. It is in these buildings where the transition from a popular architecture to an architecture that is cultured and enlightened has been observed, as evidenced by studies in archeology and architecture (Plata Montero 2007). It is possible that specialized workers, such as the Mudejar woodworkers from Guadalajara that moved to Cabezón de la Sal to work for the salt works, participated in their construction (Pérez Bustamante 1977-1978, 156).

If as of the Middle Ages we witness the evolution of the salt works in the Iberian Peninsula – the territory we have chosen as an example due to its abundance of salt lands and where there was a tendency for greater exploitation of opportunity and power by means of their salt resources – it is believed that this process was gradually being carried out from a small piece of private property during the earliest High Middle Ages (Ladero Quesada, following Pastor, 78), alongside the great and powerful of the monasteries that will form one part of the large land owners and beneficiaries, as Benoît has suggested for the salt works in Provence (the salt works in Marseilles that in the 6th century belong to the Saint Victor Abbey provide the most ancient example – Benoît 1961, 22 ff.). The Late Roman Empire, with some differences, documents this duality of property as private, religious, or public. On the one hand we know of, for example, the only '*salinae*' located in the *villa* of Albinus described by Rutilius Namatianus (*De redito suo*, 475), and on the other, the evidence of an ecclesiastical authority in charge of the elaboration and/or transport of salt. Likewise, in England a lead 'salt pan' from Shavington, Cheshire, has been found, with the inscription '*VIVENTI [ ] COPI*'. The latter term is read as '*Episcopi*' and could indicate that in the 4th century the Bishop was involved in the local salt industry (Penney and Shotter 1996, 363). This is not an isolated discovery but instead shows a more widespread and shared production in the region where the ancient toponym '*Salinae*' –Middlewich – cited in the Ravenna Cosmography (Bestwick 1975, 70) is identified. The denomination of this town may identify it with salt production, but perhaps also with a productive region, connected by means of a road network and related to possible supply sites. A similar case could be proposed with the other toponym '*Salinae*', identified with Droitwich, also cited in the Ravenna Cosmography and by Ptolemy.

Along the same lines is the case of the *Galia*, in the *Alpes Maritimae*, where we learn of the town Castellane (Alpes-de-Haute-Provence), a '*Salinae*' revealed by inscriptions (Bérard 1997, 62). The identification of the toponym has traditionally been explained by the exploitation of a number of salt springs, but it is possible to present as a hypothesis not only the exploitation of its own salt springs, but as its being the center of power of an inland salt region. To the prehistoric vestiges in the region and its nearby surroundings (Boutet and Weller 2007, 226; Morin *et al.*, 2008, 282) we must add the role played by the roads of '*Salinae*', located at an intersection of secondary roads, at the river-crossing of the Verdon River, on the way to the *Via Domitia,* and from the Mediterranean to Mont-Genèvre (Bérard 1997, 65). But it is the discovery of an inscription dedicated to the *patronus* of a corporation of *tabernarii* (CIL V 7907, ILS 6759), an inscription found in Cimiez, and dated in the 2nd century AD (a statue with the *dedicatio*: ..... *Flavio Verini filio Qu[ir. S]abino decurioni IIviro [Sa]lin(iensium) civitatis suae, IIviro [For]oiuliensis, flamini provin[c(iae)] Alpium maritimarum optimo patrono tabernarii Saliniense[s] posuerunt curantibus Matu[cis Ma]nsuet(o) et Albuci[ano i]mp. Commodo III et An[tistio] B[u]rro co[s].*), which would verify the role played by the road networks of '*Salinae*'. We can see the importance of the road network on the one hand in role played by the *tabernarii*, and on the other by the *cursus honorum* of the *patronus*, an individual from '*Salinae*', a *duumvir* who first completes his *cursus* at the *Forum Iulii,* and later becomes a *flamen* in the *Alpes Maritimae* (Gascou 1997, 89). Thus, the meaning of the term '*taberna*' can be extended to « lodging house » and not merely as « urban shop » (Kleberg 1957, 20). However, '*Salinae*' was a see according to the *Notitia Galliarum* between the years 439 and 442 that disappears in the 5th and 6th centuries, coinciding to a great extent with the periods of road renovation (a large number of milliaries in the 3rd and 4th centuries: Bérard 1997, 63-64). The ancient toponym '*Salinae*' is therefore found again in relation with salt exploitation, or its production base, articulated by a road network and in some way in relation with the creation of a see. We have already indicated this in the form of a hypothesis for *Segontia* (Sigüenza, Guadalajara, Spain), as the justification for the existence of a see in a region that had a potentially extraordinary wealth in salt, where the exit or center of production and/or taxation could possibly have existed (Morère 2007, 13), but the evidence is increasing. Here we are referring to a growing relationship between bishops and the exploitation of resources, in the Late Roman Empire, which was a period of intense exploitation of the entirety of a territory's wealth, as, for example, the case of the '*salsamenta*' (Bernal Casasola 2008, 47). In this way the second definition for the term '*Salinae*' could be confirmed: a salt warehouse, evidenced in Rome, in the 'Porta Trigemina' and proposed by Cagnat (1882, 241), an opinion taken up again with the *horrea* (Traiana 1992, 375; García Vargas and Martínez Maganto 2006, 265).

If the '*salinae*' have, therefore, taken on a toponymic and topographic meaning, especially in the Late Roman Empire (although previously in Castellane), the fact of engaging in the artificial exploitation of seawater, without it being possible to strictly specify the context, seems to be what distinguished civilized from uncivilized peoples (Plin. *NH*, XXXI, 88; Morère 2008, 374). We find salt accumulated in piles in the Homeric world, and already exploited in the Greek world (Carusi 2006, 2009), and we find it used as a raw material in the Carthaginean *salsamenta* in expansion as of the 6th and 5th centuries BC. In spite of recent proposals that beginning in the Bronze Age this technique for conserving fish could have been a practice of the Italic coastal peoples (Pasquinucci and Menchelli 2002, 181), the importance of the Punic *salsamenta* makes it possible to infer that the Italics learned this method from the Punics, as it is well known that the work of the agronomist Mago was submitted to the Romans after the 3rd Punic War and translated (Varro *Rust.* I, 1, 9; El Bouzidi 2000, 32). Further, we could also mention the very early and specific reference made by Cato to salt produced by the Carthigineans (Cato. *Orig.* II, 36).

However, it is during the period of the 1st century BC and the 1st century AD that the primary authors that speak of the '*salinae*' are dated, using this term with complete naturalness and as if it were a household word, from various points of view: descriptive, historical, and contemporaneous. There are five authors, writing at the end of the Roman Republic and at the beginning of the Empire: Cicero, Livy, Marcus Manilius, Pliny, and Columella. There are other non-Latin sources in this same historical context (Dionysius of Halicarnassus and Strabo), but they lie beyond the scope of this paper, as our focus has been on the term in Latin, as it appears in texts by the above five authors in writings on the period or in histories of Rome.

Pliny, an author from the 1st century A.C. and a great compiler of data, could represent an overview of the question related with all of the principal themes that revolve around salt. Among artificial means of salt production, he mentions the '*salinae*' together with the wells, in reference to an inland region of Hispania (Plin., *NH* XXXI, 83). He juxtaposes the two terms, contrasting them with his statements on Capadocia where he mentions the wells or springs that provide the water to the salt works (Plin. *NH* XXXI, 82). The first explanation is technical, and we can therefore propose the that the manner in which salt water (brine) is extracted from the subsoil via the construction of wells – still in use and only in the Iberian Peninsula – could indicate that the technology currently utilized in the Iberian Peninsula hails back at least to the Roman era, and may serve as a *terminus post quem*, although the technique of using wells to extract salt water is dated in the Neolithic period in Fontaines Salées (Bernard, Pétrequin and Weller 2008, 299). But apart from the technology for brine extraction, the juxtaposition of the two terms alludes to one meaning – salt production – since we know the function salt works, with a quantitative difference being possible, just as in the case of Hispania the high salinity of extracted water is specified with the term brine. In Capodoccia, the well and the salt spring are forms of water extraction and the salt works is the elaboration site, whereas in inland Hispania only the well is known. The well could be with a small unit for the exploitation, organization of production, or perhaps property(?), along the same lines as shown by the documentation from the High Middle Ages where the two terms (*salinae* / 'pozo') appear, with a possible hierarchy between the two of them. This is what could explain the multiple meanings of the term '*salinae*' as being composed of various production units, whether these units are wells for the extraction and obtaining of brine (inland salt works), man-made ponds, or areas for coastal salt works. This comparison is not coincidental since we know, based on the the descriptions that have been preserved, that the process has evolved minimally, except with respect to the mechanical advances made in relation with technological innovations.

The information on the wells identified with the 'salinae' is therefore even more significant as it only appears in Pliny, the great compiler and encyclopedist. This has led us to ask whether it is possible to discover the origins of this information, that is, the sources he used. Among the numerous compiled authors, in Chapter 31, there is a trace of Varro present in its manner of structuring the presentation (natural and artificial salt) (Morère 2008, 366). Many other chapters contain similar traces, such as the one referring to Lars Porsena's tomb (Pliny *NH* XXXI, 91-93; Jannot 2005, 633), and we believe it is quite plausible as a hypothesis that the information referring to the wells in Hispania could also have been copied from this same author (116-27a.c.), if we take his biography into account. Varro had two sojourns in Hispania: the first took place after the year 78 BC when he was legatus and accompanied Pompey in the fighting against Sertorius (Hernández Miguel 2000, 13); the second took place afterwards in the *Ulterior* Province where he was legatus propretor as of the year 50, during the Civil War when Caesar launched the Battle of *Ilerda.* Varro must have known the *Citerior* province during his first stay and the *Ulterior* during the second one. As for Cappodocia, it seems that there is doubt regarding his stay as governor in Asia since he does not appear on the list of governors, although there is some possibility. During his stay in the *Ulterior*, he could have been obtaining wealth, and for this reason he would have come into contact with the resources of the region *Corduba, Hispalis*, and *Gades*. It is not easy to specify where he could have seen the wells, since the procedure is the same for all of the inland salt works and the climate would have made solar salt production possible. Nevertheless, we are inclined to believe that he saw them in the *Ulterior* province (where there is more evidence of his plunders), especially if we take into consideration that in the *Citerior* – in accordance with the following the archaeological prospection Antonio Malpica Research Group we form part of – and, at least in the Iron Age, it has been proven that there was utilization and development of salty clay deposits that deposit salt on the ground surface (and not brine exploitation), since the technique of using wells was a very advanced phase of salt production. Further, the use of wells would amplify the technical meaning: both the use of wells and brine involves control of the manufacture, the production phase (2nd Iron Age, Roman occupation?), as compared with the previous phase which is the mere collection of muds and clays in a different economic and political context that at present we cannot specify. Based on the date provided by this data, it is possible to question whether this system of production-exploitation is specific to the peninsular indigenous world, a pre-Roman tradition, or whether it corresponds to a more intensive exploitation characteristic of the Roman world. We must not forget that tradition relates the exploitation of the mines in Hispania with Cato. Perhaps, as a form of control over salt production, the well was introduced or would be thought of at that time.

The mention of wells in Hispania and their link with Varro, that is, with the 1st century BC, and their interpretation as production sites of artificial salt, is very relevant due to the fact that in the Iberian Peninsula we do not have any other mention of '*salinae*' in all of Antiquity. We do know of citations of the mountain of salt (Morère, 1994; Carusi 2006), a specific toponym debated as *Egelasta* (Anthologie); we also have documentation of a certain *salsum flumen* in the *Hispania Tarraconensis* (Talbert 1992, 399), in the *Lusitania-Baetica* (Talbert 1992, 428); in the *Carthaginiensis* (Talbert 1992, 447, unidentified), but never documentation of the term '*salinae*', which affects the reality of inland salt exploitation and, therefore, of the wells cited by Pliny. In fact, the Iberian Peninsula, besides having a large expanse of coasts typical of peninsulas, stands out due to the high degree of salinity in soils (Figure 1 in article by Dr. Jesús Jiménez Guijarro, in this volume).

In fact, the writers of antiquity knew perfectly how to explain the procedure of how rivers were transformed into salt-water rivers, as Vitruvius does when writing about Sicilia (Vitr. *De arch.* VIII, 3, 7): '*... the other stream which runs through the other part where there are salt works has a salt flavour....*'. Salt lands, that is, lands described with specific properties, are also documented by other authors, such as Columella (*Rust.* II, 2, 15-16) when he writes of lands and their fertility. He makes reference to the '*campi salinarum*' in a semantic context of lagoons, marshlands, and their location at the ocean shore, that is, with spatial consideration as well. But the term '*campus*' with the meaning of 'plain', in line with Varro (*Ling.* V, 36), and which, by the way, is also applied to certain walled, expansive public spaces outside the city limits (Bouet 1998, 111), will be in the imperial era the terminology used for the large salt works of Ostia, known in epigraphy as the *campi salinarum*, and which could refer to both the topography of the land as well as the vast, flat expanse of the salt works. For example, there are dedicatories to the *Genius saccariorum salariorum totius urbi campi salinarum romanorum* por ejemplo *(CIL XIV 4285),* (García Vargas and Martínez Maganto 2006, 269); *conductores campi salinarum romanorum* (Morelli, Olcese and Zevi 2005, 46).

The term '*campus*', besides indicating spaciousness, also defines a topographic and geographic area, but we obtain these specifications on salt works by means of a detailed reading of the classic authors who have provided us with a vast number of small items of information that refer to their environmental surroundings and the way they function, since in the chronological context that we have chosen only the description made by Manilius of the '*salinae*' is known (*Astronomica* V, 681; Carusi, 2008). The treatise by Manilius is poetic: after providing a description of fishing techniques, tuna fishing, and so on, it then describes some salt works lyrically.

From Cicero we obtain a large amount of data. In Cic. *Nat. D.* II, 132 – a text given to me in the symposium by Carusi – he informs us about the location of the salt works, far from the ocean and up river, and about supplying water by using the tides. Nevertheless, Cicero lived in the 1st century BC, which gives greater relevance to all of the information he provides since this would mean that this data is the oldest we have from the Roman world. The mention of the movements of water, linked with the wind, tides and the effect of the wind on the lagoons, will later be taken up by Livy (XXVI, XLVI, 8) although without mentioning salt when Scipio attacked *Carthago Nova* and he was able to cross the lagoon thank to the winds that practically emptied it of water. There is no mention of the salt works, but it is a '*stagnum*' in the sense of being waterproof in accordance with the definition by Varro (*Ling.* V 26) – a term used by Caesar in Africa: '*stagnum salinarum*' (Caesar, LXXX, 1) – or by Columella who with this term designates the man-made fish tanks in relation with the constant renewal of their water due to its being moved by the wind (Col., *Rust.* 8, 17, 1).

The sources compiled from the 1st century BC to the 1st century AD have been taken from political and historical works, as well as from practical works such as those of agronomists, and from the very encyclopedia written by Pliny. Only in one of the cases will the '*salinae*' be mentioned in a poetic piece, in Book 10 by Columella (Col., *Rust.* X, 135-136), where the closeness of the salt works (of Hercules) is juxtaposed with the lake ('*palus*') of Pompey (Murano, 119). The term '*palus*' is used in the sense of shallow waters in Varro, *Ling.* V, 27. It is only found again at the end of the Empire, where, in his praise of *Narbo*, Sidonius Apollinaris mentions the '*salinae*' in a poem, in his evocative poem of the great capitals of the ancient world, with a long enumeration of the resources and goods and environs of *Narbo* (*Carm.* XXIII, 43): "*.....insulis, salinis, stagnis...*'. The case of the Iberian Peninsula is highly significant in this regard, bordered by the ocean and with a significant silence with respect to the term '*salinae*' in texts and ancient toponymy, which could, in fact, be explained due to very relevant natural salt production from lagoons, lakes, marshes, as compared with artificial production in '*salinae*', which would not be necessary as optimum natural conditions were present.

The authors under study define a historical context for the salt works and we can affirm that from the 1st century B.C to the 1st century AD, this type of facilities, the '*salinae*', are perfectly linked to their historical context. The authors know how they function and they increasingly specify more details about them. The context the authors provide us with is the Italic Peninsula, amplified by the extension of its territorial zone throughout the entire Mediterranean. Based on these dates, we have even made allusions to later time periods. With respect to historical development and social and economic organization, Cicero and Pliny make isolated references, whereas Livy, as author of '*From the Founding of the City (of Rome)*' will provide more information, although some of his references will be more about salt than about the '*salinae*'. Due to space limitations, in this paper we will not provide a complete historical analysis (Morère, 2010). However, the most relevant piece of data is the foundation-creation of the '*salinae*' under the figure of Ancus Marcius (Livy I, 33, 9- I, 34; Plin. *NH* XXXI, 89; later found in Aur. Vict. *De vir*. Ill. 5), and the creation of a *vectigal*, a tax that will be clearly identified with the *ager publicus*, which might be confirmed by the remarks made by Cicero on the *Silva Maesia* and its conversion into an *ager publicus*, also in the same time period as Ancus Marcius (*De Rep*. II, 33). The historical context of Ancus Marcius as a period of initial political power and Rome's commercial expansion towards coast makes the creation of the '*salinae*' possible (Camous 2004, 273) –although they were already in existence- even if we grant an important role in the 4th century BC to the *gens Marcia* (Rebuffat 1974, 650-651; Camporeale, 1997, 197-198), who could have controlled the resources and commerce at the mouth of the Tiber. The idea of *vectigal*, however, with its long evolution, has an impact on the public nature of all of these '*salinae*' in Rome, the '*campi salinarum*' during the Empire and takes us, as we see it, to the time these texts were written.

However, there is serious doubt regarding the historical veracity of not only certain references made by these authors to '*salinae*', but to salt as well, since we believe that the information they provide could possibly allude to events that occurred at the time the texts were written. This could be the case, for example, with the episode about Porsenna that Livy refers to the distributions of salt linked with the creation of the '*salinae*' (II, IX, 7) that Pliny also attributes to Ancus Marcius: '*King Ancus Marcius gave a largess to the people of 6000 bushels of salt, and was the first to construct salt pools*' (*NH* XXXI, 89).

References:

Alexianu, M.-T. 2008. Une catégoríe d'esclaves thraces: les halônetoi. In A. Gonzales (ed.), *La fin du statut servile? (affranchissement, libération, abolition)*, Actes du 300 colloque du Groupe International de Recherches sur l'Esclavage dans l'Antiquité (GIREA) - Besançon, 15, 16, 17 décembre 2005, *hommage à Jacques Annequin*, 487-492. Besançon, Presses universitaires de Franche-Comté.

Beltrán Costa, O. 2007. Las técnicas tradicionales en las salinas de interior. In N. Morère Molinero (ed.), *Las salinas y la sal de interior en la historia: economía, medio ambiente y sociedad. Inland saltworks and salt history: economy, environment and society*, 899-922. Madrid, Dykinson.

Benoît, F. 1961. Les abbayes du sel. L'héritage Antique du Delta au Moyen-Age. *Delta*, 17-32.

Bérard, G. 1997. *Carte archéologique de la Gaule romaine. Les Alpes-de-Haute-Provence*. Paris, Académie des Inscriptions et Belles-Lettres.

Bernal Casasola, D. 2008. El final de la industria pesquera-conservera en Hispania (ss V-VII d.c.) entre obispos, Bizancio y la evidencia arqueológica. In J. Napoli (ed.), *Ressources et activités maritimes des peuples de l'Antiquité, Les cahiers du Littoral* 2, 6, 31-57.

Bernard, V., Pétrequin, P. and Weller, O. 2008. Captages en bois à la fin du Néolithique: les Fontaines Salées à Saint-Père-sous-Vézelay. In O. Weller, A. Dufraisse and P. Pétrequin (eds.) *Sel, eau et forêt: hier et aujourd'hui*, 299-333. Cahiers de la MSH Ledoux 12 (coll. Homme et environnement, 1). Besançon, Presses Universitaires de Franche-Comté.

Bestwick, J. D. 1975. Romano-british inland salting at Middlewich (Salinae), Cheshire, in Salt. The study of an ancient industry. Report on the salt weekend held at the University of Essex 20-22 september 1974. Colcherter archaeological group. 66- 70.

Bouet, 1998. Un nouvel exemple de campus en Gaule Narbonnaise: Vaison la Romaine (Vaucluse), *RAN* 31, 103-117.

Boutet, A. and Weller, O. 2007. La question de l'exploitation du sel dans le Midi de la France durant la Protohistoire'. In Nuria Morère (ed.), *Las salinas y la sal de interior en la historia: economía, medioambiente y sociedad. Inland saltworks and salt history: economy, environment and society*, 217-239. Madrid, Dykinson.

Cagnat, M. R. 1882. *Etude historique sur les impôts indirects chez les Romains*. Paris, Imprimerie nationale.

Camous, T. 2004. *Le roi et le fleuve. Ancus Marcius Rex aux origines de la puissance de Rome*. Etudes anciennes, Paris, Les Belles Lettres.

Camporeale, G. 1997. Il sale e i primordi di Veio. *In La necropoli archaiche di Veio. Gionata di studio in memoria di Massimo Pallotino, a cura di Gilda Bartoloni*, 197-199. Roma, La Sapienza.

Carusi, C. 2006. Il sale nel mondo greco (VI a.c.-II d.c.). *Luoghi di produzioni, circolazione commerciale, regime di sfruttamento nel contesto del Mediterraneo antico*. Scuola Normale Superiore, Tesi di perfezionamento in discipline storiche classiche.

Carusi, C. 2007. Régimes d´exploitation et fiscalité du sel dans le monde grec et romain. In N. Morère (ed.) *Las salinas y la sal de interior en la historia: economía, medioambiente y sociedad. Inland saltworks and salt history: economy, environment and society*, 325-342. Madrid, Dykinson.

Carusi, C. 2008. Le sel chez les auteurs grecs et latins. In O. Weller, A. Dufraisse and P. Pétrequin (eds.) *Sel, eau et forêt: hier et aujourd'hui*, 353-364. Cahiers de la MSH Ledoux 12 (coll. Homme et environnement, 1). Besançon, Presses Universitaires de Franche-Comté.

Carusi, C. 2009. *Il sale nel mondo greco (VI a.C. - III d.C.). Luoghi di produzione, circolazione commerciale, regimi di sfruttamento nel contesto del Mediterraneo antico*. Bari, Edipuglia.

Cirer Pons, J. 1999. *Tratado de las Reales Salinas de Yviza y Formentera, ed. Mediterrània-Eivissa 1803*. Palma, Manuscrit n. 21 de la Biblioteca Central de la Diputació de Barcelona.

Delibes de Castro, G., García Rozas, R., Larrén Izquierdo, H. and Rodríguez Rodríguez, E. 2007. Cuarenta siglos de explotación de sal en las lagunas de Villafáfila (Zámora): de la Edad del Bronce al Medioevo. In A. Fíguls, O. Weller (eds.), *1a Trobada internacional d'arqueologia envers l'explotació de la sal a la prehistòria i protohistòria. 1st International archaeology meeting about prehistoric and protohistoric salt exploitation, Cardona, desembre 2003*, 111-142. Barcelona, IREC.

El Bouzidi, Saidi 2000. Les sources agronomiques au service de l'histoire: Le De Agricultura de Caton. *Florentia iliberritana* II, 11, 2000, 21-35.

García Vargas, E. and Martínez Maganto, J. 2006. La sal de la Bética romana. Algunas notas sobre su producción y comercio. *Habis* 37, 253-274.

Gascou, J. 1997. Magistratures et sacerdoces municipaux dans les cités de Gaule Narbonnaise, Michel Christol and Olivier Masson (eds.), *Actes du Xe Congrès International d'Epigraphie grecque et latine. Nîmes 4-9 octobre 1992*, 73-140. Paris, Sorbonne.

Giovannini, A. 1985. Le sel et la fortune de Rome. *Athenaeum* 63, 373-386.

Hernández Miguel, L. A. 2000. *Varrón*. Madrid, Clásicas.

Hernando García-Cervigón, A. and Alonso Sutil, MªC. 2007. Estudio lexicográfico de la terminología relacionada con la sal desde el siglo XII hasta el siglo XVII. In N. Morère Molinero (ed.), *Las salinas y la sal de interior en la historia: economía, medio ambiente y sociedad. Inland saltworks and salt history: economy, environment and society*, 757-776. Madrid, Dykinson.

Jannot, Jean-René 2005. Encore la tombe de Porsenna. *Mélanges de l'Ecole française de Rome*. 117, 633-649.

Jiménez Guijarro, J. 2009. The first exploitation of salt in Spain. Thinking about the salt used in the Iberian Peninsula during the Prehistory (in this volume).

Kleberg, T. 1957. *Hôtels, restaurants et cabarets dans l'Antiquité romaine*. Etudes historiques et philologiques. Uppsala, Almqvist & Wiksell.

Ladero Quesada, M. A. 1979. La recette de sel et son évolution dans les états de la couronne de Castille (XIIIe-XVIe siècles). In J. Cl. Hocquet (ed.), *Actes de la Table Ronde 'L'impôt du sel sel en Europe XIII-XVIIIe siècles*, 77-98. Arc-et-Senans.

Lagóstena Barrios, L. 2007a. Sobre la elaboración del garum y otros productos piscícolas en las costas béticas. *Mainake* XXIX, 273-289.

Lagóstena Barrios, L. 2007b. Columela, De Re Rustica VIII, 16-17: una fuente para el conocimiento de la piscicultura en Baetica. In L. Lagóstena, D. Bernal and A. Arévalo (eds.), *Cetariae 2005. Salsas y salazones de pescado en Occidente durante la Antigüedad*, 109-117. BAR International Series 1686.

Lagóstena Barrios, L. 2007c. Explotación de la sal en la costa meridional hispánica en la Antigüedad romana. In N. Morère Molinero (ed.), *Las salinas y la sal de interior en la historia: economía, medio ambiente y sociedad. Inland saltworks and salt history: economy, environment and society*, 301-323. Madrid, Dykinson.

Marineo Sículo, L. 2004 (reed.), *De las cosas memorables de España*. Valdemorillo, La hoja del Monte.

Minguella y Arnedo, T. 1910. *Historia de la diócesis de Sigüenza y de sus obispos*. Madrid, Imprenta de la RABM.

Morelli, C., Olcese, G. and Zevi, F. 2005. Scoperte recenti nelle saline portuensi (Campus salinarum romanorum) e un progetto di ricerca sulla ceramica di area ostiense i etè republicana. In G. Zevi Marsa and R. Turchetti (eds.), *Méditerranée occidentale antique: les échanges*, 43-55. Seminario ANSER III, Rubbettino.

Morère, N. 1994. La sal en la Península Ibérica. Los testimonios literarios. *Hispania Antiqua* XVIII, 235-250.

Morère, N. 2002. La sal en la geografía de Estrabón. In L. Hernández Guerra, L. Sagredo San Eustaquio, J. Ma. Solana Sáinz (eds.), *Primer Congreso Internacional de Historia Antigua*, 519-526. Universidad de Valladolid, Centro Buendía.

Morère, N. 2007. La sal en el desarrollo histórico de Sigüenza. Los primeros siglos. In N. Morère Molinero (ed.), *Las salinas y la sal de interior en la historia: economía, medio ambiente y sociedad. Inland saltworks and salt history: economy, environment and society*, 3-30. Madrid, Dykinson.

Morère, N. 2008. Une nouvelle approche de Pline et de l'eau salée. In O. Weller, A. Dufraisse and P. Pétrequin (eds.) *Sel, eau et forêt: hier et aujourd'hui*, 365-380. Cahiers de la MSH Ledoux 12 (coll. Homme et environnement, 1). Besançon, Presses Universitaires de Franche-Comté.

Morère, N. 2010. "Salinae", "mancipes", "conductors". Algunas observaciones sobre las implicaciones de la sal en el mundo romano. In C. Fornis, J. Gallego, P. López Barja, M. Valdés: *Dialéctica histórica y compromiso social. Homenaje al Profesor Domingo Plácido*, 1465-1473. Madrid, Pórtico.

Morin, D, Lavier, C., Guiomar, M. and Fontugne, M. 2008. Aux origines de l'extraction du sel en Europe (VIe millénaire av.JC). La source salée de Moriez – Alpes de Haute Provence. In O. Weller, A. Dufraisse and P. Pétrequin (eds.) *Sel, eau et forêt: hier et aujourd'hui*, 281-297. Cahiers de la MSH Ledoux 12 (coll. Homme et environnement, 1). Besançon, Presses Universitaires de Franche-Comté.

Napoli, J. 2007. Nouvelles réflexions sur les 'salinatores'; l'éclairage du lustrum primipili. In L. Lagóstena, D. Bernal and A. Arévalo (eds.), *Cetariae 2005. Salsas y salazones de pescado en Occidente durante la Antigüedad*, 157-161. British Archaeological Reports. International Series 1686, BAR Publishing.

Pascquinucci, M. and Menchelli, S. 2002. The Isola di Coltano Bronze Age village and the salt production in North coastal Tuscany (Italy). In O. Weller (ed.), *Archéologie du Sel: techniques et sociétés dans la Pré- et Protohistoire européenne.* Rahden/Westfalen, Verlag Marie Leidorf GmbH, 177-182.

Penney, S. and Shotter, D.C.A. 1996. An Inscribed Roman salt-pan from Shavington, Cheshire, *Britannia*, XXVII, 360-365.

Pérez Bustamante, R. 1977-1978. Las salinas de Cantabria. Aspectos económicos, jurídicos y técnicos de las explotaciones y yacimientos de sal en las Asturias de Santillana (s IX-XVI). *Instituto de Etnografía y Folklore « Hoyos Sainz »*, IX, 147-178.

Plata Montero, A. 2007. *Análisis histórico-arqueológico del valle Salado de Salinas de Añana (Älava)*, Universidad del País Vasco, Tesis doctoral.

Plata Montero, A. 2008. *Erdi aroko hiri baten sorrera. Génesis de una villa medieval*, Colección de Patrimonio Cultural, 4, Victoria.

Rebuffat, R. 1974. Tite et la forteresse d'Ostie. *Mélanges de philosophie, de littérature et d'Histoire ancienne offerts à Pierre Boyancé.* Collection de l'Ecole Française de Rome, 22, 630-650.

Reyes Garza, J. C. 2007. La innovación tecnológica y el fin de una era. El caso de Colima, Mexico. In N. Morère Molinero (ed.), *Las salinas y la sal de interior en la historia: economía, medio ambiente y sociedad. Inland saltworks and salt history: economy, environment and society*, 857-875. Madrid, Dykinson.

Rodríguez Rodríguez, Elias 2007. La explotación y el comercio de la sal en Villafáfila a través de la documentación medieval. In N. Morère Molinero (ed.), *Las salinas y la sal de interior en la historia: economía, medio ambiente y sociedad. Inland saltworks and salt history: economy, environment and society*, 499-34. Madrid, Dykinson.

Talbert, R. J. A. (ed.) 1992. *Barrington atlas of the Greek and Roman World*, Princeton, Princeton University Press.

Traiana, G. 1992. Sale e saline nel mediterraneo antico. *Parola del Passato* 47, 1992, 363-377.

****Le sel dans l'Antiquité. Anthologie et commentaires*. Paris, Les Belles Lettres (in press).

Salt in Tanning, Dyeing and Cleaning in Ancient Egypt

Virginie Delrue
Université du Littoral Côte d'Opale, Boulogne sur mer, France

Abstract
The use of salt in Antiquity is well known in various fields such as the mummification, the medicine, the glass industry, the earthenware factories or even the cooking. In this article, we shall concentrate our study on domains less informed, namely the use of alum, common salt et natron in the activities of tanning, dyeing and cleaning. We shall see, in these various uses, which product, salt, natron or alum, is used and why. Is it the closest, the most available which is privileged? The most cheaper? Or do we look on the contrary for the quality or for the specificity of a product unique as it is the case in the pharmacopoeia? We shall try to demonstrate that the choice of the product is not harmless and that it depends on several factors, the first one of them being the search for the best efficiency and for the best product. This study combines the literary testimonies and the papyri which hint at these industries of the antiquity, from Pharaonic time to Roman period, what brings to light the low sphere of influence of the industrial techniques and the perpetuity of these products.

Keywords
salt, natron, alum, Egypt, tanning, dye, wash

Definition of the products and localization of the deposits

The alum

The alum is a complex compound of potassium and hydrated aluminium double sulphate. According to the ancient authors, especially Pliny (*H. N.*, XXXV, 183), there are two main types of alum: the black and the white. In reality, we may talk of a very clear product and of a darker. In the papyruses, as for example in P. Oxy. 2567, there is another distinction, this time between the 'pure' alum and the alum 'in pieces'. For Pliny (*H. N.*, XXXV, 184) and Dioscorides (*De mat. Medic.*, V, 122), the best alum is that of Egypt, and then that of Melos. Incidentally, we found exploitation traces going back to the Roman in the Egyptian oases where the alum was produced (Wagner 1987, 306-310). Nevertheless, we find it also in Macedonia, Sardinia or even at Lipari. The best one, the Egyptian alum, is produced in the oases of Dakhleh, Khargeh and Baharieh, but it is not the most commun in the Antiquity.

The alum had been known in the pharaoh Egypt at least since the New Empire. The term designing it, *abennou* is attested for the first time in the Ebers papyrus, around 1500 BC (Jéquier 1922, 106-107). We are talking here of a mineral product and not of a plant, as some have stated (Loret 1893). But the fact that the word did not exist before doesn't necessarily mean that the product was not known. G. Jéquier suggests there was a single term for both the natron and the alum. Nevertheless, it would be surprising that the Egyptians should have mistaken one salt for another, given the fact that they were experts in salt, and the fact that they had been using especially natron for mummification since early Antiquity. Also, the use of a single term for two very distinct products as regards their chemical components and their collection places seems very unlikely.

The natron and its derivates

The natron is a carbonate and sodium bicarbonate compound, with water added. The 'pure' product always has impurities, giving it the distinct properties of its extracting place. It is mainly found in Egypt, in Ouadi Natroun, in the North, or at Elkab, in the South. For a description of the different types of natron, see Delrue and Napoli (2008). We should also mention that the translators of the ancient authors used the term 'nitre' to designate the natron, even if we are talking of two different products. The nitre, also named today saltpetre, was probably not known by the ancients, and they could talk about the 'natron'.

The foam of nitre or *aphronitrum* is a product of the natron efflorescence. When the natron crystals lose their water, they crack off and it results a very soft product, often compared to 'fleur de sel' because of its texture.

The salt

The common salt, or sodium chloride, is the most frequent product and the simplest, chemically speaking. We can find it in most of the regions in Egypt.

We may thus see that Egypt has three types of salt and that they constitute the subject of this article. It is thus interesting to know whether they privileged one type of salt and why.

The use of salt in leather tanning and mordanting

The phases

Our purpose here is not to describe the whole process of ancient tanning, but to simply emphasize the phases involving salt. For more information regarding tanning, see Jemma-Gouzon (1972), who offers a large description of the contemporary tanning, whose methods have not changed much since the Antiquity.

During the first phase, called the river work, the hides are either 'green', meaning non-treated, either dried out, coated with salt to be conserved for a couple of days, to avoid the decomposition. The use of salt during heating (when we remove the residuals off the hides) is not attested during the Antiquity, even though it appears that the natron was incorporated in heating baths to accelerate this process (Leguilloux 2004, 22). In the second phase, the tanning, the hides are soaked in tanning agents baths. Among the latter, we may mention the use of alum to prevent the decomposition and to soften the hide. The last phase, the finishing works, does not involve the use of salt.

The products used: salt and alum

The leather may only be tanned with salt, as Cato states (*De Agri.*, 135, 3). The salt dries out and prevents the decomposition of hides (Jemma 1972, 32). We are talking here about dry tanning, without tanning soak. It is a rudimentary operation (Leguilloux 2006, 18), but it could be practiced almost anywhere, with no special equipments. The hides are simply covered with salt (Talmud *Schab*. VII and XI), on the flesh part, and then sun-dried. The leather is rigid, but fragile. In this state, they are mostly used as carpets. For other uses they should be oiled.

The salt is also used for the preliminary phase, for the tanning process itself, instead of alum or beside it. During the tanning process itself, they associate the salt and the alum in order for the fibres to be better impregnated with the tanning product.

While salt may be used for 'coarser' operations, the alum is used for the finishing touches. The leather tanned with alum is slender, white, smooth and delicate, but rather water-fragile; that is why it is used for small objects, such as purses. The alum is, thus, the best for this kind of operation, even though we are not sure of when exactly it began to be used. Pliny, Ovid and Juvenal seem to make certain allusion to this type of alum-tanning (Pliny *H. N.*, XXXV, 190; Ovid *A.A.*, III, 271; Juvenal VII, 192), which has been officially attested since the Mesopotamian period (Leguilloux 2006, 17). But, for the Pharaonic period, even though the tanning was depicted in the graves (Figure 1), we know nothing about the products used, because of a lack of sources.

Consequently, the alum seems to have been used for the tanning operations, and the salt, more aggressive, to make the leather more rigid.

The preliminary stage to dyeing

The leather was sometimes coloured, naturally or with tanning products. The skins were soaked, partially or totally, according to the purpose, in a bath of mordant agents. Pliny (*H. N.*, XXXV, 190) mentions leather dyeing using alum. Maybe he mistook it for copper or iron sulphate (Croisille 1985, 285-286, note 4). Alum soaking gives a whitish colour to the leather. Nevertheless, it is not the only product used, as the natron, well known for its property of fixing the colours, also had a role in the tanning process (Leguilloux 2006, 21).

Salts within the tanning process and its stains

We shall now identify the products used in the dyeing and mordanting.

The case of salt

The introduction of salt in a dyeing soak has as purpose fixing the dye on the tissue, even though this operation was not indispensable for the murex (Macheboeuf 2007, 387). Of course, the salt is an agent that ensures a high-quality dye for a long period, but it might not have been largely used, given the concurrence of the natron and the alum. Anyway, we may say it is hardly ever mentioned in our sources.

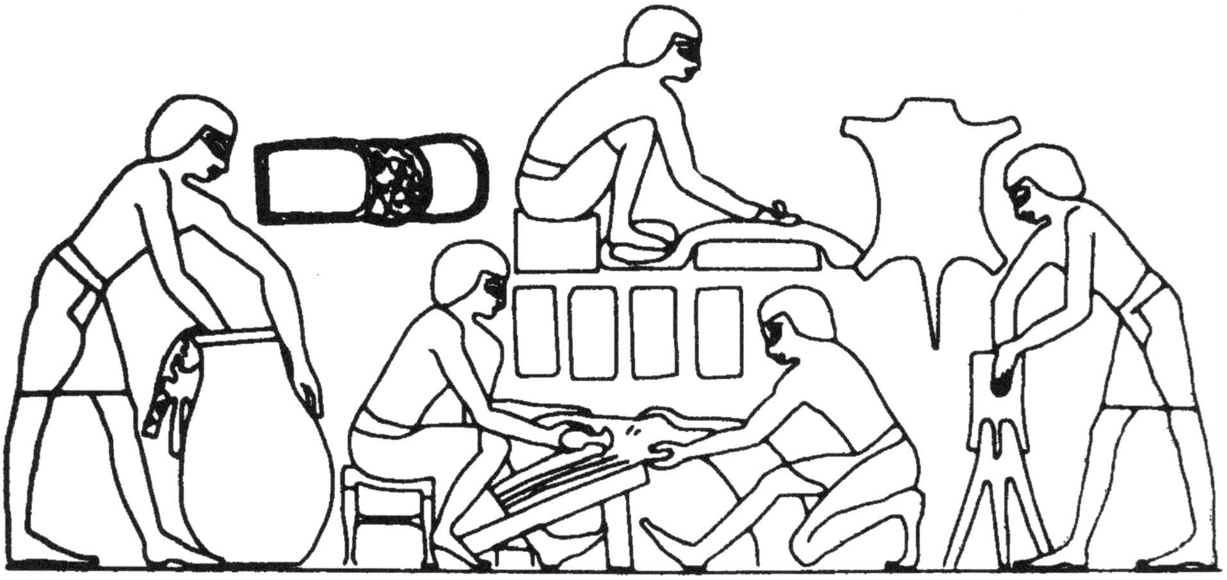

Figure 1. Scene of tanning in the grave of Rêkhmirê (Leguilloux 2004, 55).

Figure 2. Scene of wash in the grave of Baqt (Beni Hasan) (Marsh-Letts 2002, 26).

The privileged alum

Today we know that the alum has the property of fixing the colours (concerning the chemical action of the alum as stain in dyeing, see Delamare and Monasse 2005) and many modern authors, such as Diderot and D'Alembert (1782, 312) or Jéquier (1922, 109, who quotes Pauly-Wissowa, *Real-Encyclopädie*, I, 1297), explain that one of the main uses of alum is in the leather dyeing process. The ancient testimonies give the same assertion. According to Scribonius Largus (*Compositiones, ad dentiunt dolorem*, LVII, see Borgard 2001) it is alum that is 'used by the most famous dyers'. Pliny *(H. N.* XXXV, 183; XXXIII, 88) shows that the white and liquid alum of Cyprus is 'very much used to dye the wool in order to lighten it'. It may be used as stain or directly as colouring agent. The 'black' alum was also used to dye 'dark-coloured' fabrics.

A Roman papyrus (P. Cair. Zen. 59326 bis, l. 26), which shows a list of things that a dyer should buy, mentions the alum, another confirmation that this product was used during the dyeing process. In the P. Oxy. 467 (Wagner 1987, 310), around 129 dC., we see that the alum allows 'to fix mineral dyeing to ochre, red or brown, among others' (Wagner 1987, 310; Borgard 2001, 56-58).

In these references we find the two main types of alum: the black and the white. Nevertheless, we do not find the distinction between the pure alum and the alum broken in pieces.

The use of natron

According to our sources, the alum was not used during the pharaoh era. The natron, that had the same properties as the alum in the dyeing process, probably replaced it successfully. The alum probably became far more popular, and people got more used to this product when the Greeks and the Romans came along. In fact, it came to be used largely in dyeing after the Greek era. Nevertheless, according to Pliny (*H. N.* XXXI, 110), the 'impure [natron] is good for dyeing', which shows that its use had not perished despite the popularity of the alum. The fixing properties of the natron allow an effective mordanting (Goyon 1996, 21-22), and Plutarch confirms it (*De defic. Oracul*, c 41).

But, during the following eras, Greek and Roman, the alum totally replaces the natron for dyeing, as it disappears from our sources. And it is certainly more common and cheaper than the natron.

The use of salt in cleaning

They cleaned not only the dirty clothing, but also the wools and materials before the fabrication, as they were usually oily from the animal fat and they got dirty after the manipulation of the fabric. Thus, there are two operations, that of bleaching and that of cleaning. They are often represented in the pharaoh iconography (Figure 2).

Concerning this use, we can't find any mentions for the salt or the alum. Only Aufrère (1991, 607) talks about the dry-cleaning and lightening properties of the alum: 'the natron (alum)-treated fabric, (...) became white' as both of them trimmed the fat out, giving the fabric a lighter appearance. But, as far as we know, in the ancient documents there is no mention of another product besides the natron used for these operations in Egypt.

The omnipresence of natron

In Egypt, the natron was a monopole product, as the salt and the alum (Préaux 1939, 253), belonging to the clergy during the pharaoh era, then to the State during the following eras. Certain documents suggest this type of monopole – those which mention the nitriké tax, for example, and allow us to say that the natron was used in the bleacheries, to get the fabrics washed and cleaned.

Thus, the ostracon DeM 314, dating from the New Empire, evokes the 'rough' natron (*hsmn* in Egyptian), indispensable during the washing process (Davies and Toivari 2000, 73). The Tebt. 703 papyrus (Figure 3) also mentions the natron among the substances used in bleaching: 'Visit also the washing-houses where the flax is washed and make a list, and report so that there may be a supply of kastor-oil and natron for washing.' This document, far more recent, around 210 BC, shows that the natron was still used during this period. The Cair. Zen. III 59304 papyrus (Figure 4), since 31 August 250 BC, shows that a bleachery is about to close as it was out of natron.

The tax receipts nitriké, meaning the tax on the natron, indicate that the bleachers are the ones to pay it. The 'Theban bank receipt for the tax on the natron' (Figure 5) edited by Kayser (1991) states that a certain Héracleidès paid the 'tax on the natron of the bleachery'. This document, from 162 BC, evokes a fuller that cleans wool and linen clothing (Kayser 1991, 221). This ostracon is not the only document mentioning this tax for this particular case; there are also other papyruses that make reference to it, as for example P. Tebt I 40.

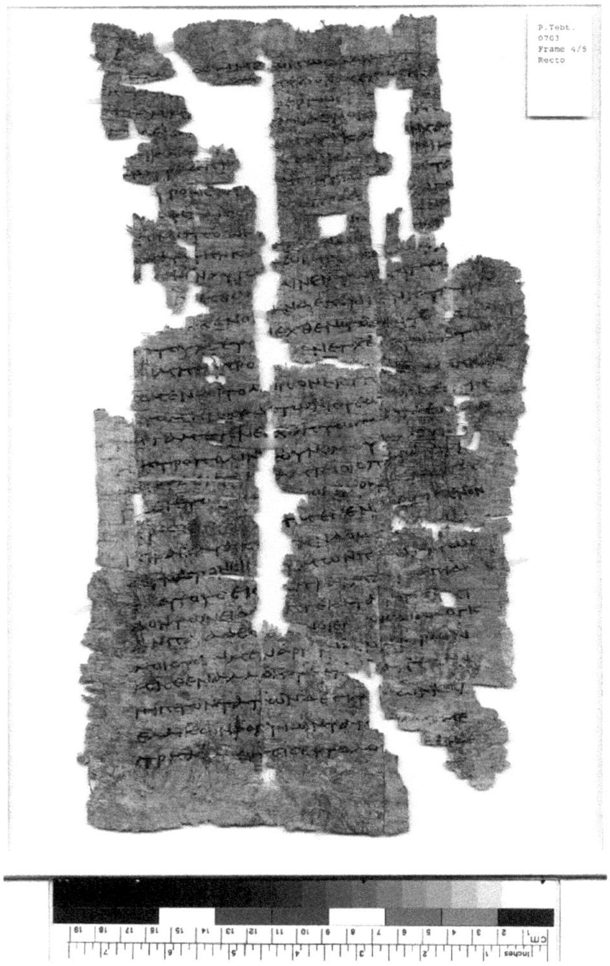

Figure 3. P. Tebt. III 1.703 (http://digitalassets.lib.berkeley.edu/apis/ucb/images/AP01538aC.jpg).

The ancient authors also evoke the natron as washing product. Strabo (XI, 529) and Isidore (*Orig*., XVI, 2, 7) state that waters containing natron are efficient in washing. The two authors are quoted by Jacob (1963, 86). The use of natron has a history of several centuries, as modern authors show it as being effective within the washing process, authors such as Leclerc de Buffon (1799, 81), who refers to the *Voyage de Thévenot*, 1664, t. 1.

The quality of natron

Chassinat (1968, 784-785) evokes the foam of nitre as washing product. This product, also called aphronitrum, was rare and pricy. We are talking about a high quality product, which did not seem to be largely used, even for the priciest clothing. Moreover, the interpretation of the document that E. Chassinat refers too is still under discussion. Thus, it seems that we may not take into account this specific product.

The above mentioned Deir el-Medineh ostracon uses the Egyptian term *hsmn*, which designated a rough natron. We have already shown, in a previous article (Delrue and Napoli 2008, table 1, 366), that the ancient Egyptians knew very well the differences, such as in the case of medical use, among the various types of natron, as the 'impurities' within the natron turned it into a unique product, specific to each production site. We may also suppose that the *hsmn* natron was the one used in bleacheries, as it was little refined, common and cheaper. As it lacked natron, the bleachery evoked in the Cair. Zen. III 59304 papyrus chose to get the necessary supply from another nome in order to avoid bankruptcy. The supply for monopole products is usually done through within the nome, meaning the administrative division. Once the product couldn't be found anymore in a nome, the laundry workers were forced to get the supplies from a surrounding nome. This suggests the fact that the geographic origin of the natron was only secondary compared to the pursuit of the action. Maybe the stocks of the surrounding nomes came from the same nitre bed. But, most probably, the quality of the natron did not have a notable influence upon cleanness degree of the washed fabric.

Figure 4. P. Cair. Zen. III 59 304 (http://ipap.csad.ox.ac.uk/4DLink4/4DACTION/IPAPwebquery?vPub=P.Cair.Zen.&vVol=3&vNum=59304).

While the natron was with no doubt privileged, we may assume that, besides its obvious efficacy, it also had a predominant role in various purification rites because of the almost sacred character given by the ancient Egyptians.

* * *

The use of the alum during the Pharaonic period is anecdotic within these uses. We may assume that the natron easily replaced it, which would explain that it was little used, despite the fact that the product was known. The Egyptian term designating it, as we mentioned before, has appeared since the New Empire. Amasis, pharaoh of the 26th dynasty (around 550 BC), gives a substantial gift of a thousand talents of alum for the construction of the temple of Delphi. It was thus really known and massively collected. Nevertheless, it seems not to have had a general daily utility.

But during the Greek and Roman eras, the use of alum generalized in Egypt, according to the papyruses of the period, probably because the same process took place in the West. The alum became more used within quality leather tanning, while the natron was mostly used for washing purposes. Nevertheless, these conclusions might be influenced by the sources we have.

Figure 5. Ostracon on the tax on the natron (Kayser 1991, 220).

We may note that the sources did not make the difference among the varieties and qualities of the products in their uses, unlike what we may see in the pharmacopeia.

In return, we may note a slow evolution of the uses, given the incontestable influence of the traditions. Thus, during the Pharaonic period, the natron was the traditional referential product, mostly because of its sacred character. Then, after the Greeks and especially the Romans arrived, the mentalities changed, and the habits, too. The natron did not seem to be used anymore on a daily basis, but it still remained prestigious in medicine, while the alum started being used in everyday life. In consequence, the cultural traditions, more than the chemical properties, influenced the use of salt.

References:

Aufrere, S. 1991. *L'Univers minéral dans la pensée égyptienne*. Cairo, IFAO.

Borgard, P. 2001. *L'Alun de l'occident romain, production et distribution des amphores de Lipari*, Thèse de doctorat, Aix-Marseille 1.

Chassinat, E. 1968. *Le Mystère d'Osiris au mois de Koiak II*. Cairo, IFAO.

Croisille, J.-M. 1985. Edition de Pline l'ancien, *Histoire Naturelle* XXXV, Paris, Les Belles Lettres.

Davies, B. G. and Toivari, J. 2000. A Letter of reproach (O. DEM 314) Corruption in the administration of the washing service at Deir el-Medina. In R. J. Demaree, A. Egberts (éd), *Deir el-Medina in the third millennium AD*, 65-77. Leiden, Nederlands institut voor het nabije Oosten.

Delamare, F. and Monasse, B. 2005. Le rôle de l'alun comme en teinture; une approche par la simulation numérique; cas de la teinture de la cellulose à l'alizarine. In *L'Alun de méditerranée*, 277-290. Colloque international, Naples 4-6 juin 2003, Lipari 7-8 juin 2003. Paris, De Boccard.

Delrue, V. and Napoli, J. 2007. L'utilisation du natron dans la pharmacopée médicale antique. In N. Morère Molinero (ed.), *Las salinas y la sal de interior en la historia: economía, medio ambiente y sociedad. Inland saltworks and salt history: economy, environment and society*, 351-375. Madrid, Dykinson.

Goyon, J.-C. 1996. Le lin et sa teinture en Egypte. In *Aspect de l'artisanat du textile dans le monde méditerranéen*, 13-25. Collection de l'institut d'archéologie et d'histoire de l'antiquité, Lyon 2, Vol. 2.

Jacob, A. 1963. Nitrum. In Ch. Daremberg, E. Saglio, *Dictionnaire des antiquités grecques et romaines*, 85-86. Akademie druck-u. Verlagsanstalt, t. IV.

Jemma-Gouzon, D. 1972. *Les tanneurs de Marrakech*, Mémoires du C.R.A.P.E. XIX.

Jequier, G. 1922. Matériaux pour servir à l'établissement d'un dictionnaire d'archéologie égyptienne. *BIFAO* 19, 106-109.

Kayser, F. 1991. Un Reçu bancaire thébain pour la taxe sur le natron. *BIFAO* 91, 219-223.

Leclerc, de Buffon 1799. *Histoire naturelle générale et particulière*, tom 10.

Leguilloux, M. 2004. *Le cuir et la pelleterie à l'époque romaine*, Paris, Errance.

Leguilloux, M. 2006. *Praesdia du désert de Bérénice*, vol. 3, *Les objets en cuir de Didymoi*, Ifao, Sodis, 10-21.

Loret, V. 1893. Le nom égyptien de l'alun. In G. Maspero, *Recueil de travaux relatifs à la philologie et à l'archéologie égyptiennes et assyriennes*, vol 15, 199-200.

Lutz, H. 1923. *Textiles and costumes among the peoples of the ancient near east.* New York, G. E. Stechert and Co.

Macheboeuf, C. 2007. Le sel et les coquillages à pourpres. In L. Lagostena, D. Bernal et A. Arevalo (eds.), *Salsas y salazones de pescado en occidente durante la antiguedad*, 387-390. Colloque international 7-9 novembre 2005, British Archaeological Reports, BAR Publishing.

Marsh-Letts G. S. 2002. *Ancient Egyptian linen, The role of natron and other salts in the preservation and conservation of archaelogical textiles. A pilot study*, University of Werstern Sydney.

Preaux, C. 1939. *L'Economie royale des Lagides*. Brussels, Editions de la Fondation Égyptologique Reine Élizabeth.

Wagner, G. 1987. Les oasis d'Egypte à l'époque grecque, romaine et byzantine daprès les documents grecs. *Bulletin d'Egypte* 100.

Part IV. Historical Approaches

Salt Production in Mediterranean Andalusia in the Transition from Late Antiquity to the Early Middle Ages

Antonio Malpica Cuello
Universidad de Granada, Spain

Abstract
The coasts of both the Andalusian Mediterranean and its neighbour North Africa are replete with archaeological remains evidencing the existence of a great deal of fishing activity and later preparation of salted fish. Important traces of these garum (fish sauce) factories have been found, but this is not the case with the Roman salt works, which must also have existed, and were probably concentrated on the same coasts. These salted fish factories seem to have largely halted production by the 4th century. The existence of fully formed salt works with obvious production capacity dating from the Middle Ages has been detected in the Mediterranean. We do not know when the Roman operations changed over to become medieval ones. There may have been a time of crisis in the former and they may have used salt wells near the sea, the various dry rivers in the Mediterranean and some marshes that also exist. Only an analysis of the archaeology, especially landscape archaeology, could clarify the issue, which is a fundamental starting point, above all now that we are in danger of losing archaeological remains and traditional landscapes.

Keywords
salt, Mediterranean Andalusia, Late Antiquity, Early Middle Ages

Salt in Mediterranean Andalusia and its trade during the Roman period and Late Antiquity
Despite the relatively feeble tides in the Mediterranean basin, there were plenty of salt marshes along the coast of the Mediterranean Sea. Due to the increasing industrialization, they have decreased in number time passing by; nowadays only a few saltworks still exist, and are located within the same areas. In the province of Almeria there were once seven salt marshes between Adra and Pulpí, but only those in the area of Cabo de Gata still exist. These saltworks are located east from the capital city extending on across a wide area that nearly reaches the centre of the national park of Cabo Gata. This situation is quite common along the Mediterranean coast of Andalusia, and as a mutter of fact other saltworks, like the very well known one in Torrenueva (a hamlet in the province of Motril) have disappeared in the 1950s.

Putting things in perspective, we realise that salt production in the Mediterranean area underwent different changes that not always can be documented in a satisfying way. It is only possible to outline a general overview that, considering our field of specialisation, will be limited to the Middle Ages, and more precisely to the early Middle Ages, taking into account the Roman period as well.

Focusing on the coast of the present province of Almeria, the work by Lorenzo Cara, Jorge Cara and Juana Mª Rodríguez should be taken into account. Firstly, they analyse a fish salting installation dating between the early 1st and the mid 3rd century AD, describing into details how similar installations along the coast were organised and related to salt production in Roman times (Cara, Cara and Rodríguez 1988). Nowadays there is no archaeological evidence left (Cara, Cara and Rodríguez 1988, 53), but thanks to this study it was possible to reach a broader understanding of fishing and salt use along the Almerian coast during the Roman period.

Fishing was an important resource in this area. There are two different fishing areas along the coast, one of which opposite Aguadulce, 10km away in the northeast. It might be possible that the inhabitants of the nearby Ribera de Algaida, or maybe those from the ancient village of Turaniana (Cara and Cara 1994) started to exploit it. The other fishing area is located between Punta Entinas and Torrenueva, 11km away in the southeast.

In both areas different kind of fishes, like tuna fish, mackerel, and other ones were fished. Sealing and mooring conditions are excellent in spring and summer time, when fishing was carried out extensively. Moreover, there was plenty of fresh water in the nearby.

Likewise, salt production was of vital importance to supply this fish salting installation; in fact, not far away from there, some saltworks already working in the Middle Ages (and probably even earlier) have been found. We are talking about the salt marshes located 3,5km away to the southeast, and known as Xata and Mudagüara (Cara, Cara and Rodríguez 1988, 65). These place names refer to the topographical position of the two sites. Xata means 'by the shore', evidently the seashore; Mudagüara means 'round', but it probably refers to the fact that this place is usually surrounded by water and located on a small hilltop. Anyway, the *Catastro* (land registry) of the marquis of the Ensenada states that there were 'three ponds close to the sea, where during the summer a lot of salt was curdled' (Cara, Cara and Rodríguez 1988, 65). We are aware of the existence of these salt marshes thanks, as well, to the important 19th century *Diccionario* (Dictionary) by Pascual Madoz (Madoz 1845-1850, s. v. Roquetas).

Thus, the relationship between *garum* production and saltworks has been previously pointed out, but it is clear that while farms and other installations left marks in the landscape and can be archaeologically studied, salt marshes have nearly disappeared leaving no evidence (Ponsich 1988). This is certainly related to the fact that, as only a few spots were suitable for making salt, most salt marshes, especially those along the Mediterranean coast, have been exploited more or less continuously. It should be taken into account that because of the feeble tides in the Mediterranean basin, seawater should be carried on the dry land in different ways. This was made possible thanks to the existence, on one hand of lagoons, on the other hand, of lowlands protected from waves by sandbanks.

Roqueta is not a unique case in this respect. The area of the province of Almeria is certainly the best-known one: there were several farms producing *garum* and at the same time there was the greatest concentration of saltworks, some of which have kept on working until recently. We are going to mention the farms identified (Cara, Cara and Rodríguez 1988, 68-69) along the Almerian coast, located from west to east as follows: Adra, Guardias Viejas, Roquetas and Ribera de Algaida, Almería, Torre García and Villaricos. All these farms, whatever their size and importance was, were related to saltworks either in the nearby or further off.

The province of Almeria is quite different from the Mediterranean coast of Andalusia, where evidence of salt marshes are hardly known[1], even if all along the coast there are remains of small pools used for making *garum*.

Further research centred on coast changes is needed to prove this relationship; so far it is possible to state that the exploitation of salt marshes was mainly located in certain spots and that salt was sent on to fisheries and to *garum* production centres. Salt moved around following fishers and people who worked in fish salting. For this reason it is not correct to state that there was a saltworks in each fishery or production centre related to fish conservation. Salt was obtained where possible and added to fish on the same spot if it was a good fishing place, if not, it was traded where needed. We know that it was a seasonal activity and that lot of people moved around because of this work.

Thanks to the work by L. Cara. J. Cara and J. M^{a}. Rodríguez, already mentioned above, we could reach a better understanding of salt exploitation and trade during the Roman period. It is worth mentioning that the three scholars show how a remarkable change occurred at that time, and already pointed out by Ponsich and Tarradell (Ponsich and Tarradell 1965, 116-117), could be applied to the province of Almeria (Cara, Cara and Rodríguez 1988, 69).

During the first phase, starting in the 1st century BC, large production centres were created in towns controlled by the Phoenicians and once founded by them. Later on, after the second quarter of the first century AD, several secondary productive structures specialised in processing one kind of fish, were established. Lastly, small productive centres that supplied the local and hinterland market with few different kinds of goods were established. These production centres were settled during the second half of the 1st century AD, and lasted for a different length of time; the one studied by the scholars from Almeria was working until the 3rd century.

Along most of the Mediterranean coast of Andalusia, activities related to *garum* production ceased around the 4th century. One of the best know and of the most important production centre 'El Majuelo', in Almuñécar (Granada), stopped working between the end of the 4th and the beginning of the 5th century, or maybe, as it has been suggested, during the second half of the 5th century (Gómez 1998, 365-367).

Anyway, it has been proved that from the 3rd century onwards, the trading system that connected all the Mediterranean area, as well as the political ideas behind it, gradually faced a crisis (Wickham 1988).

At the time when the production system described above underwent major changes, it was already affected by the crisis, or at least the crisis had forced the production centres to adjust to the new situation. To transport the *annona* meant to move ships and goods all other the lands and seas of the Roman Empire. As Michael McCormick states, 'volume and, to a certain extent, variety characterized goods shipped around the late antique Mediterranean' (McCormick 2001, 86).

The loss of Cartago (440) in Northern Africa, from then on controlled by the Vandals leaded by Genseric, meant if not the collapse, a deep crisis of this system in western Mediterranean, while in the eastern part of the Empire trade went on until immediately before the Arab conquest. Because of the decay of this system what the Arab found at their arrival was a serious crisis in the urban areas.

To make it clear we are going to quote once more Michael McCormick:

> 'In terms of communications and exchange, massive, centralized, and homogenous movements of goods between a few places on opposite shores of the Mediterranean were finished. In their stead stood what had probably

[1] We can mention only the saltworks in the province of Motril (Granada), in the hamlet of Torrenueva. This site was very important for salt production during the *nazarí* period, as well as after the Castellan conquest (Malpica, 1981). During Antiquity and in the Middle Ages there was a settlement in the nearby (Gómez, 1992). There were probably some more settlements in the Palmones River, in the surroundings of Algeciras, maybe in the mouth of the Guadiaro River. This theme is discussed in Pérez Hurtado de Mendoza 2004, 160-161) and in the article by Ojeda (2004, 160-161), about the possible presence of saltworks in the Palmones River.

always been there, operating on the margins of the state-subsidized transport system: local and not so local shippers who bought and sold many different things on a small scale, connecting a multitude of minor centers of production and consumption. The most exotic goods they had to trade probably passed though countless hands, slowing their movement and boosting their place. But that mattered little, for along the northern Mediterranean at least, demand cannot have been very strong' (McCormick 2001, 117).

As a consequence, the production of *garum* decreased considerably and was probably reduced to a small-scale one, so to satisfy local needs. Nevertheless, we cannot assume that maritime activities or fishing ceased. The fall of the Roman Empire did not mean the end of the world. Even if local aristocracy had to face a deep crisis, daily life kept on going. This means that 'exotics goods', as defined above by McCormick, were not common anymore.

Salt was not an exotic good, but something necessary, a vital resource for ancient societies. Because of the collapse of the trading and production system, big saltworks disappeared or were reduced to small production centres, but salt was still produced.

The period between the final phase of crisis of the Roman Empire and the Arab landing in Hispania, from then onwards al-Andalus, is the one we know less about concerning the organisation of production and trade. We do not know exactly how salt marshes started to be exploited again. It is not easy to reconstruct this phase and the only way to do it, is to undertake a thorough archaeological research. The increasing number of building has changed deeply the coast aspect, leading to an irreparable loss of archaeological remains. This makes it difficult to undertake a research, not only on salt production, but also on related themes, like coastal navigation (for example it is difficult to locate place for watering, anchorages and harbours) or the economic organisation of farmed lands and its relationship with different resources, like the maritime ones.

Unfortunately, the remains we aim to identify left few traces and this makes our research even more difficult.

Salt during the Early Middle Ages

There is little evidence of saltworks in al-Andalus before the 10th century. As stated above, archaeological sites show few remains and nowadays it is not easy to preserve them, so it is not possible to undertake a detailed research that could provide reliable results over this period.

Despite all these problems we can – and should – raise some questions. As far as we know, the salt (known in Arab as *milh*) used in al-Andalus was obtained from seawater or from salt water present in the hinterland. Considering their position and their interactions within the surrounding area, it is possible that the saltworks far away from the coast date to the *andalusí* period (Malpica 2008a), while the ones along the seashore already existed before. This does not mean that before the Arab arrival salt was not collected or produced in the hinterland. It is likely that salt water wells were used for this purpose. These kinds of wells have been used in some areas until recently, showing a small-scale production strictly controlled by neighbourhood. Reyna Pastor (Pastor 1963) has collected some written source from Castile and Leon dating to the early Middle Ages that prove this use since then. This point will be discussed below.

There is evidence that in al-Andalus salt was obtained from seawater in different spots along the seashore in the nearby of Cadiz, Almeria, Alicante and Ibiza (Vallvé 1980). Far away from the coast, in the environs of Zaragoza, there were rock salt mines that can be easily identified with the ones mentioned by the geographers Ibn Galib (Vallvé 1975, 377) and Ahmad al-Razi (Catalán and De Andrés 1975, 55; see also, Lévi-Provençal 1953).

As discussed above, fishing and salt production were strictly related in Antiquity. If we consider the crisis started in the 3rd century, reaching later on a point of no return, it should be beard in mind that it had consequences on the economic system and on the regional structure. These changes are quite evident if we analyse, even if not into details, life along the coast.

Sailing changed and long distance trade nearly disappeared. It has already been underlined how McCormick (McCormick 2001, 117) has clearly pointed out in his work that in the Mediterranean sea trade went on without interruption, even if on a smaller scale. Probably he is right, but a thorough study of the Andalusian coast based on archaeological evidence is lacking[2] and the limits of written evidence dating to the 8th-10th centuries are well known.

It seams like life along the coast underwent some changes. Sailing changed and was then on a smaller scale than at the time of the Roman Empire, when large amount of goods were regularly traded across the Mediterranean because of the *annona*. Sea trade continued on a different scale. The Arab invasion, or better, their arrival in Hispania shows this clearly. After the first landing in 711 other landings or – depending on the point of view – arrivals of settlers, followed. 'Abd al-Rahman al-Dajil, known as the Immigrate, and first Omayyad of al-Andalus, clearly states in his account that the route between northern Africa and the Iberian Peninsula was covered on a regular base. As analysed by Virgilio Martínez Enamorado in his work on the landing place of the man who was to become Emir, Arab written sources state this quite clearly (Martínez 2006).

[2] It is worth quoting some works that cannot be included in the Andalusian reference frame, but are nevertheless important reference points. One of these is Azuar Ruiz (2004).

Activities along the coast continued on a local basis: fishing and to a certain extent salt production left historical and archaeological evidence. These activities did not disappear, even if they were undertaken on a smaller scale, and a smaller amount of goods were traded. Of course it is not possible to calculate the total amount of the production, as it is not for earlier periods, when it took place on a larger scale. Thus, we have to focus our research on different kind of data, which are not quantifiable.

It is important to point out that some events mentioned in written source are, as well, documented by archaeological evidence. All these data should be discussed in regional context, which is the best way to explain and present them within a more general historical debate.

This is the case of the sailors from Pechina, who founded this village located north from modern Almeria, in the nearby of the Andarax river. They founded settlements on both the southern and northern side of the Mediterranean or at least were settled in there.

From the 8th century onwards, but mainly during the 9th century, there is evidence of regular landing from northern Africa to Mediterranean Andalusia in spots far away from Gibraltar.

These journeys tell a lot not only about the ability needed for undertaking them, but also about the skills in sailing, the good knowledge of the coast and of the regional organisation of this area.

Moreover, boats with a different structure than those normally used until then made this kind of sailing possible. As these boats had a shallow draught, they could go extremely close to the coast and could be easily beached on the shore. They were extremely manageable and approximately four people were needed to steer and dock them. They measured only between 12 and 15m in length, served different purposes and could be used for costal as well as for open sea fishing. Furthermore, these boats could move along accessible coast with a few cove and anchoring grounds, carrying goods as well as a medium number of passengers. With or without making stopover, their effectiveness lied behind the use of the Latin sail that made possible to make the most of the Mediterranean winds. Nevertheless, a deep knowledge of the coasts was needed in order, if necessary, to know where to collect water, to be able to find every single ravine where to hide, to chose in which cove or wharf take shelter from a strong wind depending on its direction. As stated above, only a small amount of goods and a few people could be carried on this kind of boats, but it was possible to go around this problem by sharing load between several boats.

These boats could be used for several activities, among which fishing and trading salted fish, as for example the sailors from Pechina did; indeed Lorenzo Cara (Cara 1998) has shown how this hypothesis is very close to reality.

These boats sailed constantly and regularly between the two shores of the Mediterranean Sea, carrying out different and sometime complementary activities. Moreover, they stayed on each shore at different seasons. Al-Bakri, writing in the 11th century, but describing previous facts, said: 'the sailors we talked about used to leave al-Andalus to spend the winter in the harbour of Ténès' (Bakri 1913, 128).

This text is extremely clear. They spent the winter in the Algerian harbour, while they stayed in al-Andalus when the weather was better in there. It can be assumed that in both sites they carried out economic activities that could be either related to production or to trade. Obviously, there was not a strict division between these two kinds of activities, but they show seasonal changes related to particular occasions.

These seasonal movements might be related to fishing migratory species moving from one side to another of the Mediterranean, like tunas that were caught using special nets known as tunny-fishery. Their constant journeys followed tunas and other fishes. Moreover, it might be possible that they produced salt during the summer, so to trade salted fish on the other shore of the Mediterranean and eat or sell it during the winter while getting other goods. Thus, it is possible to recognise small unloading spots from where it was easy to get access both to the hinterland and to maritime resources like salt.

Economy was centred on a variety of different productions that could rely on trade and, among different things, on salt trade. Salt production kept been carried on as it was necessary for fishing as well as for livestock, an activity we hardly know anything about, even though the majority of salt marshes exploited in a later phase, from the 10th century onwards, were related to local seasonal migration of livestock (Cara and Rodríguez 1989).

As salt resources were not fully exploited until then, they started to be used to advantage. We know that the production of salted fish ceased around the 4th and/or the 5th century, not being any longer the major cause of salt consumption. As a consequence, salt production fell to minimum levels and the extension of big saltworks was considerably reduced. From this period onwards, large amounts of salt were not necessary any longer. During the Roman period small saltworks were established while the big ones were still working. This does not mean that the first ones survived, while the other ones disappeared. We believe that priority was given to produce salt on a regular basis, but without investing too much into it.

It is important to underline what Reyna Pastor has pointed out concerning salt production in Castile and Leon during the early Middle Ages (Pastor 1963). This activity was mainly controlled my local farmers, who produced, as well, most of the salt in al-Andalus.

Probably at that time salt exploitation was not as complex as in big saltworks. It was possible to use salt ponds and salt marshes, but also wells that reached deposits of underground salt water from which enough salt for peasants' villages could be obtained. Problems arose when natural resources slowly started to be controlled and exploited by feudal lords. Ladero underlines this too, showing that:

> 'From the 10th century to the reign of Alfonso the 7th [12th century] small saltworks' owners gradually disappeared, leaving space to bigger ones, among which the sovereigns themselves can be listed. Moreover, a royal tax known as *alvará* was introduced, implying more generally that royal privileges were granted' (Ladero 1987, 822).

Certainly until the 10th century, when the Omayyad State was established in al-Andalus, the caliphs did not control natural resources, salt included. As a consequence written sources, most of which refer to the centralised fiscal registry in Cordoba, mention salt only from that time onwards.

In these documents clear references to salt are lacking: as it is mentioned only exceptionally, we can suppose that its exploitation was controlled by the state. Documents show that probably only salt mines were used as resources, while the establishment of saltworks next to salt marshes might have started as a process of enlivening.

Salt production was deeply related to the needs of each local community living next to this resource. As it shows up clearly in the following centuries, from the 11th century onwards, due to the increasing importance of towns, it was associated to trade (Malpica 1997).

Thus, it is necessary to define this research as related to the evidence of salt marshes exploitation dating to the first period of al-Andalus, with special reference to Mediterranean Andalusia, the area on which our research is centred.

Firstly, the presence of feebly tides in the Mediterranean Sea made our research easier, as this make it more difficult for salt water to reach the hinterland. This reduce the number of spots where it was possible to produce salt, but at the same time it identify a limited number of place where to start this activity, making it lasting in the same place throughout time. In other words, it can be said that if we have to find where, at different times, this resource was exploited, we have firstly to search saltworks where it is morphologically possible to establish them. Why we have to do this 'firstly' will be explained below. Despite all the changes that this production underwent through time, salt can also be made in extremely simple ways, so simple that normally they are not taken into account.

Concerning this point we cannot avoid quoting a scholar who summarises the advantages and problems in producing salt along the Mediterranean coast. It is worth reading the following passage:

> 'From the Rock of Gibraltar to the boundary of the province of Murcia the Mediterranean coast of Andalusia shows natural features that are slightly different from Atlantis shore: the mild climate and the average salinity, which is higher in the Mediterranean than in the Atlantic, prove to be extremely favourable to detect and exploit salt marshes. However, tides are extremely feeble in the Mediterranean Sea and changes of water depth at high and low tide seldom reach one meter (microtidal sea). Moreover, the Betic Mountain Chain is quite close to the coast and the rivers that drain off water from there, reaching this part of the coast, are small ones, with a marked longitudinal slope and torrent-like hydrological regime that characterise mainly the eastern area (summer drought, sudden rise, etc.). For all these reasons it is not easy to find a large estuary in this region, and extended level lands that could be naturally flooded by the sea are associated to the deltas of the major rivers and to the presence of lagoons formed by the sea where sediments accumulate in large amount because of river-sea interaction. The situation described above is the current one, but throughout the last few millenniums the coastal zone underwent relevant changes. Therefore, its paleogeographic evolution can throw new light on the location of spots where salt marshes were exploited in the past and that are nowadays far away from the present coastal profile' (Ojeda 2004, 161).

Paleogeographic research on the Andalusian coastal profile has shown that relevant changes occurred, but most of them, especially those on the Mediterranean coast, are quite recent ones (Hoffmann 1988).

Moreover, it should be taken into account that alluvial land around deltas like the ones of the Guadalfeo River (Malpica 1996) or of the Adra River (Sermet 1950), was farmed and drained as much as possible, but the flow of salty and fresh water was preserved.

It was more convenient to start a saltwork next to sandbanks or to lagoons.

The research undertaken on the geomorphologic formation of the Mediterranean coast, with special reference to saltworks and to their location, tells a lot about this (Ojeda 2004).

It is worth taking it into account for our research, but we are going to summarise the conclusions only.

We are going to focus on two areas where salt production was very important along the Andalusian coast of the Mediterranean: the countryside in the environs of Dalías and Cabo de Gata. Nevertheless, we are going to take into account as well the eastern extremity of the province of

Almeria, in the extreme east of Andalusia (Pulpí), in particular San Juan de los Terreros.

Dalias is the best place where to start salt production without facing many problems; in fact, there are depressions close to the seashore. On the inner side, a cliff dating 6000 years ago marked the boundary between the dry land and the sea, outlining the shore. This reef made possible the formation of sand deposits, thanks to interactions occurring in the costal zone and resulting into the deposition of detritus. As a consequence small inlets and island formed, and thanks to the gentle slope of the continental platform, seawater could flood the area between these islands and the dry land. These areas could be overflow by the sea, or due to percolation, so that salty lagoons formed. The interactions on the coastal zone went on, so to let seawater reaching these depressions it was necessary to create channel and sometime underground water channels; lately, pumping systems have been used as well (Ojeda 2004, 166-167).

In Cabo de Gata (Cape of Gata) the coastal platform is broader and the coastal interaction make possible the deposition of debris originated in the bay of Almeria, where the Andarax River forms a delta. In this cape the debris form a natural barrier, creating a saltwater lagoon. These lands were not particularly fertile and could be regarded as marginal. We are talking about lands close to the area of coastal interaction, where debris has been accumulated by wind as well; in other areas these lands have been farmed but occasionally and with low production[3]. The low fertility of this kind of soil, as well as the weather conditions with high temperature and little rain, made it more reasonable to use this area as a salt resource close to the seashore of the Mediterranean Sea (Ojeda 2004, 167).

It is clear that things have been changing time passing by, and that recently it has been necessary to pump seawater in the hinterland to keep on working the last salworks still existing in Andalusia.

It is necessary to mention the saltworks of San Juan de los Terreros, for which we have no evidence in Ancient and Medieval times, but they have been working until a few decades ago. Its location is related to the presence of marshes or wetlands next to the watercourse named 'los Pérez', which was an ancient creek that was gradually separated from the sea by natural barriers created by debris.

It can be assumed that these lagoons, containing enough salt water, could be easily used to obtain salt, especially if in relatively small amounts. Salt could be obtained with little technical knowledge, for example by lifting water to a higher level and collecting it into ponds or small pools. This technique was used in small scale in the countryside, where saltwater well where used. The works published at different times on Montejícar about the so-called stream Salado, turns to be useful in this case to explain what we wish do describe (Malpica 2008b). Thanks to a pulley, water was drawn out from a rock lined well, and then poured into a small circular deposit or into a pond to make brine; finally, it was collected into small pools to make it crystallise. This structure need a few space and can be used for salt supply.

It might be suggested that this process was used in other spots along the coast, in the mouth of some watercourse, getting into some aquifers enough unstable so to make it possible seawater intrusion that is salty. For a limited production of salt, like the one needed for limited needs, it was not necessary a large amount of work and/or of money.

Lorenzo Cara, analysing sea fare in the offshore of Almeria, points out that in that area it is not easy to find fresh water wells as water is usually salty in there (Cara 2005, 126). This shows clearly that small amount of salt could be obtained from wells. Salinity is a measure of salt concentration in water and is the ratio between freshwater and saltwater aquifer. When the salinity level decreases, salinity intrusion is the immediate consequence and is not easily reversible.

Thus, it is likely that salt production continued in the sites mentioned above, without disappearing completely.

Therefore, as Reyna Pastor pointed out for Castile and Leon (Pastor 1963), and as already mentioned above, before the 11th century salt production was part of a basic economy related to a rural society that exploited as well maritime resources, among which salt production.

Thus, further research should be carried out on the points we have underlined above, focusing on specific areas that should be studied into details before evidence disappear or the remains of these structure would be unrecognisable, once transformed for good.

Acknowledgements
We would like to acknowledge the Research's Projet: *Análisis de los paisajes históricos: de al-Andalus a la sociedad castellana*, Consejería de Innovación, Junta de Andalucía (Spain) for their support.

References:

Al-Bakri. 1913. *Description de l'Afrique Septentrionale*. Translation De Slane, M. G., Alger. A. Jourdan.

Azuar Ruiz, Rafael (ed.) 2004. *Fouilles de la Rábita de Guardamar*. Madrid, Casa de Velásquez.

Cara Barrionuevo, L., Cara Rodríguez, Jorge and Rodríguez López Juana M^{a}. 1988. Las Cuevas de la

[3] An extreme example of this is what we have seen in a hamlet of Tjønuvík in the northeast of the island of Estremoy, in the Faroes Islands. Nevertheless, this should be regarded as something exceptional that might have not occurred in the area we are talking about.

Reserva (Roquetas) y otras factorías pesqueras de época romana en la provincia de Almería. *Boletín del Instituto de Estudios Almerienses* 8, 53-72.

Cara Barrionuevo, L. and Rodríguez López, Juana Mª. 1989. El ámbito económico del pastoralismo andalusí: grandes aljibes ganaderos en la provincia de Almería. In *El Agua en zonas áridas. I Coloquio de Historia y Medio Físico*, Almería, 631-653.

Cara Barrionuevo, L. and Cara Rodríguez, J. 1994. *Roquetas de Mar, arqueología e historia: desde la prehistoria hasta inicios de la Edad Moderna*, Almería, Instituto de Estudios Almerienses.

Cara Barrionuevo, L. 1998. *La ciudad de Pechina y su territorio, siglos IX y X*, Unpublished PhD dissertation, Granada.

Cara Barrionuevo, L. 2005. Ciudades portuarias, alquerías y comercio en el Sudeste peninsular. El ejemplo de Almería en la Edad Media. *Arqueología medieval* 9, 125-137.

Catalán, Diego and De Andrés, Mª Soledad. 1975. *Crónica del moro Rasis*, Madrid, Seminario Menéndez Pidal / Gredos.

Gómez Becerra, A. 1992. *El Maraute (Motril): un asentamiento medieval en la costa de Granada.* Motril, Ayuntamiento.

Gómez Becerra, A. 1998. *El poblamiento altomedieval en la costa de Granada.* Granada, Tharg.

Hoffmann, G. 1988. *Holozanstratigraphie und Kustenlinienverlagerung an der Andalusischen Mittelmeerkuste.* Bremen, Fachbereich Geo wissenschaften der Universität Bremen.

Ladero Quesada, M. Á. 1987. La renta de la sal en la Corona de Castilla (siglo XIII-XVI). In *Homenaje al profesor Juan Torres Fontes* vol. 1, 821-837. Murcia, Universidad de Murcia.

Lévi-Provençal, É. 1953. La 'Description de l'Espagne' d'Ahmad al-Razi. *Al-Andalus* XVIII, 51-108.

Madoz, P. 1845-1850. *Diccionario geográfico-estadístico-histórico de España y sus posesiones de Ultramar.* Madrid, Establecimiento literario-tipográfico de P. Madoz y L. Sagasti.

Malpica Cuello, A. 1981. Las salinas de Motril. (Aportación al estudio de la economía salinera del reino de Granada a raíz de su conquista). *Baetica* 4, 147-165.

Malpica Cuello, A. 1996. *Medio físico y poblamiento en el delta del Guadalfeo. Salobreña y su territorio en época medieval.* Granada, Universidad de Granada.

Malpica Cuello, A. 1997. El tráfico comercial de la sal en el reino de Granada en época medieval. In S. Pira (ed.), *Storia del commercio del sale tra Mediterraneo e Atlantico,* 83-112. Cagliari, AM&D Edizioni.

Malpica Cuello, A. 2008a. Techniques et aménagements des salines médiévales de l'intérieur d'Andalousie. In O. Weller, A. Dufraisse and P. Pétrequin (eds.) *Sel, eau et forêt: hier et aujourd'hui,* 433-450. Cahiers de la MSH Ledoux 12 (coll. Homme et environnement, 1). Besançon, Presses Universitaires de Franche-Comté.

Malpica Cuello, A. 2008b. El medio físico y la producción de sal. Propuesta para el análisis de las salinas granadinas desde una perspectiva arqueológica. In J. M. Martín Civantos (ed.), *Medio ambiente y arqueología medieval,* 145-161. Granada, Editorial Alhulia.

Martínez Enamorado, V. 2006. Y al-Dajil arribó a al-Andalus... En torno al desembarco de 'Abd al-Rahman I en la playa de Burriana/Bitruh Riyana. *Al-Qantara* XXVII, 199-210.

McCormick, M. 2001. *Origins of the European Economy. Communication and Commerce, AD 300-900.* Cambridge, Cambridge University Press.

Ojeda Zújar, J. 2004. Salinas marítimas: marco geográfico y ambiental. In *Salinas de Andalucía*, 147-168. Sevilla. Consejería de Medio Ambiente de la Junta de Andalucía.

Pastor, R. 1963. La sal en Castilla y León. Un problema de la alimentación y del trabajo y una política fiscal. *Cuadernos de Historia de España* XXXVII-XXXVIII, 42-87.

Pérez Hurtado de Mendoza, A. (Coord.), *Salinas de Andalucía.* 2004. Sevilla, Consejería de Medio Ambiente de la Junta de Andalucía.

Ponsich, M. and Taradell, M. 1965. *Garum et industries antiques de salaison dans la Méditerranée occidentale*, Paris, Presses Universitaire de France.

Ponsich, M. 1988. *Aceite de oliva y salazones de pescado: factores geo-economicos de Bética y Tingitania.* Madrid, Universidad Complutense de Madrid.

Sermet, J. 1950. La costa mediterránea de Málaga a Almería. *Estudios Geográficos*, 4 (1943), 15-29, y «La vega de Adra», *Estudios Geográficos* 11, 695-710.

Vallvé, J. 1975. Una descripción de España de Ibn Galib. *Anuario de Filología de la Universidad de Barcelona* I, 369-384.

Vallvé, J. 1980. La industria en al-Andalus. *Al-Qantara* 1, 209-241.

Wickham, C. 1988. L'Italia e l'Alto Medioevo. *Archeologia Medievale* XV, 169-206.

Land Organisation and Salt Production in Region of the Salado River (Sigüenza, Province of Guadalajara, Spain): Ancient and Medieval Times. Results of the First Campaign 2008

Antonio Malpica Cuello
Universidad de Granada, Spain

Nuria Morère Molinero
Universidad Rey Juan Carlos. Madrid, Spain

Adela Fábregas García
Universidad de Granada, Spain

Jesús Jiménez Guijarro
Universidad Rey Juan Carlos. Madrid, Spain

Abstract
In this paper, the provisional results of the first year of this research project are being presented after having completed the archaeological and laboratory work and corresponding documentary analyses. Through this, we could determine a whole process, starting from the simple use of salt in prehistoric times to a closer relation between salt as a resource and including the habitat during the Iron Age and the control of this resource within the comprehensive rule of the territory under the Roman Empire. During the Middle Ages, this relation persisted in form of small settlements and large defensive centres, and it still did during the Christian repopulation of the whole territory, when the king took hold of salt production.

Keywords
land organization, salt production, ancient and medieval times

Introduction
In recent years, the use of salt since prehistoric times has become one of the most researched topics. Salt is important because herbivores experience a deficit in salt due to natural causes, sweating, etc. (Meyer 1982). In prehistoric times, human beings could reduce their own deficit by eating the meat of game, but this deficit increased when they started domesticating animals and plants, not only because of an increased proportion of vegetables in their diet, but because they started boiling and roasting meat.

As a consequence, it was in the Neolithic period when humans needed to increase their salt intake and therefore started to extract and process it. This has been confirmed by archaeological findings on sites near salt brooks in Europe (Weller *et al.*, 2008).

After this long period, there is evidence to show that salt played a primary role in the first urban societies and in the first states, for instance Mesopotamia and Egypt Carusi, 2007a), the ancient Greek cities, and Rome (Carusi, 2007b; Morère, 2007), which started to manage salt extraction –creating the concept of *saltworks*–, salt transportation, and trade. Since ancient Greek times, but especially under Carthaginian and Roman rule, more and more salt extraction sites appeared on the coast and large-scale production soared due to increased demand from the thriving salting industry.

However, contrary to the widespread notion that coastal saltworks greatly conditioned both production and trade in ancient times, Plinius provides precise descriptions that the salt production sites in land were equally as important in spite of being smaller (Plinius, *HN*, XXXI, 73-105).

Research on salt production during the Middle Ages has improved greatly due to documentary analysis from this period and has given birth to salt historiography. Research on the Antiquity has also developed in recent years, but these efforts have not crystallised into integral programs for the study of territories, neither with regard to these periods nor to prehistoric times. For this reason, we decided to start a research line on salt production in relation to the genesis and the diachronic evolution of the territory. After completing the first year of our research, we present some initial results in this paper.

Research project
Our research project was submitted to the Spanish Ministry of Culture, where it was awarded a R+D grant with code HUM2007-66118. The title of the project is *Land organisation and salt use from late ancient times until the emergence of feudal society in the region around the Sistema Central: The Provinces of Madrid and Guadalajara*[1]. The study was based on the strategic reconnaissance of the territory and took into account the evolution of salt use as well as the diachronic formation of saltworks. The methodology we used was based on

[1] *Organización del territorio y explotación de la sal desde la Tardía Antigüedad a la formación de la sociedad feudal en el área del Sistema Central: provincias de Madrid y Guadalajara.*

the principles of landscape archaeology (Malpica 2007), and our project was divided into three phases which we will describe in this paper.

The area studied

One of the two primary variables that determine any historiographical research is space. Our activities were developed in one of the most important salt extraction areas in the interior of the Iberian peninsula, namely the basin of the Salado river, near the town of Sigüenza, in the province of Guadalajara, located on the fringes of the Sistema Central (Figure 1). In fact, the Salado river is part of one of the most important river systems in Spain. It flows from North to South into the Henares river, which is a tributary of the Tagus. The Tagus flows westwards thanks to the slope of the Peninsula towards the Atlantic ocean. Its source is next to La Laguna, and the river flows through terrain formed by marl, plaster, and clay, the last remains of the ancient sea that once covered the interior of the Peninsula during the Late Triassic period. The high salt concentration in the marshes of La Alcarria, is intermittent and heterogeneous and conditions the halophile vegetation, which is small in size and not very diverse in species, as well as the fertility of the soil. However, it does not hinder agricultural production. Presently, the lands are made up of a meadow used for growing cereals and the upper grazing peneplanes. Moreover, this area is an important stockbreeding region, which is shown by the presence of paths and sheep tracks. All of this must have influenced the ancient settlement patterns.

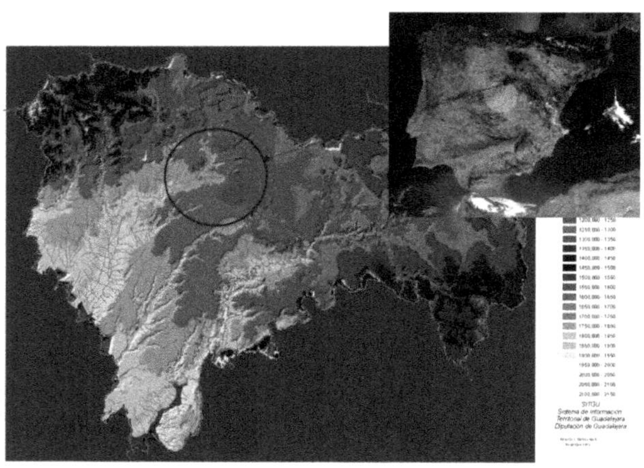

Figure 1. Overall view of the area of our study. Basin of the Salado river (Sigüenza, province of Guadalajara).

For our study, we chose the stretch from the river source (next to La Laguna) to the point where it flows into the Henares river, in the town of Baides, after flowing through the rough karst terrain of the Santamera region, where the massive limestone from the Tertiary which sheltered some of the oldest human settlements in the valley surfaces (Valiente Malla 1984). In addition, we decided to mark a whole area that included also the main tributary currents of the Salado river on both banks, some of which are salty, because we needed to configure the territory not only with regard to the use of the salt marshes, but also to the fresh water sources that make settlements viable in the long term. In view of this, the sector considered during the archaeological works of the year 2008 was located in the municipality of Sigüenza, in the province of Guadalajara, in 17 of its districts, as well as in other municipalities, like La Olmeda, Paredes, and Baides itself, as the final location.

The Project will be carried out in 3 phases. The first phase began in 2008 and is almost completed. In this phase, developed a prospection method and drew up a systematic prospection programme, based on the total coverage models and the postulates of the Site Catchment Analysis (Jarman *et al.* mainly aimed at 1972) finding the largest possible amount of archaeological entities or sites, as well as of dispersed elements, which cannot be ascribed to a specific site, but are related to various cultural and economic activities and are therefore useful for carrying out the integral study of a territory. The aim of the first phase was to obtain thorough knowledge about the historical development of an area. This approach is related to Zedeño's notion (1997) and it makes it possible to study the territory in terms of an economic factor of high importance, i.e. salt. In this phase, the written documents kept in the Diocesan Archive of the Cathedral of Sigüenza[2] have been analysed in their entirety and are currently being interpreted.

The second phase of our Project started in 2009. It involves the continuation of prospections and isolated archaeological interventions, which are aimed at improving our understanding of salty marshes as an archaeological entity. The knowledge available is admittedly poor, as it is limited to the results of the study of the lakeside area of Villafáfila, in the province of Zamora (Delibes *et al.*, 2007a; Delibes *et al.*, 2007b). The last phase of our Project (deadline 2010) will consist of a general data analysis, drawing conclusions, and finally a presentation of the results, therefore complying with the research-outreach cycle expected from any scientific activity.

Earlier research

In spite of the shortage of archaeological studies available, the territory we have studied is far from having been neglected by fellow researchers. Indeed, the area around Sigüenza has always attracted historians, archaeologists and scholars. As far back as ancient times, they tried to trace back Sigüenza to ancient *Segontia*. Whether this fact is true or not (Abascal Palazón 1986; Morère Molinero 2007), the surroundings of Sigüenza have always been of archaeological interest. It is in this context where the work of Enrique de Aguilera y Gamboa, Marquis of Cerralbo (1916a, 1916b) took place, focusing on finding and excavating the Celtiberian necropoleis of La Olmeda, Carabias, El Atance, Valdenovillos and other sites (Valiente Malla 1997). At a later time, when these findings where analysed again in a systematic

[2] We would like to thank Father Felipe Reces y Rata, archivist of the Cathedral and a great expert on the Cathedral's treasures, for his friendly and generous help.

way, the territory around Sigüenza was rediscovered (Cerdeño 1981). Further archaeological work took place in El Perical and Sigüenza (Cerdeño 1976), new archaeological sites of interest like the Roman *villa* of La Torrecilla were discovered (Sánchez Lafuente 1982) and the territory was studied globally for the first time (Morère 1983). This study lead to the first historical research on salt production in Sigüenza (Morère 1991) and, years later, to the first International Salt Conference, which was held in Sigüenza (Morère 2007). This conference laid the foundation for our current Project, as it highlighted the great potential of the region around Sigüenza for salt production studies.

The first phase: Prospections (2008)

The first phase of our Project, which is ongoing, started with the definition of prospection units in the area. To this purpose, we analysed the basin of the main river (the Salado river), determined subsidiary areas (streams) and search for all kinds of salt sources systematically: brooks, salty terrain, water courses, wells, etc. From June 2008, in order to improve our understanding of the social and economic importance of this area, sites of interest were determined and a preliminar diachronical analysis of the development of the use of the territory was carried out.

This phase was intended to set out the foundations of the following phases. At the same time, drawing up a provisional map of all the archaeological sites involved will allow us to analyse population density and its relation or not, as the case may be, with salt sources throughout different historical and cultural periods.

Salt sources

In our first approach, we were able to find an important group of salt sources irregularly dispersed throughout the area (Figure 2). The salt emergences can be categorised into:

1. Salty brooks and emergences, like in El Salobral.
2. Endorrheic lakes with a large salt concentration, as in La Laguna and in La Fuente de las Praderas.
3. Salty spaces that have been anthropised, that is, salt production sites based on artesian wells, boiling containers and salt pans from which salt was extracted by evaporation, like Santamera, Riba de Santiuste, Imón and La Olmeda. All of these sites are related to the *salinas históricas*, which have been studied extensively (García Grinda 1983; Cruz García 1989).

There is evidence of different patterns in how the area was used with regard to salt storehouses and to the regulation of settlements according to cultural and chronological factors.

This allows us to put forward our first hypothesis pertaining to the development of the use of this salty basin. Although our hypothesis reflects only the first impressions derived from the preliminary analysis of archaeological findings, a diachronic process of salt use can be determined in consideration with different sequences of events that become apparent in this territory.

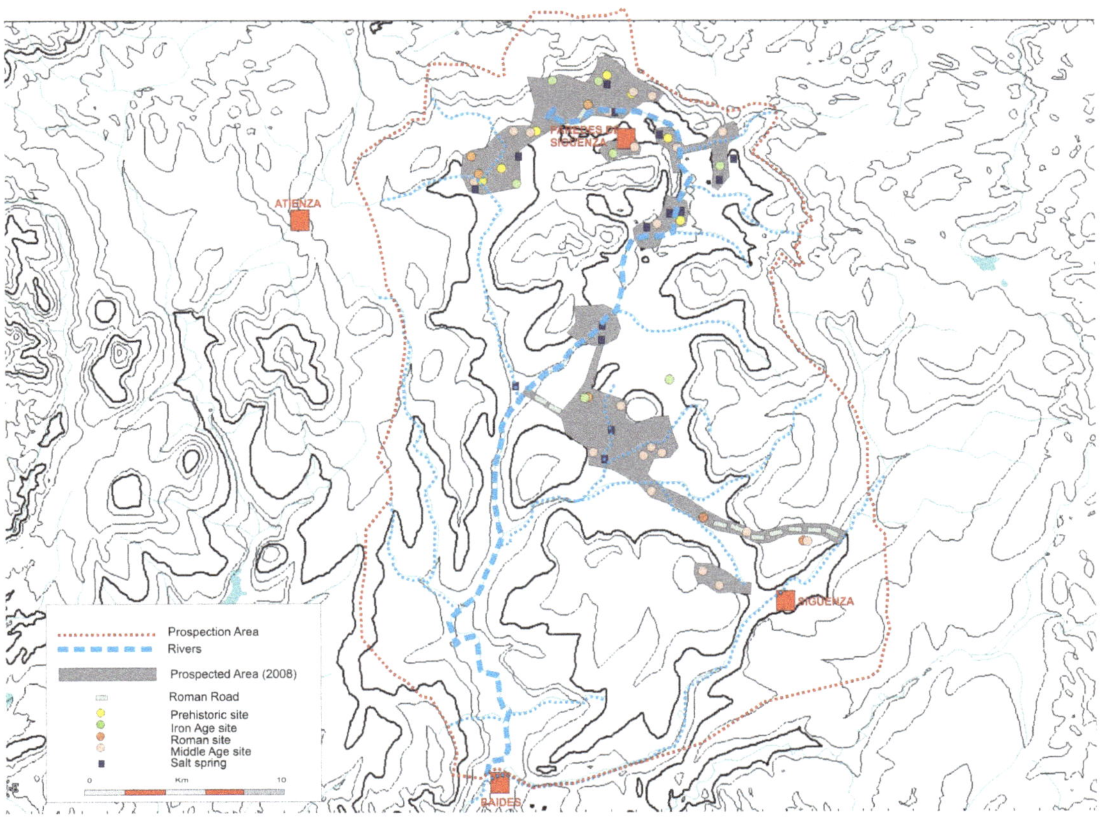

Figure 2. Working area. The areas of prospection during the first phase of the Project (2008), as well as archaeological sites and salt sources are marked.

Materials were found that can be ascribed to settlements from the Calcolithic period and the Bronze Age, in particular from the middle and final Bronze Age. These settlements are dispersed in the surroundings of salt sources, especially around sources of type 1 and 2, although not necessarily located next to them. There have been many settlements concentrated around La Laguna since the Calcolithic period and the late Bronze Age, but to date there are remarkably few findings from the Neolithic. During this period, settlements might have been seasonal, small and dispersed, often located on high areas, at about 1,000-1,100 metres over the sea level and always near fresh water sources. These settlements suggest occasional use of salt for stockbreeding. In fact, the most frequently encountered archaeological entities may be related to patterns conventionally thought to coincide with the breeding of cattle with a certain mobility (like in the groups Cogeces and Cogotas I), with some presence of elements from the Beaker culture. So far, only remains have been found from the Ciempozuelos horizon, which is characterised by incised pottery. Other materials, especially lithic remains, presumably from the time before the Neolithic, have been found dispersed and do not seem to be related to salt sources. All of this suggests that occupation of the territory followed a pattern of integral use of the environment. Strategies may have been specialised on hunting in the marshes around the salty lakes, which were frequently visited by all kinds of animals, especially large-sized herbivores. For these animals, licking the salty mud was the easiest way to satisfy their daily need of salt (Jiménez Guijarro, 2007). Moreover, to date no evidence of *briquetage* has been found. This suggests that salt might have been produced by evaporation and by precipitation of mud obtained from the edges of salty marshes and salt sources.

Several *oppida* from protohistoric times have been found. These are enclaves on higher ground with a high strategic value, since they provided a good view over the valley or the junction of valleys. These findings have allowed us to identify more and more settlements and draw a map of the territory during the Iron Age depicting the settlements, visual elements and dissuasive and defensive elements. Until now, only the *castrum* (hill fort) of the Salido river had been studied extensively (Talavera Costa, 2007). These new *oppida* were impregnable and fit well into the defensive architecture of the region, shown in the Salido river area or in Los Castillejos de Pelegrina (Talavera Costa, 2005). The villages were equipped with a complex wall system towards the most vulnerable hillsides, made up of several circles and enclosing several rooms. The inner walls and the highest point of the hill were equipped with small towers. The entrance to the village (Salido river) and the approach path (Los Castillejos de Pelegrina) were built with huge stone boulders roughly fitted together (Cyclopean masonry). Studying the areas of Paredes de Sigüenza, Val del Cubo, La Olmeda, Ures and El Atance has facilitated mapping out the strategic settlements. These settlements can be ascribed to the Second Iron Age, since large amounts of Celtiberian wheel-thrown pottery have been found. For now, it cannot be asserted that it corresponds to the First Iron Age, given that the fragments of large hand-made containers could not be ascribed to any culture. Even though the containers appear to be very ancient, we would like to put forward a hypothesis about their function. In fact, we determined an almost direct relation between the raised settlements (*castra* and *oppida*) and the existence of salt sources (mostly of type 1), since there was either a salt brook, sometimes a salt river or a stream in the surroundings or it was easily accessed from the settlement. This apparent physical and visual connection shows that brine was indeed used at the time. But how was brine actually used? There is little evidence of salt use in the Celtiberian period, and the evidence that is found refers to igneous salt (Morère, 2007). This suggests that, at least initially, salt water and mud were captured but that this resource was tapped only very sparsely. Again, there is no evidence of *briquetage* and this fact becomes more and more apparent, so this could suggest that the strategies applied for salt production and use must have been different than those determined in most of Central and Eastern Europe (Hees 2002), although this should be taken with caution until there is more data available. At the same time, the presence of thick-walled pottery from medium to large size suggests that they might have been used to wash salt clay and to purify salt. However, this probably did not happen next to the salt brook, but within the *castrum*. Nevertheless, some of the evidence we collected suggests a previously undocumented method of salt extraction by evaporation, which consisted in digging canals and bowls into the rock, and which may have been used during the Iron Age.

The notion that open-air wells were used for salt extraction in the interior of the country in Hispania, which was described by Plinius, may not correspond to the Celtiberian region, it may refer to a recent (maybe contemporaneous) description of the first Roman occupation of Hispania and of both the reorganisation of the administration and the use of this territory (Morère Molinero, in press), during the 2nd and 1st centuries BC, when Roman production methods started to take hold.

While we have a certain knowledge about the settlement pattern in Roman times in the Henares valley, we do not know much about the basin of the Salado river, since no archaeological prospections have been carried out to date. However, it can be ascertained that most settlements on high ground from the Celtiberian period were abandoned. Only the *oppidum* of La Olmeda remained inhabited from the Iron Age until the end of Classical times. Its strategic location and the fact that a Roman road went through the plain kept it as a defensive and control outpost. Even though there is no evidence so far of dense population of this area in Roman times, the main road from Sigüenza to Tiermes, a town in the province of Soria, has indeed been identified and it had been outlined in earlier research (Arias, 1989). However, a number of subsidiary roads (*diverticula*) have also been found in this area, making up a complex infrastructure which was integrated into the road system of the Empire. Considering the structure of the Imperial administration, these roads must have covered the whole region. The existence of such a system would prove that the land was used intensively and in a specialised way for fiscal or trading purposes (ensured or controlled trade), by tapping the main resources of the area, in particular the salt sources, which can be found everywhere in the region. The dense road system would have facilitated the control of the area and of the salt extraction sites, which were dispersed and were documented by us in the surroundings of the *villae* located on the plain. The main road deserves the name *via salaria* because of its specialised use, and it is highly probable that production took place following a model based on large farms devoted to extensive stockbreeding and some agricultural activity. One of these farms, which has been called Palazuelos, must have been especially wealthy, given the differentiation of the materials on its surface (Sánchez Lafuente, 1982). Therefore, it might have had a higher rank among other farms, playing an administrative, fiscal and productive role in a large territory, namely in the whole salt production site between Carabias and La Olmeda, an area which nowadays is highly anthropised. Other settlements on the plain, which are also close to salt brooks and wells, like the brook in Paredes de Sigüenza, may have played a rather local role. This settlement and control pattern (road, settlement on the plain as the power and spatial planning centre), is completed by way of reoccupation and use of some old Celtiberian *castra*, but this time for more powerful functions: road control, outposts, garrisons, etc.

As far as the late Antiquity is concerned, we have not found any evidence that could be directly related to new Visigothic settlements as such, but certain formal traditions can be traced back in some chapels and churches of the area. These are cave habitats whose prototype could be Torre Morenglos, near the village of Tordelrábano (Daza Pardo, 2005-2006). We believe that some the Christian archaeological elements that are more difficult to ascribe could have their origin in this period. This would mean that the settlements would have been relocated from the meadow, a typical habitat in Roman times, to a higher point on the plain, at its meeting with the hillside, but without getting too far from the salt brooks. These settlements remained until the High Middle Age.

In fact, since the Islamic period, the pottery findings become more abundant and we have determined that the settlements were unstructured, maybe due to a process that started in the late Antiquity. The habitat was located on the hillside near fresh water sources — mostly brooks— and near salt sources, as in La Olmeda, which were clearly used in this period. The Islamic settlement pattern, based on small farms and corresponding to a decentralised political model, is introduced and consolidated. Prospections have yielded evidence of small farms and some large centres of population (Valdecubo), which is confirmed by early Christian documents. The military settlements are located on hillocks which were highly valuable for defense and strategic control of a large territory. The archaeological findings suggest that around these settlements a kind of suburb developed in which the main production system was located. In this area, apart from these forts, which control several valleys, like Riba de Santiuste, many of the elevations and hills with strategic value have provided us with Islamic findings, although the origins of some of them can be traced back to the Iron Age. These may have been small control outposts dispersed throughout the land. In this period new roads were built, some of which integrated into the old Roman road system. The purpose of these roads may have been connecting the largest military centres of Islamic Spain. In the case of our study, the connection between the Sigüenza region and today's province of Soria through the Altos de Barahona and more specifically, to the outposts of the fortress in Gormaz, in the province of Soria, as well as new roads connecting the medieval settlements must have been especially important.

The Late Middle Ages are characterised by the existence of a habitat initially located in caves and used to control flocks and herds and the spread of settlements in caves on the medium-upper part of the hillside. In this area there are many examples of tombs dug into the rock around the church. People lived in the hills located opposite to the former Islamic centres of population, whenever there had been such, like in Imón and Riba de Santiuste. The large military premises of the previous period were occupied by feudal lords, around which the population would slowly concentrate, often on the basis of the high medieval suburbs. However, there are remarkable examples of abandonment and disappearance of prominent settlements that had been inhabited during the first years of Islamic rule. In the late High Middle Ages, the increased control over the management of the territory facilitated by a monarchic system becomes perceptible, but generally speaking, in this period the patterns of settlement still existing today were introduced and the number of settlements increased dramatically (Figure 3). In most cases, it is not possible yet to differentiate clearly between settlements that originated in the Early Middle Ages from those founded in the Late Middle Age.

The most recent archaeological findings in this area can be dated back to the Modern Era. The first industrial sites devoted to salt extraction from deeper levels, by way of artesian wells, seem to have originated during this period. This system was probably used by the Romans (Plinius, *HN*, XXXI, 96; Morère, 2008), in order to take advantage of the higher salt concentration in deeper layers. Some of the most remarkable archaeological findings of industrial character come from this period (García Soto and Ferrero Ros 2007), which mostly consist in devices destined for natural evaporation of brine (waterwheels, pipes and salt pans). In this period, production and even rural life was centred around salt storehouses and specialised workers appeared. We have also found evidence of well-developed roads built according to the guidelines of the Enlightenment of the 18th century, when salt production boosted in the area around Sigüenza. It was not until the arrival of absolute monarchy that salt became subject to large-scale transportation.

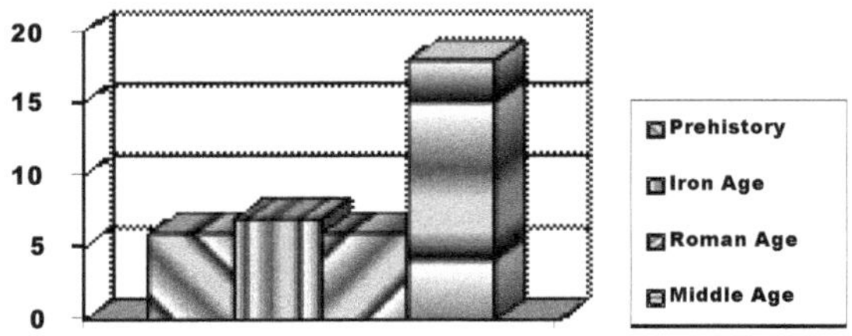

Figure 3. Graphic representation of the number of archaeological sites in terms of chronological and cultural factors.

References:

Abascal Palazón, J.M. 1986. En torno a la promoción juridical de la Segontia de los Arévacos. *Gerion* 4, 213-224.

Aguilera y Gamboa, E. 1916a (inédito). *Páginas de la Historia Patria por mis excavaciones arqueológicas*. 5 Vol.

Arias, G. 1989. La calzada Tiermes-Segontia. *El Miliario Extravagante* 22, 18.

Aguilera y Gamboa, E. 1916b. *Las necrópolis ibéricas*. Madrid.

Carusi, C. 2007a. Recherches sur le sel dans la Méditerranée orientale pendant l'Âge du Bronze. In A. Fíguls, O. Weller (eds.), *1a Trobada internacional d'arqueologia envers l'explotació de la sal a la prehistòria i protohistòria. 1st International archaeology meeting about prehistoric and protohistoric salt exploitation, Cardona, desembre 2003*, 257-280, Barcelona, IREC.

Carusi, C. 2007b. Régimes d´exploitation et fiscalité du sel dans le monde grec et romain. In N. Morère Molinero (ed.), *Las salinas y la sal de interior en la historia: economía, medio ambiente y sociedad. Inland saltworks and salt history: economy, environment and society*, 325-342. Madrid, Dykinson.

Cerdeño Serrano, Mª.L. 1976. Excavaciones arqueológicas en Alcolea de las Peñas (Guadalajara). Informe de la campaña efectuada en el cerro 'El Perical'. Julio 1974. *Noticiario Arqueológico Hispánico. Prehistoria* 5, 9-18.

Cerdeño Serrano, Mª. L. 1981. Sigüenza, enterramientos tumulares de la Meseta Oriental *Noticiario Arqueológico Hispánico. Prehistoria.* 11. 189ff.

Cruz García, O.1989. Norias de tradición mudéjar en Imón. *Revista de Folklore* 107, 147-166.

Daza Pardo, E. 2005-2006. *Formación de ocupación rural en la Sierra Norte de Guadalajara (siglos V-XII). El despoblado de Morenglos, Alcolea de las Peñas*, 31-32, 195-218.

Delibes de Castro, G., García Rozas, R., Larrén Izquierdo, H., Rodríguez Rodríguez, E. 2007a. Cuarenta siglos de explotación de sal en las lagunas de Villafáfila (Zamora): de la Edad del Bronce al Medioevo. In A. Fíguls, O. Weller (eds.), *1a Trobada internacional d'arqueologia envers l'explotació de la sal a la prehistòria i protohistòria. 1st International archaeology meeting about prehistoric and protohistoric salt exploitation, Cardona, desembre 2003*, 111-144. Barcelona, IREC.

Delibes de Castro, G., Fernández Manzano, J., Rodríguez Rodríguez, E., del Val Recio, J. 2007b. Molino Sanchón II: Un salín de época campaniforme en las lagunas de Villafáfila (Zamora). In N. Morère Molinero (ed.), *Las salinas y la sal de interior en la historia: economía, medio ambiente y sociedad. Inland saltworks and salt history: economy, environment and society*, 47-72. Madrid, Dykinson.

García Grinda, J.L. 1983. Las salinas de Imón, en Guadalajara: parte de un conjunto salinero. *Q. Revista del Consejo Superior de los Colegios de Arquitectos* 64. 28-35.

García Soto Mateos, E., Ferrero Ros, S. 2007. Estudios de algunos complejos salineros contemporáneos de las comarcas de Atienza y Sigüenza: Cercadillo, Santamera, Bujalcayado-La Olmeda de Jadraque, El Atance y Cirueches. *Wad Al Hayara* VIII. 23, 85-120.

Hees, M. 2002. Production et commerce du sel à l'Âge du Fer en Baden-Württemberg (Allemagne). In O. Weller (Ed.), *Archéologie du sel: techniques et sociétés.* Internationale Archäologie, ASTK 3. Colloque 12.2, XIVe Congrès UISPP, Liège, 209-215.

Jarman, M., Vita-Finzi, C., Higgs, E. 1972. Site catchment analysis in archaeology. In Ucko, P., Tringham, R., Dimbleby, G. (Eds.) *Man, settlement and urbanism*. London, 61-66.

Jiménez Guijarro, J. 2007. ¿Aprovechamiento o explotación?: reflexiones acerca de la minería y uso de la sal durante la Prehistoria. In N. Morère Molinero (ed.), *Las salinas y la sal de interior en la historia: economía, medio ambiente y sociedad. Inland saltworks and salt history: economy, environment and society*, 185-216. Madrid, Dykinson.

Malpica Cuello, A. 2007. Análisis de las salinas medievales desde la arqueología del paisaje. In N. Morère Molinero (ed.), *Las salinas y la sal de interior en la historia: economía, medio ambiente y sociedad. Inland saltworks and salt history: economy, environment and society*, 469- 498. Madrid, Dykinson.

Meyer, Ph.1982. *L'homme et le sel*. Paris, Fayard.

Morère Molinero, N. 1983. *Carta Arqueológica de la región Seguntina. Institución Provincial de Cultura "Marqués de Santillana"*. Guadalajara.

Morère, N. 1991. L'exploitation du sel dans la región de Sigüenza. *Alimenta. Gerion. Homenaje a Dr. Michel Ponsich*, 223-236.

Morère Molinero, N. 2007. Hommes et sel: Archéologie et Histoire jusqu'au premier millénaire av. Notre ère. In A. Fíguls, O. Weller (eds.), *1a Trobada internacional d'arqueologia envers l'explotació de la sal a la prehistòria i protohistòria. 1st International archaeology meeting about prehistoric and protohistoric salt exploitation, Cardona, desembre 2003*, 97-110. Barcelona, IREC.

Morère Molinero, N. 2007. La sal en el desarrollo histórico de Sigüenza. Los primeros siglos. In N. Morère Molinero (ed.), *Las salinas y la sal de interior en la historia: economía, medio ambiente y sociedad. Inland saltworks and salt history: economy, environment and society*, 3-30. Madrid, Dykinson.

Morère Molinero, N. 2011. Historical development of the 'salinae' in Ancient Rome: from technical aspects to political and socio-economic interpretations (in this volume).

Sánchez Lafuente, J. 1982. Nuevos yacimientos romanos en la provincia de Guadalajara. *Wad-al-Hayara*, 9. 103ff.

Talavera Costa, J. 2005. *El poblado arévaco de Los Castillejos de Pelegrina (Guadalajara, España). Evolución de sus fases*. Oxford. British Archaeological Reports, BAR Publishing. International Series.

Talavera Costa, J. 2007. La sal en la comarca de la Sigüenza Arévaca: ¿Riqueza natural-riqueza social?. In N. Morère Molinero (ed.), *Las salinas y la sal de interior en la historia: economía, medio ambiente y sociedad. Inland saltworks and salt history: economy, environment and society*, 241-262. Madrid, Dykinson.

Valiente Malla, J. 1997. *Guía de la Arqueología en Guadalajara*. AACHE. Tierra de Guadalajara, 15. Guadalajara.

Valiente Malla, J. 1984. El Abrigo de Peña Corva en Santamera, Riofrío del Llano, Guadalajara. *Wad Al Hayara*, 11, 289-312.

Weller, O., Dufraisse A., and Pétrequin P. (eds.) 2008. *Sel, eau et forêt: hier et aujourd'hui*. Cahiers de la MSH Ledoux 12 (coll. Homme et environnement, 1). Besançon, Presses Universitaires de Franche-Comté.

Zedeño, M. N. 1997. Landscapes, Land Use and the History of Territory Formation: An example from the Puebloan Southwest. *Journal of Archaeological Method and Theory* 4(1), 67-103.

Sea Salt and Land Salt. The Language of Salt and Technology Transfer (Portugal since the Second Half of the 18th Century)

Inês Amorim
Universidade do Porto, Portugal

Abstract
The study of salt landscapes, from a multidimensional approach based on the evolution of the wetlands, which bear the marks of human action over time and space, could be developed at different levels. We defend a comparative study of technological terminology as a way of reflecting on the use of a specific language and perhaps understand technology transfer systems since the 18th century. These issues stem from an elementary observation around the definition of the unit of production of Portuguese saltpans, **Marinhas** (also Salinas or Salterns), generally used in all Portuguese data terminology with regard to both seaside salt-marshes and inland salt works.

Keywords
Marinha, saltern, salt landscape, salt architecture, technology transfer, environmental history

1. Introduction: Portuguese historiography – ancient and new approaches

Since the 1960s research on Portuguese salt production has been practically ignored. The works of Virgínia Rau at the time represent the dominant research in this domain not only because she participated in historical works produced by 'La Commission International de Histoire Maritime Française' but also as an expert on the international historiography of Portuguese Salt, with her participation in the well-known work 'Le rôle du sel dans l'histoire' (Rau 1968). These publications proved the strategic position of Portuguese sea salt in international trade, particularly that produced in the Aveiro (16th century), Setúbal and Lisbon (17th and 18th centuries) saltpans and exported to the Netherlands, England and the North of Europe (Rau 1984).

More recently, three International Seminars on Portuguese Salt organised in 2004, 2006 and 2008 gathered programmes and projects to restore multidisciplinary and interdisciplinary scientific research and improve the viability of salt-related historical heritage. The first one identified documental data funds in Portugal and in archives abroad with interest for the History of Salt Production in Portugal, comparing methodologies according to specific research objectives developed in other European salt-making regions (Amorim 2005).

It was an opportunity to justify future research ventures, as a convergence point for several projects (Portuguese and International) opening a debate which could serve as a platform for interdisciplinary and thematic studies, intersecting several chronological moments (Classical Antiquity to the 21st century), multidisciplinary research (History, Social Environment, Biology, Chemistry, Geography, Cartography, Archaeology, Museology), and to establish national and international geographical contacts sharing a common historical past (Portugal, Spain, France, Italy, Sweden, Holland, Russia).

In this First Seminar, Jean Claude Hocquet presented a review of current historical research, and the need to move scientific research from salt trade (dominant studies) to production, which depends not only on human or technical factors, but also on the inevitable changes in the ecosystem. As he wrote, the history of climate and sea level variations can be evaluated by the challenges that require men to make saltpans and salt, while offering sound indicators for ecology and biotechnology (Hocquet 2005, 15).

Moreover, it became clear that many shared the idea of the urgency in identifying and evaluating the heritage associated to the salt industry (public and private documental and cartographic collections, tangible or intangible heritage, iconography, etc.), thus contributing to an architecture of salt (in some cases as industrial heritage), through a contextualised reconstitution of a collective memory and history characterised by salt. As a result, an application was submitted to the Portuguese Research Foundation, for a project named SAL(H)INA - *Salt History - nature and environment, from the 15th to 19th centuries*. It was approved, and the Portuguese Scientific Foundation was to provide further support to the organisation of the two following international seminars (in 2006 and 2008).

The organisation of the Second International Portuguese Salt Seminar under the title 'The adaptation of Portuguese salt to the world markets – old and new uses for salt' took place from 19th to 21st October 2006, exploring multidisciplinary issues, not only the long-term reconstitution of production centres, covering both the dimension of traditional production and the introduction of new sustainable products, capable of sharing markets; but also the different nuances of the everyday consumption of salt, in the conservation of foodstuffs, in industry and also as a tourist product.

There is the need to evaluate the increasing importance of new consumption patterns achieved in the identification of heritage linked to the saltpans (private and public documents, cartography, tangible or intangible heritage, iconography, etc.) conducive to the development of an 'architecture of salt' (in some cases as industrial heritage), with a view to the re-establishing memory and history, on the basis of the traces left by the salt industry.

Traditional approaches to the salt trade intersected with new findings was the object of thorough debate which took into account: the issue of salt trade and the nature and intensity of land and sea routes, taking into account periods of war and peace; the evaluation of the importance of monopolies that were set up, breaking down preconceived ideas of these markets in the presence of competition, the hierarchy of commercial areas and the development of trade systems; structural factors that influence production costs, which, in turn, not only shape production but also the consumer market, leading to the existence of differentiated markets.

The uses and consumption of salt deserved special focus, particularly new consumptions leading to the immediate renewal of the saltpans: a focus on the need to regulate 'other' consumptions taking into account the evolution of tastes and social and cultural changes that have modified the nature and the contents of people's shopping bags, in terms of quality and quantity, according to diachronic criteria of priority or posterity in the consumption of salt: a focus on the role of the scientific system as a development agent in the salt sector, by stimulating new products deriving from production (product innovation, certification and packaging); the preservation of biodiversity, the safeguarding of heritage and culture (as eco-tourism), the setting up of interpretation centres that foster research in all sectors (history, biology and biotechnology, chemistry, architecture, tourism, etc.) (Amorim 2008).

Conclusions envisaged a further study proposed for the Third Seminar which took place in October 2008, thus fulfilling the plan laid out in 2004, under the title 'The salt landscape - tradition and innovation'. It focused on a humanised construction of the landscape, analysing, among other aspects, techniques as an element of constant adjustment, namely to the local ecological circumstances, where the labour force is an unavoidable strategy. This led to a 'culture of work' which we today strive to restore and study, disseminated in cultural intervention centres. In the meantime, the Seminar also focused on the reconstitution of salt landscapes which represent a challenge and an incentive to finding forms of environmental reconstruction, regenerating the areas of production with new producers, in an attempt to renew and revitalise traditional techniques, as well as revive memories through the creation of cultural centres, the success of which we will try to appraise.

The study of salt landscapes, from a multidimensional approach based on the evolution of the wetlands, which bear the marks of human action over time and space, could be developed at different levels. In this paper we defend a comparative study of technological terminology as a way of reflecting on the use of a specific language and perhaps understanding technology transfer systems and social adjustments to environmental evolution. These issues stem from an elementary observation around the definition of the unit of production of Portuguese saltpans, 'Marinhas' (also Salinas or Salterns), generally used in all Portuguese data terminology with regard to seaside salt-marshes as well as marshes built from salt wells in the interior of the country.

2. Portuguese salterns: location and distribution

The tradition of Portuguese coastal salt-making is well known. At least from medieval times, sea salt production was carried out extensively in the coastal marshes where solar evaporation was the dominant technique: seawater was run or pumped (using the tidal system) into a series of tanks with different levels, where its brine content was progressively concentrated by evaporation under sunny summer conditions. Indeed, Portuguese salt marshes are assembled in five large groups: Aveiro (River Vouga and Ria de Aveiro (the Aveiro estuary), Figueira da Foz (River Mondego), Lisbon (River Tagus, on both the north and south banks), Setúbal (just at the mouth of the River Sado) Alcácer do Sal (further up in the course of the same River Sado) and the Algarve (from west to east, the main locations: Lagos, Alvor, Portimão, Faro, Olhão, Tavira and Castro Marim) (Silva 1956, 15). Each one has its own characteristics in terms of operations, organisation of work and quality of salt, even if it is all harvested from sea water in salty estuaries. The climate, the soil, the situation vis-à-vis the sea and the communication system with water introduce variables that affect and determine the architecture of the marshes and their technology.

A few exceptions are found in the interior of the country: some characterised by a system of water evaporation from salt wells, and others by the exploitation of rock mines. In this case, with the establishment of the Portuguese Geological Services, engaged in surveys for oil, and the Chemical and Fertilizer Industry in the second half of the 19th century and beginning of the 20th, the location of rock salt mines expanded. Maps drawn up in the first half of the 20th century, locate and distinguish various land salt regions (in the interior of the country the existence and exploitation of rock salt and salt water wells) and sea salt coastal areas (see Figure 1 and Figure 2).

In a context of economic change at the turn of the 20th century, evaluating the quantity and quality of the salt produced and distributed to salt markets, based on export data, there is a clear decline in the Portuguese position on the international markets ranking. This tendency becomes all the more evident from the middle of the 20th century, which is corroborated in reports (surveys on the salt industry) and statistical data comparing the main salt production locations. The methodology used attempts to compare several sources of information from the 18th to the mid-20th century (Amorim 2005, 111-125), produced by royal academic members (Lobo 1793; Lobo 1789-1815), Public Works and Statistics Administration (1850-1860), chemical research (Lepierre 1936), Department of Chemical Services (1954-1959) and more recently by official statistics (Neves 2005, 127-134).

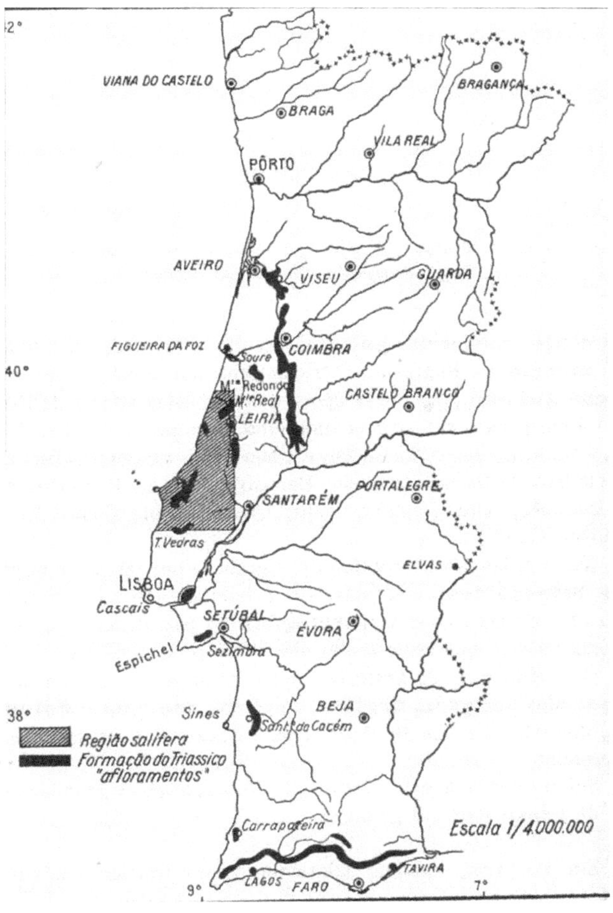

Figure 1. Inland salt location (Sá 1951, 84).

As can be seen in Figure 3, when comparing the most important centres of production and their units of production from the 18th to 20th centuries, there is a decline in the number of units ('marinhas' and 'talhos'), especially over the last fifty years, with one single exception – the inland saltern of Rio Maior. Located 75km north of Lisbon, 30km from the sea, in the Natural Park of Serra de Aire and Candeeiros, it is an extensive rock salt mine (between Leiria and Torres Vedras), fed by a well-located saltwater undercurrent at its centre. The water emerges at the surface seven times more salty than the Atlantic ocean (see Figure 1 and Figure 2).

From the quantitative point of view these statistics reveal the dominant positions in sea salt production (Aveiro, Figueira da Foz, Lisbon, Setúbal, Alcácer do Sal and Algarve), but this predominance disappears in the second half of the 20th century with the exception of Rio Maior, an original example of an inland saltern which applies the evaporation system, the only inland salt works still in operation today in Portugal, a unique example which contrasts with the 43 units running in Spain (Carrasco Vayá and Hueso Kortekaas 2008, 34-41). The system of evaporation of seawater is the common denominator.

3 – A language of salt – inland /sea land – the same language?

Our aim is not so much to observe the reasons behind that decline and the maintenance of the Rio Maior inland saltpan but rather to discuss and compare the language used in different centres of production in terms of units of production, and also the social actors involved, such as the workers. As mentioned previously, the methodology used is a comparative analyse of technological terminology in each location and a specific language as a reflection of the technology transfer system.

Figure 2. Sealand salt exploitation (Lepierre 1932, 18).

Briefly, the Portuguese salt evaporation techniques are characterised by several steps in the salt water evaporation process. A saltpan ('marinha' in Portuguese) contains a set of tanks that are intended to receive seawater, from which salt is obtained through evaporation. Each tank serves a specific purpose in the process, as water content decreases and salinity rises:

- preparation surface: it functions as a storage reservoir, with a variety of designations depending on the region, which concentrate salt water and the deposit of clay and sand particles carried in salt water;
- evaporation and in some cases also concentration: the water flows through successive compartments,

allowing for gradual evaporation, which removes moisture and allows the formation of salts
- crystallisation surface: to collect and obtain a concentrated solution in the tanks, as final salt deposits.

The difference between salt marshes procedures of making salt and the Rio Maior procedures of inland exploitation is derived from the origin and the quality of the water. In the first case, the driving tidal seawater requires several levels of tanks until the final level of crystallisation, whereas, in the second, with a higher degree of water concentration, obtained directly from the stream with some regularity, only one crystallisation tank is required.

Supplementary to this whole process, another common working operation of great importance is carried out in all centres:
- water circulation and movement (hydraulic system) and networked canals for the flow of water;
- roads and land divisions or walls separating the compartments requiring a very precise technical efficiency;
- deposits and final storage facilities for salt packaging or assembly work instruments, transport systems (boats, cars or railway) providing connection with ports and distribution warehouses.

Place/Year	1790-1791	1857	1932	1960s	2000
Aveiro (River Vouga and Ria)	178	259	276	270	15
Figueira da Foz (River Mondego)	1,150 (*)	-	400	229	50
Rio Maior (inland salt works)	350 (*)	-	400	385	400
Lisbon (River Tagus)	245		212	230	1
Setúbal and Alcácer (River Sado)	352		170	300	6
Algarve (Faro, Tavira, Castro Marim)	146	94	227	136	15

*in these cases, they are named 'talhos'. Sources: Lobo 1789-1815. Inquérito à Indústria do Sal 1869. Lepierre 1932; Silva 1954; Silva 1956; Silva 1957; Menezes 1956; Virgolino 1957; Neves 2005.

Figure 3. Units of production ('marinhas').

Based on the three steps taken in salt production (preparation, evaporation and crystallisation), we organised all the data related with the different names of each division and subdivision, and compared parallel terms. Figure 4 presents the information in a comparative way, highlighting common terminology as well as their variants.

We can draw some preliminary conclusions:
- the term 'marinha' is a common expression for all the centres of production, as a general concept. The Portuguese origin of this word is 'mar', which means the sea, but, as we have already seen, Rio Maior is located at some distance from the sea (30 km). Moreover, in one of the cases, that of the Aveiro coastal salt-marshes, the final part of crystallisation is named 'marinha' as the real and final stage of production;
- the preparatory reservoir which receives the first salt water, before it is distributed to the other tanks, has different names, such as 'viveiro', 'pejo', 'tejo' and 'praia', and, in some cases, with variants. The same applies to the evaporation and crystallisation surfaces;
- some of the production units have more than six tanks, from the evaporation to crystallisation surfaces;
- the specific crystallisation tanks vary from place to place: 'meio' is exclusively used in Aveiro, while the more generalised terms are 'talho' followed by 'peça'.

It should be noted though that the terms 'talho' and 'talhão' are common to almost in every location with the exception of Setúbal and Alcácer do Sal, where it is replaced with the expression 'peça'. We have no real reason to explain these differences but it could possibly have to do with the fact that, in the 1930s, there were three methods of sea salt exploitation: the same procedures in Aveiro and Figueira da Foz, a different one in Lisbon and Setúbal, and the third resulted from a combination of the other two techniques in the Algarve.

The different systems of production were pointed out by authors at the end of 18th century (Lobo 1793; Lobo 1789-1815) and in the 19th century (Alcoforado 1877; Silva and Moura 1873), particularly by the French scientist Aimé Girard (1872) who had visited the international exhibition in Portugal in 1865 (Almeida 1873, 27). He had the opportunity to study the original manufacturing process and the quality of Portuguese salt (with great interest in Aveiro salt), going as far as to advising the French salt producers to import Portuguese technology to the French marshes. He stressed the distinction between the three methods based on the presence or absence of a specific deposit at the bottom of the tanks of the different regions (Setúbal, Alcácer do Sal and Lisbon). It was a compact lining 2 to 10 mm thick, produced with marine algae, *microcoleus Corium,* known in Setúbal and Alcácer do Sal, called 'casco' in Portuguese. It lined the bottom of the saltern like a carpet, settling and purifying the salt, acting as a dialyser by allowing the more rapid formation, by osmosis, of salts of magnesium. Finally, it enhanced the better absorption of light, raising the temperature. In contrast with those salterns of Setúbal and Alcácer, in Aveiro, Figueira da Foz and Algarve, the bottom of the crystallisers was not protected by the felt plant ('casco') but rather a precise human technique was employed in the construction of the bottom of the saltern, workmanship which the Frenchman, Aimé Girad, praised highly.

These procedures used in the North, as praised by Girad, are a consequence of climatic and geo-morphologic circumstances. In Aveiro and Figueira da

Locations:	Aveiro		Figueira da Foz		Rio Maior		Lisboa-norte		Lisboa-sul		Setúbal		Alcácer		Tavira		Faro		Castro-Marim	
Tanks and subdivision	1790-1791 (Lobo 1793, 130)	1930-1932 (Lepierre 1936, 66-70)	1790-1791 (Lobo 1793, 131)	1930-1932 (Lepierre 1936, 84)	1790-1791 (Lobo 1793, 153)	1930-1932 (Lepierre 1936, 48)	1790-1791 (Lobo 1793, 128)	1930-1932 (Lepierre 1936, 48,135)	1790-1791 (Lobo 1793, 128)	1930-1932 (Lepierre 1936,48,121)	1790-1791 (Lobo 1793,129)	1930-1932 (Lepierre 1936, 92)	1790-1791 (Lobo 1793, 129)	1930-1932 (Lepierre 1936, 92)	1790-1791 (Lobo 1793, 131)	1930-1932 (Lepierre 1936, 48,138,157)	1790-1791 (Lobo 1793, 131)	1930-1932 (Lepierre 1936, 48,138,157)	1790-1791 (Lobo 1793, 131)	1930-1932 (Lepierre 1936, 48,138,157)
a. Preparation tanks	viveiro	viveiro	viveiro	viveiro	?	poço	pejo	pejo	viveiro	viveiro	pejo	pejo	praia	pejo	viveiro	1º viveiro	viveiro	tejo	viveiro	viveiro
	algibebe	algibé	vasa	vasa							caldeirão	caldeiro	caldeirão	caldeiro	caldeira	2º viveiro ou governo	caldeira	contra-tejo	caldeira	caldeira
b. Evaporation tanks	ante-caldeiro	caldeiro	entrebanho	entrebanho	?		caldeirão	caldeirão	caldeirão	caldeirão (or contra-caldeira)			caldeira	caldeira				governo		
	sobre-cabeceira	sub-cabeceira	cabeceira	cabeceira de cima			caldeira	caldeira	caldeira											
	talho	talho		cabeceira de baixo			cabeceira	cabeceira	cabeceira	cabeceira										
	cabeceira	cabeceira		sertões																
			talhão	talhão																
c. Crystallization tanks	meio de cima	meio de cima	talho	talho da praia do meio	talho	talho	talho	talho	talho	talho	caldeira	caldeira	peça	peça grande	talho	talho	talho	peça	talho	talho
	meio de baixo	meio de baixo		talho da praia de baixo							peça	peça grande		peça pequena						
												peça pequena								
Workers names	?	marnoto	?	marnoteiro	?	marinheiro	marroteiro	marnoteiro and marroteiro	marroteiro	marnoteiro and marroteiro	marroteiro	marroteiro, contramestre, carregador	marroteiro	marroteiro?	?	encarregado	?	capataz	?	mestre and sanheiro
Number workers	178	700	1150	500 man and 2000 women (more 1300)	350	70	1860"	240		1500	1760*	1000 a 1500*		650 to 700	#437	570#				
Number of saltern	178	276	1150 talhos	400	25	450 talhos	245"	20		194	352*	170*			#146	227#				

Legend:
*** including Alcácer do Sal**
" including Lisbon South
 # all saltpan of Algarve

Figure 4 - Tanks names in a comparative view - 1790-91 and 1930-31

Foz, the saltern is composed primarily of a tank ('viveiro') which stores seawater during high tides. Then a series of basins successively strip materials and insoluble sulphate of lime from the water, and then drop it into a salt crystalliser. The saltworker ('marnoto' in Aveiro and 'marroteiro' in Figueira da Foz, as can be seen in Figure 4) must harvest the salt formed every two days, while in Setúbal and even in Lisbon, harvesting is only required three times a year (in the summer) to obtain the crystals. In this case, the entire volume of water is introduced almost at once, left to evaporate during the season and leaving a layer of salt about 4cm thick. Once the first harvest is completed, a small amount of water-mothers remains. Once the first harvest is completed, a small amount of mother water remains and, immediately following, about 15cm of new water flows in. Evaporation of these waters leads rapidly, within twenty days, to the second harvest, resulting in a 2cm layer. The season runs until September and is concluded with a third harvest (Lepierre 1932, 66-82).

4 – Technology transfer

In the years following 1932 (see Figure 3), an evident change in saltern distribution can be seen. In Figueira da Foz and Algarve the number of saltpans declines in contrast with Setúbal and Alcácer. However, this did not reflect on production, as can be seen in Figure 5, with total production almost duplicating in global terms, and trebling in Setúbal and Algarve.

	Aveiro	Figueira da Foz	Lisbon	Setúbal and Alcácer	Algarve	Rio Maior	Total
1933-34	50,000	30,000	110,000	21,000	21,000	1600	233,600
1954	90,000	40,000	100,000	60,600	60,000	1000	351,600

Sources: Lepierre 1932, 47; Sá 1954, 27; Silva 1954, 41;Virgolino 1957, 36; Silva 1957, 67

Figure 5. Portuguese Salt Production 1933-34 and 1954-1957.

In the 1960s, a National Commission to Regulate Chemical and Pharmaceutical Products was created, adopting several measures in light of the difficulty of maintaining regular consumption of Portuguese salt. Statistics account for only 266,900 tonnes consumed in contrast with the 351,600 tonnes produced (Silva 1956, 21-24). Almost half of the production remained in storage and the solution was to prohibit the construction of new saltpans, to impose fixed prices and to reform old ways of production (Silva 1956, 26-31).

In those thirty years, between 1932 and 1960 (see Figure 3), we assume that some changes took place which explain the higher level of production in Setúbal, Alcácer and even in Algarve. In this latter case, climate conditions and well-applied technology could be the explanation. But in the other cases, the possibility that some technological processes applied in salterns could be the real reason. Surveys in the 1950s indicate a technology transfer from the Northern salt production centres (Aveiro and Figueira da Foz).

The description is common to the salt production centres of Alcácer and Setúbal: 'for the exploitation of the saltpans two different techniques are adopted - the Sado and northern techniques. The conception of the fixtures, with reference to the saltpan itself, the reservoir where salt is produced, varies according to the technique adopted. The typical saltpans of the River Sado are essentially constituted by a 'pejo' - reservoir which receives water directly from the source of supply, - one or various 'caldeirões' -reservoirs from where water is distributed to the ponds, - and the box of the saltpan -where the crystallisers are located. The regional exploitation technique is very rudimentary and for the harvest, the salters (the saltpan workers) limit themselves to waiting for the water from the previous year's flooding to evaporate, leaving the salt deposited in the crystallisers. For the subsequent harvests the crystallisers receive the solution from the 'caldeirões', which by that time has attained a reasonable graduation. And a similar procedure is adopted for the remaining harvests, the number of which never exceeds 3 or 4 per year'.

And describing the northern technique implemented in both places the technological language is common: 'the saltpans are constituted by: a 'viveiro' which corresponds to the 'pejo' of the Sado; one or various 'algibés' - corresponding to the 'caldeirões' and the box of the saltpan. It is in the interior of the box that the structure is completely different, because, apart from the crystallisers, it incorporates a series of ponds which form part of the preparatory area and constitute the 'mandamento'. When complete, the 'mandamento' comprises the following rows of ponds: 'Caldeiros', 'Sobre-cabeceiras', 'Talhos' and 'Cabeceiras''.

Continuing with the description (comparable in Figure 4) 'the preparatory area also includes a further one or two series of ponds - lower 'meios' - intercalated with the crystallisers - upper 'meios'. The saltern is known as Single or Double, according to whether there are one or two rows of crystallisers. Sometimes the 'cabeceiras' are also prepared to produce salt, there being several with 3 rows of crystallisers. According to the northern technique the exploitation works are always preceded by preparatory works which consist in the cleaning of the saltpan, repairs to the canals and partitions and preparation of the bottom of crystallisers. These preliminary works are of great importance, as the proper functioning of the saltpans during the production period depends on them' (Virgolino 1957, 548-549).

5. Conclusions

In conclusion there were specific reasons to introduce efficient modes of production and the northern technological process seems the more appropriate choice. Figure 6 compares the production and number of salterns in each location. Setúbal, with 34 'marinhas' of the Aveiro type, reached 29,7 percent of total production and represents already 44,55 percent of local type production.

	Type North (Aveiro)	%	Local type (Sado)	%	Type Tejo (only in Setúbal)	%	Total	%
Setúbal number	34	20,1	132	78,1	3	1,8	169	100
Setúbal production	9937	29,7	22, 302	66,8	1140	3,5	33, 379	100
Alcácer do Sal	11	9	110	91	-	-	121	100

Source: Virgolino 1957, 18; Silva 1957, 67

Figure 6. Number of salterns, production and percentage by typology of saltern ('Marinhas') – 1957 – Alcácer do Sal and Setúbal.

The reasons behind this change have never been explained. The inexistence of studies about migratory movements of salt workers prevents any conclusion. But we can advance some hypothesis based on academic discussion around the 1940s with regard to the pollution of salt products, or the decay of the 'casco' (the natural algae that covered the bottom of the saltpans in Setúbal and Lisbon), which explained the complaints of a higher percentage of dirty salt from the Setúbal unit (Lepierre 1936; Silva 1957, 73). It could be this context which justifies the care taken to apply technology carefully when preparing the saltpans in Aveiro and Figueira da Foz. The operation applied envisaged the substitution of an ancient system used in the north of building an artificial floor covered with a thin layer of deposits in which the salt could be collected (Lepierre 1936, 66-82).

Another hypothesis to explain the changes in methodology and the transfer of techniques is based on the new uses of coastal marshes when rice cultivation intensified on the banks of the Sado, and this economic change could explain the workforce transfer from salt exploitation to rice plantation. We have no numbers to prove this idea but the characteristics of the local workers, as undifferentiated groups, applying simple operations as describe in 'limit themselves to waiting for the water of the previous year's flooding to evaporate, leaving the salt deposited in the crystallisers (see Figure 6 and Figure 7), predisposed them to changing their way of life and taking up rice cultivation. The lack of saltmen could explain their replacement and migration movements from the northern salt centres. In conclusion, the change in the nature of the saltern bottoms, with the degradation of the 'casco', explains the application of techniques from the north and the need for workers could be a complementary reason for the employment of a new language of salt. The Alcácer do Sal study also distinguishes between the tools used in Aveiro and Alcácer (Virgolino 1957, 32-33). There was a tendency in the organisation of salt production in Aveiro from the beginning of the 20th century, to change saltpans into fish aquaculture (Nobre, Affreixo and Macedo 1915). Moreover, a shift in State economic policy was operated through the Commission for Chemical and Pharmaceutical Products, created in 1952. It was required to evaluate the state of production conditions at each location considering also comparative costs, and to describe and evaluate the technical characteristics of the various Portuguese salt production centres and the markets they served (Silva 1956, 24-26).

Figure 7. Rio Maior Summer 2008 (Photo João Catalão 2008).

Changes were not exclusively felt in salt-marshes but also impacted on the inland salt production areas, as is the case of Rio Maior, which had come to act as a 'laboratory' for growth and innovation, as reported in a 1957 study (Silva 1954). Even though the salt it produced had the highest prices at national level, as mentioned in the report, local production was always sold to companies and farmers around Rio Maior, given its high salinity, to be used in the meat canning industry and iron craftwork industry. Tourist interest seemed an important reason for the Commission to hire two men to look after and repair the well and the motor regulating water flow, and to guard the saltpans.

We can indicate two main changes: the well system exploitation (turning the mechanical process into a motorised system, as from 1953 (Silva 1954, 12) and the developments in the construction of the 'talhos' (see figure 8). Around 400 'talhos' of different sizes and depth, were coated, since 1944, in cement or gravel, and since 1954, they were covered in stone slabs (Silva 1954, 12-15). This system is not Portuguese in origin, when compared with innovations at the inland saltpan of Alava, Biscay, in the north of Spain. In all those cases the innovation was intended to combat degradation.

Figure 8. Rio Maior Summer 2008 – 'talhos' (Photo João Catalão 2008).

Using salt technology terminology to measure knowledge transfer is, perhaps, a delicate methodology. Hocquet highlighted the critical and debateable method of allocating things to names in a pioneer study of 1974. Indeed, he recommended that first one needed to clarify the local meaning (in each place) and then compare technologies (Hocquet and Hocquet 1974, 528). However, in this case, there is evident data about technology transfer with consequences on the terminology adopted. The reasons for these changes are unclear but the aim of higher production rates and revenue could be a good explanation for worker migrations, reasons which were also known in other foreign realities (Bouhier 2006, 151) and also in earlier centuries in Portugal (Amorim 2006, 141; Costa and Cleto 2008, 73).

References:

Alcoforado, M. da M. 1877. *A indústria do sal.* 'Museu Technologico', 1°Anno, Setembro.

Almeida, D. A. 1873. A indústria salina em Portugal, Revista *de Obras Públicas e Minas*, T. IV, n° 37, Lisboa, 117.

Amorim, I. 2005. *I Seminário Internacional sobre o Sal Português*. Porto, IHM-UP.

Amorim, I. 2005a. Os inquéritos sobre o sal Português nos séculos XVIII a XX. In Inês Amorim (ed.), *I Seminário Internacional sobre o Sal Português*, 111-125. Porto, IHM-UP.

Amorim, I. 2006. Propriétaires et regime d'exploitation du sel dans les marais salants d'Aveiro, Portugal (fin XVIIe –XVIIIe siècles). In *Le Sel de la Baie et Ses Concurrents à travers les Âges, Colloque international,* 6e Congrès de la Commission Internationale d'Histoire du Sel 16, 17, 18 septembre 2006, 137-147. Nantes, Presses Universitaires de Rennes.

Amorim, I. 2008. *The articulation of Portuguese Salt with worldwide routes. Past and new Consumpption Trends*. Porto, IHM-UP.

Bouhier, C. 2006. Le transfert de technologie saunière de Guérande à Noirmoutier au début du XVIII siècle. In *Le Sel de la Baie et Ses Concurrents à travers les Âges, Colloque international,* 6e Congrès de la Commission Internationale d'Histoire du Sel 16, 17, 18 septembre 2006, 149-156. Nantes, Presses Universitaires de Rennes.

Carrasco Vayá, J. and Hueso Kortekaas, K. 2008. *Los paisages ibéricos de la sal. Las salinas de interior*, 25-42. Guadalajara, Asociación de amigos de las salinas de interior.

Costa, Patrícia and Cleto, Joel 2006. O sal do esquecimento. Salinas e comercialização de salgados na foz do rio Leça. In Inês Amorim (ed.) *II Seminário Internacional sobre o sal português - A Articulação do Sal Português aos circuitos mundiais · antigos e novos consumos*, 65-78. Porto, IHM-UP.

Girard, A. 1872. Étude sur les marais salants et l'industrie saunière du Portugal. Annales du Conservatoire des Arts et Métiers, Paris 1872. In Charles Lepierre (ed.), *Inquérito à Indústria do sal em Portugal,* XLII-LI. Lisboa Universidade Técnica de Lisboa, 1936.

Hocquet, J. C. and Hocquet, Jaqueline. 1974. Le vocabulaire des techniques du marais salant dans l'Adriatique au Moyen âge. *Mélanges de L'École Française de Rome. Moyen Âge Temps Modernes* 86, 527-552.

Hocquet, J. C. 2005. L'actualité de l'Histoire du Sel. In I. Amorim (ed.), *I Seminário Internacional sobre o Sal Português*, 15-28. Porto, IHM-UP.

Inquérito à Indústria do sal, VIII volume: Salgado do Algarve. 1959. Lisboa, Comissão Reguladora dos Produtos Químicos e Farmacêuticos.

Inquérito à Indústria do sal. 1869. Manuscripts in Archives of the Minister of Public Works.

Lepierre, C. 1936. *Inquérito à Indústria do sal em Portugal*. Lisboa, Universidade Técnica de Lisboa.

Lobo, C. B. L. 1789-1815, 1991. Memória sobre as marinhas de Portugal. In *Memórias Económicas da Academia Real das Ciências de Lisboa 1789-1815,* v. 4, 127-152. Lisboa, Banco de Portugal.

Lobo, C. B. L. 1793. Memória sobre a História das Marinhas de Portugal. In *Memórias da Literatura Portuguesa*, v. 5, Lisboa, Academia das Ciências de Lisbos.

Menezes, J. C. 1956. *Inquérito à Indústria do sal, IV volume: Salgado de Aveiro*, Lisboa, Comissão Reguladora dos Produtos Químicos e Farmacêuticos.

Neves, R. 2005. Os salgados portugueses no séc. XX- que perspectivas para as salinas portuguesas no séc. XXI? In Inês Amorim (ed.), *I Seminário Internacional sobre o Sal Português*, 127-134. Porto, IHM-UP.

Nobre, A., Affreixo, J. and Macedo, J. 1915. *A Ria de Aveiro, relatório oficial do regulamento da Ria de 28 de Dezembro de 1912*. Lisboa, Imprensa Nacional.

Rau, V. 1959. Sources pour l'étude de l'économie maritime portugaise. In Michel Mollat (ed.), *Les sources d'histoire maritime en Europe du Moyen Age au XVIII siècle, Actes do 4ºColloque International d'Histoire Maritime*, 255-266. Paris, SEVPEN.

Rau, V. 1968. Les courants du trafic du sel portugais du XIV au XVIII siècle. In Michel Mollat (ed.), *Le rôle du sel dans l'histoire*, 53-72. Paris, Presses Universitaires de France.

Rau, V. 1984. *Estudos sobre a História do Sal Português*. Lisboa, Presença.

Sá, M. V. 1951. *Sal comum: I - Sal de mar e sal de mina.* Lisboa, Sá da Costa.

Silva, J. F. 1954. *Inquérito à Indústria do sal, II volume: Salgado de Rio Maior*. Lisboa, Comissão Reguladora dos Produtos Químicos e Farmacêuticos.

Silva, J. F. 1956. *Apontamentos para um curso de salineiros*. Lisboa, CRPQF.

Silva, J. F. 1957. *Inquérito à Indústria do Sal. V volume, Salgado de Setúbal*. Lisboa, Comissão reguladora dos produtos Químicos e Farmacêuticos.

Silva, S. A. P. and Moura, A. M. 1873. *Breve Notícia sobre as Marinhas da Ria d'Aveiro. Notícia da Exposição Universal de Viena de Áustria em 1873*, Bruxelas.

Virgolino, F. B. M. 1957. *Inquérito à Indústria do sal, vol. VI-Salgado de Alcácer do Sal*. Lisboa, Comissão Reguladora dos Produtos Químicos e Farmacêuticos.

A short Overview on the Main Salt Production in Italy from the End of the Middle Ages up to the Modern Period

Valdo D'Arienzo
Università degli Studi di Salerno, Italy

Abstract
The salt production in different Italian countries between the Middle Age and the Modern Period is very differentiated. This historical overview presents a description of the main sources.

Keywords
salt production, Italy, Middle Ages, Modern Period

Salt production in various Italian regions provides a particularly differentiated system, although, most often, the system of the evaporation of the marine waters seems the most common even from ancient times. The Mediterranean environment of the Italian salt mines, like the main 'marine structure', is a long-term characteristic covering a period of more than one millennium.

The difficulty in outlining a common history is partially and only partially overcome by the common denominator which can be traced in the whole Mediterranean basin: the crisis encountered during the passage from the Middle Ages to the Modern Period, a crisis which is not only a production-, fiscal policy- and trade-related one, but it concerns most importantly the technical and structural transformations, as well as investments. Only beginning with the 16th century did the owners, both public and private or religious authorities, introduce an active policy at the same time as the market revitalization. It is with this meaning that the interpretative model suggested some decades ago by Jean-Claude Hocquet (Hocquet 1981, 4-5 and *passim*) still appears of topical interest: it is equally possible, by means of this interpretative key, to understand the silence of the documentary sources which had been 'silent' throughout the High Middle Ages.

As I already had the chance to defend some years ago, when I drew up the Italian Bibliography on 'salt', the fragmentary nature of the studies published up to the present moment, the 'strong local character' as well as the absence of a thorough work cause the impossibility of outlining an exhaustive picture of this matter (D'Arienzo 2001, 60).

Still, aside from these considerations, it must be said that the history of the Italian salt mines in the Middle Ages is marked by a strong continuity in relation with the reality of the previous centuries, especially in what concerns the working production and techniques, as well as the correlations with both road and sea transport systems. For this period of time, even without the quantitative data which are so hard to be found especially for the most recent centuries, it is practically impossible to suggest numbers and hypotheses to explain the high contraction of the general production level between the 5th and the 10th century: the various typologies of the documentary sources and the peculiarity of salt which, being imposed everywhere a very high taxation, inevitably lead to 'silences' and 'false' numbers.

Still, one may notice the extremely obvious bond connecting the Roman and late imperial period (Giovannini 1985; Moinier 1997; Carusi 2008), both with the Early and Late Middle Ages, crossing even the so-called 'monastic' period, in which the salt mines property is reorganized, being transferred, outside the central authorities' control, from the private ownership into the religious one and thus taking part in the foundation of the ecclesiastical patrimony which is being shaped during the centuries previous to the year 1000 and after. Nevertheless, it must be underlined, how a historiographic debate on this historical process developed in the previous years, where not all authors, especially Michel Mollat, Jacques Le Goff and Jean-Claude Hocquet, agree with certifying this process's value, temporal division if not even its existence (Le Goff 1959; Mollat du Jourdin 1993; Bergier 1984; Hocquet 1990, 2001, 12-15). The fact that between the 12th and 13th century, politics put its mark in the exploitation of the salt mines, that is the absolute power of the monarch or of the local authorities, where they existed and were functional, interferes by reorganizing the entire fiscal system in view of the salt property, production and trade regulation; in 12th century Italy, emperor Frederick II started a rigorous and accurate reorganization of all that concerned the matter, even coming into conflict with the papal authority itself (D'Arienzo 1996, 44-46). The monopoly's reorganizing stage begins and, to some extent, is consolidated during these decades, generating then the conditions for the production and trade expansion in the Early Modern Period and the following centuries.

Beginning with the North-East of Italy, an area in which Venice wielded for many centuries a strategic influence not only at a local level, the salt mines of Chioggia can be found. This productive structure seems quite old (Fazzini 2006[1]), and proves that the lagoon area has always been an area meant for production, given the proximity of the neighbouring Comacchio and Cervia. It is Cassiodoro who provided the first exact evidence when, between 537 and 538 d.C., addressing the *tribunis maritimorum,* he urged the people to do their best in order to increase the salt production and trade, underlining what the earlier authors had claimed – its irreplaceableness: 'In salinis autem exercendis tota contentia est: pro aratris, pro falcibus cylindros volvitis: inde vobis fructus omnis enascitur, quando in ipsis et quæ non facitis possidetis. Moneta illic quodammodo percutitur victualis. Arti vestræ omnis fluctus addictus est. Potest aurum aliquis minus quærere; nemo est qui salem non desideret invenire' (Cass., XII, 24).

Still, it is during the Middle Ages, between the 11th and the 12th century (Hocquet 1991, 13), that Chioggia reaches its highest splendour, but then, little by little, it loses its importance, especially compared to the other salt mines of the Adriatic coast, such as Pirano, under the control and influence of the Serenissima which privileges one or the other, depending on the production costs and a clever competitive strategy between them. 'Nel 1200, dopo due secoli di espansione, la produzione del sale aveva raggiunto il suo culmine collocando la laguna al primo posto tra i produttori mediterranei. Chioggia, diventata un'autentica capitale del sale, era in grado di sopperire ai bisogni di un mercato che si estendeva dal versante meridionale delle Alpi alle valli dell'Appennino... La concorrenza, tuttavia, diventava sempre più vivace su tutto il circuito dei mercati del sale lagunare: nuovi centri di produzione apparivano e si sviluppavano sulle rive dell'Adriatico o nel cuore delle Alpi. La concorrenza, d'altronde, era conveniente per i Veneziani stessi... Venezia ricevette ben presto importazioni provenienti dalla Puglia, poi da Alessandria e, alla fine del XIII secolo, da tutto il Mediterraneo... Le saline di Chioggia entrarono in crisi e molte scomparvero' (Hocquet 1991, 24-25[2]).

Thus, the luck of the salt mines of Chioggia decreases quickly during the 14th century, consequently to the clever, but ruthless policies of the Venetians, though connecting its own fame with the expansion and increase of the salt traffic during the next centuries of the modern period in the whole Mediterranean.

Figure 1. View of Chioggia in the mid-20th century.

Similarly located in the lagoon area of the Veneto coasts, the salt mines of Comacchio also play an important role, even though, before Venice expanded even there its control on the production areas of the high Adriatic, they seem to subject to the domination of both Ravenna and Ferrara, without ever succeeding to produce a social class of owners who could invest and capitalize in this field. Even the relationships with the ecclesiastic property, on one hand, and the controversies with the bishops' rights, on the other hand, seem to influence the lasting development of the Comacchio salt mines for a great part of the Middle Ages, in a supremacy game never solved entirely in favour of any one competitor (Dondarini 1995, 274-275). And one must not forget that in the 'wars' which came one after the other along all the Adriatic coasts in the Late Middle Ages, the salt mines of Comacchio are repetitively attacked and destroyed between the 9th and the 10th century – a famous Venetial attack dates to 932 d.C. – without ever succeeding to reach again the level of the previous centuries, if not for short periods of time (Dondarini 1995, 275).

The transfer of the Comacchio salt mines into the possession of the Estense family, lords of Ferrara, characterizes the time, during the Early Modern Period, in which they tried to relaunch the salt mines' activity and still carry on the conflict with the Venetian interests, as always, hostile to the salt production in that area of the Po's Delta. The subsequent transfer of Comacchio into the property of the Papal States at the end of the 16th century marked the history of this centre, especially in what concerns the trade of its production, with the development of contraband, which had already characterized the previous period.

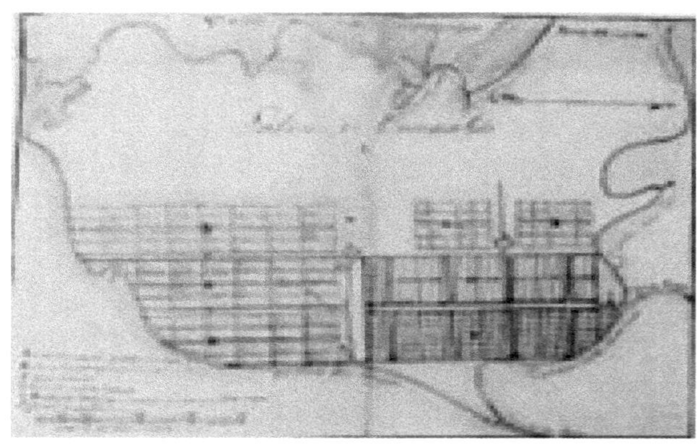

Figure 2. G. Farinelli, *Plan of Comacchio Salt Pan*, 19th century.

The salt mines of Cervia seem to have very old origins, connected to transhumant sheep farming, the main activity of the Umbrian populations who needed great quantities of salt; other populations, throughout the centuries, had exploited the waters of this area, from the Greeks to the Etruscans and Romans. The salt mines of Cervia were in some way connected to an well developed production system integrating them with the ones of Ravenna, closed in 1441, and of Cesenatico, buried in 1775. This stage is followed by a period of silence in what concerns the sources which, eventually, provide again information on salt production in Cervia in 965 d.C., and it is well known that in the 11th century Cervia supplied the important Bologna market, using the Reno's waterway.

Throughout the centuries, salt production in Cervia appears particularly developed as much as to satisfy not only the local markets of Romagna but also the ones in Lombardia, the already mentioned city of Bologna and the Marca Anconetana; then, its annexation to the Papal States, in 1509, transforms this centre into a strategic one from the point of view of the Adriatic policies of Rome (D'Arienzo 2007a, 219-235), mainly as far as the fishing and brine industry are concerned, even though the sources of the early 19th century often complain about the fact that the salt of Cervia, due to its chemical and organoleptic qualities, seems little suitable for this use, and it is for this reason that the salt of Comacchio is preferred (D'Arienzo 2007a, 229-230[3]).

Figure 3. Aerial view of the Cervia salt mines.

Moving on the Tyrrhenian seaside, we must mention the salt sources of Volterra, in Toscana, gathered in the so-called *lagoni* ('great lakes', basins and natural lakes), but it is not mentioned whether they were exploited or not in ancient times, as it happened with the lead, silver and copper mines (Borelli 2000, 8[4]). The sources testify about this important production for the second half of the 10th century by means of the salt water boiling for obtaining the brine in the so-called 'salt pits', the lead boilers. This technique was obviously favoured by the ease of supplying the wood necessary to start up and feed the boilers, thanks to the abundant woodlands of Val Cecina.

The period comprised between the 10th and the 13th century is marked by the legal disputes between the bishops and the Tuscany community, until through a policy of slow enfeoffment of the salt pits, the city community succeeds to set up the customs for salt which, *de facto*, represents a real monopoly on the salt production and trade.

In the Modern Period, the salt mines, transferred into the property of Great Dukedom of Tuscany, represent a very important source of income, taking into account also the widespread contraband which the Medici family attempt to contain, until the promulgation, on 2 December 1701, of the 'General Law on Salt' by Cosimo III, for the reorganization of the entire matter. 'La legge generale del 1701 non aveva caratteristiche innovative tali da risolvere gli annosi problemi legati al sale, era stata promulgata per riordinare organicamente la vecchia legislazione, adattandola alle esigenze dell'epoca; in pratica si trattava di una sorta di 'legge quadro' che si inseriva in un tessuto amministrativo variegato con la finalità di semplificare l'opera di governo. La legge di Cosimo III non abolì le incongruenze causate dalla tassa sul sale, che insieme al sistema daziario costituiva una rendita importante per lo Stato, ma costituiva anche la causa principale del commercio illegale che gli amministratori cercavano in ogni modo di estirpare' (Borelli 2000, 26[5]).

Figure 4. Salt pit of Santa Maria (Romegialli 1636).

Figure 5. Salt pit of Sant'Antonio (Romegialli 1636).

The Barletta salt mines have been and still are the most important productive structures in the Italian Peninsula. Their extension, from Canne to the lake of Salpi, is among the largest, and their name has changed through the centuries, from Santa Maria de Salinis to Barletta and finally to Margherita di Savoia. Somehow connected to them, there were the salt mines of Manfredonia, which were closed during the Modern Period.

The origins seem to be very old, previous to the Roman domination, and already at that time they were different from the other salt mines active in the Mediterranean for their productive capacity and because they were solely dedicated to export, a characteristic they kept for many centuries. Their importance can somehow be appreciated considering the conspicuous investments made in the imperial age in order to connect the salt mines to the harbour of Barletta, where the salt was probably shipped to other Mediterranean centres: 'verso il 115 dopo Cristo l'imperatore Traiano fece costruire sull'Ofanto il magnifico ponte, crollato nel 1850, per unire queste saline a Barletta' (De Luca 1926, 15-16[6]).

Figure 6. Running boiler at the end of the 19th century.

The salt mines of Barletta, like many others in the Italian Peninsula, are not mentioned anymore after the fall of the Empire but only around the year 1000 more sources become available, usually quoting concessions, feudal laws and prerogatives often related to bishops or ecclesiastic structures. These salt mines, particularly, are related to events connected to the figure of Frederick II, who saw for them an essential role in the reform of the imperial assets because of the income they produced and to the possibility to accumulate considerable sums of money thanks to exports. The controversies with the papacy, which lead directly to the excommunication of the emperor, prove the quantity and the variety of the interests which gravitated around the salt mines, which, especially in the early Modern Period, through the achievement of monopoly of the state, will turn out to be essential not only to satisfy the internal demand, but also for the budget itself thanks to the profits derived form exports. Exports which are subject, between the 15th and 16th century, to the political logic of the crown of Aragona following commercial routes wanted by the authorities, while during the following century the favourable international circumstances mark the fortune of the Apulian salt (D'Atri 2001, 54-64), especially starting in the late 1650s and at least until the end of the next decade, according to statistical data regarding the ships docked at the wharf of Barletta for loading salt (D'Atri 2001, 60). Throughout the 18th century the exports-related data (from 1723 to 1739) indicate a relatively constant trend (Russo 2001, 78).

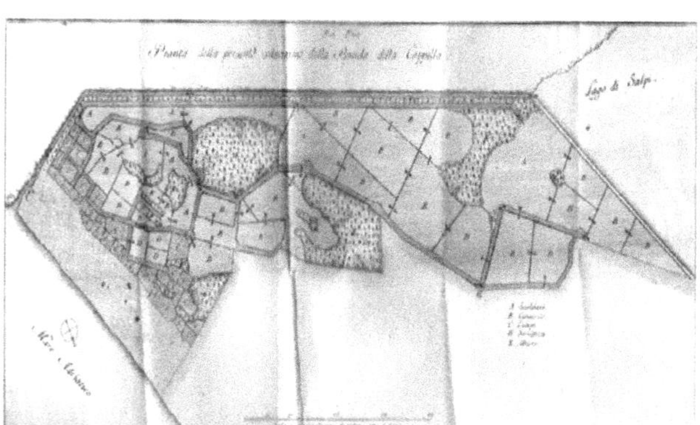

Figure 7. 19th century map of the salt mines of Barletta.

The exploitation of the salt mines of Barletta is distinguished by two important aspects both in Middle Ages and in the Modern Period. The first one is that the salt mines, even if subject to the monopoly of the State, are privately owned and not State property (D'Atri 2001, 35); the second one is that for a long time they represent a reference point for Venetian commercial traffic, taking into consideration in particular the Venetians' subtle market strategies and the fact that they benefited from various privileges and tax exemptions as early as the 13th century (D'Arienzo 2000/2001, 81); the Venetians indeed took provisions from Barletta only occasionally or for contraband (Hocquet 1982, 103).

Lastly, it must be mentioned that the sovereigns of the Kingdom of Naples intervene decisively to restore the salt mines of Barletta using, for example, the ability of the great architect Luigi Vanvitelli in 1754 (Russo 2001, 46), and also the precise study formulated by the general administrator Vincenzo Pecorari (Pecorari 1784; De Stefano 1981, 51-67), who continues the improvements already under way. Starting with those changes, the salt mines of Barletta, even within a great reduction of the international trade of salt, become and remain up to the present moment the most important salt mines in Italy for finished product.

Italy - particularly rich in marine salt mines - has only one rock salt mine, apart from the Sicilian ones, active until the beginning of the 20th century and situated in the Calabrese Apennines, in the mountains of Orsomarso in the National Park of Pollino, at Lungro.

The abundant production and the fine quality of this rock salt have been known since ancient times thanks to the description made by Pliny the Elder. Both the inhabitants of the Greek colonies of Sibari and Thurii and the Romans knew and exploited this rock salt mine, exporting large quantities of salt, despite the difficulties related to the inaccessible ways of communication, most of the times formed of narrow paths, gorges and steep mule tracks (Frega 2003).

Figure 8. Entrance to the mine of Lungro.

The exploitation of this rock salt mine, together with those of silver at Longobucco and of iron at Stilo, continues also in Middle Ages, both in Norman Age and Suebic Age amidst the overall backwardness of the mineral field all over Europe (D'Arienzo 2007b, 595-596). But, over the centuries, the state of neglect and the continuous landslides and floods of the mine's gallery and at the same time the lessening of the repair and renovation works determine its slow decline. Since the beginning of the 17th century and throughout the 18th century, the expenses undertaken by the Royal Court of Naples and by private contractors in the attempt of restoring the mine to its previous level of rock salt extraction are very abundant, but always with ephemeral results (D'Arienzo 2007b, 600 and *passim*). It is only during the 19th century that a significant renewal is registered, a renewal which will, however, collide with the rise of fierce unions, and with their struggles aimed at making the miners' work conditions more human, up until 1970 when the mine is definitively closed.

Figure 9. Gallery of the mine of Lungro.

Only a short note about Serpico, near Avellino, where a salt water well was exploited near the Salso river. The well, mentioned in few medieval documents, has a certain fortune in the early Modern Period, but it is often in the centre of disputes for the continuous and unsolved phenomenon of contraband which leads several times to its closing down and to important maintenance work in 1697 (D'Arienzo 1996, 91-92). In the late Modern Period the wood crisis all over Europe leads to a slow decay of the small structure of Serpico; because of the increase in the price of wood, the necessary raw material for the boiling of salt water to extract the brine, similarly to what was happening in Volterra where, however, a process of innovation was started, the well will be exclusively exploited by the local population until the final shutdown in the mid-20th century.

Lastly, there is just one piece of information about the well of Capriglia, a few kilometers from Serpico, at the beginning of the 17th century (D'Arienzo 1996, 91).

The salt mines of Cagliari are among the oldest in Italy and, because of the thousand-year exiguity of the demographic curve of the Island, have always been devoted to exports, also thanks to the favourable position of Sardinia along the Mediterranean commercial routes. If Phoenicians and Punicians started and gave impulse to the salt mines of Cagliari, the Romans enlarged them both by entrusting hem to private contractors (the *Societates Publicanorum*) and, later on, during the Imperial Age, by managing them directly through the *Corpus Salariorum* which run all the active salt mines in the Italian Peninsula (Manca 1966, 42).

After the fall of the Roman Empire, the salt mines of Cagliari, like most of those in the Mediterranean, fell into a progressive state of neglect accentuated by the decrease in population and by its moving towards the inland zones to the detriment of coastal areas. In Sardinia, it is only with the coming of the Giudicati in the 9th century that a resettling of the salt mines that go back to the State property is registered (Manca 1966, 43), feature that will last for a long time, like the 'obbligatorietà dell'organizzazione servile' [tr.: compulsoriness of the feudatory organization] (Manca 1966, 44[7]).

Starting with the confiscation of the salt mines from the Pisans, the Crown of Aragona, through their expansionist policy, places the production of salt mines and the commercialization of salt in the centre of its own political and economical strategies. The Sardinian salt mines, set in the Catalan-Aragona context and within the unfavourable economical circumstances of the 14th century, among which notably the bubonic plague of the mid-14th century, are at the top of their glory exactly at that time (Manca 1966, 321-325).

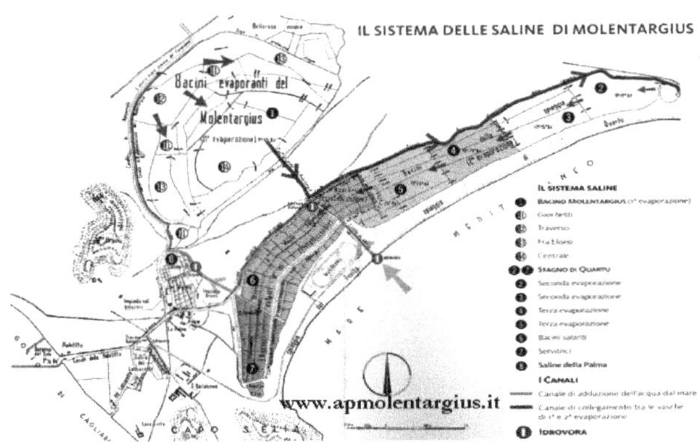

Figure 10. 20th century map of the salt mines of Molentargius, Cagliari.

The following centuries register a slow decline followed by a renewal, a 'second era', but within a completely different market; if between the 16th and the 17th century the levels of the early of the 15th century (Anatra and Carboni 1997, 149 and *passim*) are not reached, despite the fact that Cagliari supplied in party the wealthy Neapolitan market (D'Arienzo 1997, 144-146), during the 18th century, the arrival of cargo ships from Northern Europe revitalizes the field for some decades: 'The international requests of Sardinian salt had in the 18th century direct consequences on the whole productive structure of the salt mines of Cagliari and on the mechanism of forced hiring of workers, who had to produce more and more in order to satisfy a request coming from north European and Atlantic markets: Great Britain, Denmark, Sweden, Russia and also North America' (Pira 1997, 180).

The salt mine of Trapani also appear to be among the oldest Italian structures, while no documentary confirmation can be found for the information offered by the historian Giuseppe Mondini, who says that the first concession for the construction of salt mines dates back to 1440, but admits they must have existed for a very long time before then (Mondini 1999, 5).

Similarly to the situation in the rest of Italy, also at Trapani the production of salt marks a certain continuity between Middle Ages and Modern Period, regardless of the long 'silence' that characterizes the ancient times and the centuries before the year 1000. In the 12th century the well-known Arabian geographer Al-Idrisi mentioned the Sicilian salt mines, while between the second half of the 14th and the whole 15th century, the Aragona family started a policy of enfeoffment of the salt mines which can be interpreted as a turnround in comparison to the monopoly wanted by Frederick II, before, and by kings of Angioni family, after, until the Vespro War (Bianchini 1841, 220; Ruocco 1958, 32; Cancila 1999, XV). 'Molti documenti ufficiali del secolo XIV e XV, infatti, dicono che le saline trapanesi furono oggetto di concessione in feudo da parte dei sovrani aragonesi, sotto i quali, per altro, furono infeudate le Saline della Curia e la Salina grande; nel '400 furono infeudate parecchie altre saline. Proprio questa conversione di regime, da demaniale in feudale, può aver nuociuto alla produzione, per difetto, da parte dei feudatari, d'una organizzazione unitaria, adeguata al razionale sfruttamento dei giacimenti' (Manca 1966, 237[8]). The activity of the salt mines in that period, however, does not seem to be particularly significant in spite of the important requests from the manufactories of salting the tuna present in the whole north-west of Sicily; actually the resort to imports *extra regnum* (Ibiza and Cagliari) and the opening of new salt mines (Cancila 1999, XIX[9]), purer and therefore more suitable for the working of the fish hauls, indicate, probably, the inadequacy of the old medieval structures.

In the first half of the 16th century, therefore, the salt mines of Trapani are deeply restructured, new ones are founded and the capacity of production increases considerably even if the exports-related data show fluctuations both for that century and for the following one (Cancila 2001, 174-175). The political scenery changes deeply as well: Genoa restocks with high continuity from Trapani, Milan seems to prefer the Sicilian salt, cheaper than the Spanish one from Ibiza (Cancila 2001, 174-175) and the continental South, especially during the 17th and 18th century, is supplied with Sicilian salt (D'Arienzo 1996, 129 and *passim*).

In conclusion, and considering what initially anticipated, one can say that the period between the fall of the Roman Empire and the 11th-12th century is marked by the rarefaction of the documentary sources. Additionally, the production of all the salt mines is affected by a series of negative coincident elements: the emigration of the population towards the inland areas; the continuous raids of pirates who ravage the deserted structures along the coast; the standstill of the mineral extraction activities all over Europe; the lack of capital to invest in restoration of the structures; the uncertainty in what concerns the property which does not guarantee a constant production when the central authority - prominent figure in the management of the monopoly - is absent; the disappearance of old tested commercial routes throughout the Mediterranean.

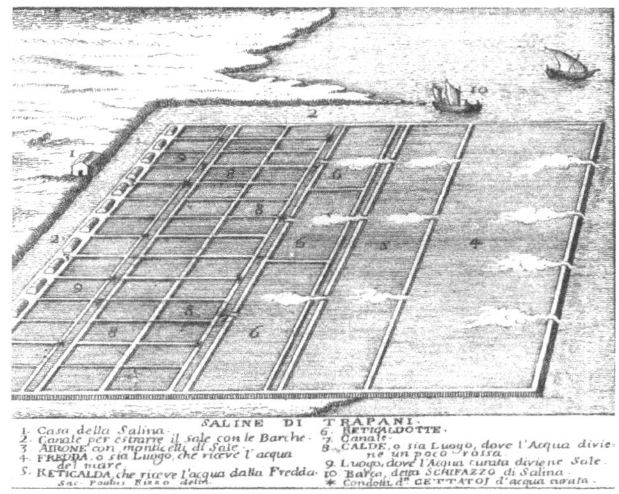

Figure 11. 18th century map of the salt works of Trapani.

This phase is followed by a new cycle when the monarchies try to restore the State property and in consequence the monopoly of salt; but above all, there is a re-launching of the investments in this field, investments which, because of the dynamics of the capitalism, find in salt mines the primary field for exploitation; in connection with the few work interventions and with the costs of a non specialized labour force – one must keep in mind that often the farmers and the labourers seasonally abandon their work in the fields and move down to the coasts for the picking of salt, so that one can refer to salt works as to an agricultural activity - the profits, thanks to the monopoly, are guaranteed and the dynamics allow a quicker accumulation of capital.

With the rise and the strengthening of various reigning families, there comes also a first form of economical policy, like in the case of the Aragona family, which allows setting up production and market strategies that exalt the local and regional realities, like in the case of the salt mines. And this is why at least until the 16th century the salt mines resume the production of bigger and bigger quantities of salt, even considering the trends of the economic cycles such as the negative one in the 14th century; and this is also why, throughout the 16th century, almost everywhere in Italy the old establishments are renovated and new foundations built. During this transition a lot of salt mines disappear, others are drastically reorganized and others are forcefully re-launched thanks to the strong internal and external demand. The market marks in the Modern Period a new phase of growth, still within different economical cycles, until the final crisis marked by the strong competition with salt produced in different European areas outside the Mediterranean, and the progressive replacement of salt for the preservation of food.

---

[1] 'Many archaeological testimonies support the hypothesis that also our lagoon had already been used in the Roman age for the management of the salt mines. The existence of a Roman salt mine seems to have been documented in S. Giorgio, in the S. Marco basin, and many other findings, even recent ones, lead in the same direction.'

[2] 'In 1200, after two centuries of expansion, salt production had reached its highest point, putting the lagoon on the first place among Mediterranean producers. Chioggia, having become an authentic capital of salt production, was able to provide for the requirements of a market expanding from the Alps' southern slope to the Apennines' valleys... Nevertheless, the competition became more and more dynamic in the whole circuit of the lagoon salt markets: new productions centres appeared and developed on the Adriatic shores or in the heart of the Alps. On the other hand, the competition was good for the Venetians themselves... Venice soon received imports from Puglia, then from Alessandria, and at the end of the 13th century, from the entire Mediterranean... The salt mines of Chioggia were affected by the crisis and many of them disappeared.'

[3] 'The possibility to exploit the production of the local salt mines, is referred to in a March 1st [1828] letter by the Inspector of Rimini, in which he writes about the arrival in his office of three samples of salt from Cervia and Comacchio and, after various experiments, the latter are chosen as they seem to be the most indicated for the salting of sardines, mackerels and mullets.'

[4] 'Documentary searches have provided more important information compared to the archaeological studies; the first historical news allowing verification of the existence of systems for the extraction of salt in the district of Volterra date back to the Early Middle Ages.'

[5] 'The general law of 1701 did not have innovative characteristics insofar as to solve the old problems connected to the production of salt, it had been promulgated in order to reorganize organically the old legislation, adapting it to the requirements of that age; in fact, it was sort of a "framework-law" inserted in a variegated administrative network in view of a simplification of the government's activity. The law of Cosimo III did not abolish the incongruences caused by the tax on salt which, together with the excise system, represented an important income for the State, but was also the main cause of the illegal trade which the administrators were trying in any way to eliminate.'

[6] 'Towards 115 d.C., the emperor Trajan had a magnificent bridge built across the Ofanto river, to connect these salt mines to Barletta, but it collapsed in 1850.'

[7] 'The State ownership, of old Roman origin, expanded and affirmed itself around the year 1000, when, following complex political and religious events, in which directives of the Gregorian reform had a decisive role, almost the totality of Byzantine assets was added to the giudicati's property, the *rennu*. After that, this characteristic, which never changed, was partially transferred to the advantage of religious institutions, like the case of the salt works granted to the monks of San Vittore and to the archbishop. When the Pisans took possession of all the salt works, taking them from the Victorians in 1216 and then submitting the Giudicato of Cagliari forty years later, they re-established completely this prerogative, favored by the reunification in their hands of the complex of salt ponds, which before had been granted in a fractioned way.'

[8] 'Many official documents from the 14th and 15th century, indeed, say that the salt mines of Trapani were the object of enfeoffment from the Aragonese kings who took control over the Salt mines of the Curia and the Big Salt mine; in the 15th century many other salt mines were subjugated. This system of conversion, from the State to feudalism, possibly harmed the production, for lack of a unitarian organization by the feudatories, suitable to the rational exploitation of the deposits.'

[9] 'In the 1430s, authorizations for the opening of new salt mines were more frequent and at the beginning of the second half of the century - probably because of the increase in the population of the island, and even more because of new tuna-fishing establishments - the enfeoffment for the construction of new marine salt works started to intensify, so that in the following decades until 1507 it reached – especially along the coastline from Trapani to Marsala - the dimensions of a real boom in favour of the representatives of the urban aristocracy and of the high bureaucracy, like it had never happened before or in the following centuries. At the end of the Middle Ages, the salt mines were mostly located inside the island, at the convergence of the three 'Valli' in which Sicily was divided at that time, and more precisely in the quadrilateral between Petraia, Cammarata, Sutera and S. Filippo d'Argirò, branching out until Caltabellotta and Naro in the south-east and until Modica and Noto in the south-west; while the marine salt works were spread along the southern coast from Mazara to Terranova, in the Syracusan (Augusta), in Messina and mostly along the short coastline between Marsala and Trapani. With the new system - equipped with windmills for mine dewatering - the island was accomplishing self-sufficiency and was getting ready to conquer the foreign market, while the city of Trapani was conquering the monopoly of production, which for several decades during the 16th century was bound to a large extent to satisfy the requests of the many tuna-fishing establishments from its own coastline.'

References:

Anatra, B. and Carboni, F. 1997. Sale in Sardegna nella prima età moderna. In S. Pira (ed.), *Storia del Commercio del Sale tra Mediterraneo e Atlantico*, 147-173. Cagliari, AM&D Edizioni.

Bergier, J. F. 1984. *Una storia del sale*. Venezia, Marsilio.

Bianchini, L. 1841. *Della storia economico-civile di Sicilia*, vol. I. Napoli, Stamperia Reale.
Borelli, F. 2000. *Le saline di Volterra nel Granducato di Toscana.* Firenze, Leo S. Olschki.

Cancila, O. 1999. Saggio introduttivo to G. Mondini, *Le saline della provincia di Trapani*, XII-XLII. Trapani, Banca del Popolo (anastatic reprint edition, 1881 Trapani, Tipografia Giuseppe Gervasi-Modica).

Cancila, O. 2001. *La terra di Cerere*. Caltanissetta-Roma, Salvatore Sciascia Editore.

Carusi, C. 2008. *Il sale nel mondo greco (VI a.C.-III d.C.). Luoghi di produzione, circolazione commerciale, regimi di sfruttamento nel contesto del Mediterraneo antico.* Bari, Casa Editrice Edipuglia.

Cassiodoro, F. M. A., *Variarum libri XII*, cura et studio A.J. Fridh. *De anima*, cura et studio J. W. Halporn, 1973.

D'Arienzo, V. 1996. *L'Arrendamento del sale dei Quattro Fondaci. Struttura, organizzazione, consumi (1649-1724)*. Salerno, Elea Press.

D'Arienzo, V. 1997. Le fonti di approvvigionamento del Regno di Napoli e il sale sardo tra Medioevo ed Età Moderna. In S. Pira (ed.), *Storia del Commercio del Sale tra Mediterraneo e Atlantico*, 129-146. Cagliari, AM&D Edizioni.

D'Arienzo, V. 2000/2001. Corfù e il commercio del sale in età angioina. In C. D. Litchfield, R. Palme and P. Piasecki (eds.), *Le Monde du Sel. Mélanges offerts à Jean-Claude Hocquet*, 73-84. Halle, Berenkamp Verlag.

D'Arienzo, V. 2001. Bibliografia Italia 1990-1999. In *CIHS Bibliographie. Bibliographie zur Geschichte des Salzes und des Salinenwesens*, 59-69. Herne.

D'Arienzo, V. 2007a. Industria ittica e incentivi statali nello Stato Pontificio nella prima metà dell'Ottocento. In L. Palermo, D. Strangio and M. Vaquero Piñeiro (eds.), *La pesca nel Lazio. Storia economia problemi regionali a confronto,* 219-235. Napoli, Editoriale Scientifica s.r.l.

D'Arienzo, V. 2007b. La produzione di sale nelle aree interne del Regno di Napoli. L'arrendamento dei "Sali di Monte di Calabria" nella seconda metà del XVIII secolo. In N. Morère Molinero (ed.), *Las salinas y la sal de interior en la historia: economía, medio ambiente y sociedad. Inland saltworks and salt history: economy, environment and society*, 593-608. Madrid, Dykinson.

D'Atri, S. 2001. *Il sale di Puglia tra marginalità e mercato: monopolio e commercio in età moderna.* Salerno, Edizioni del Paguro.

De Luca, V. 1926. *Il Comune di Margherita di Savoia (Già Reali Saline di Barletta). Spunti storici.* Barletta, Papeo.

De Stefano, R. 1981. Le saline di Barletta nella seconda metà del XVIII secolo e l'opera del Pecorari. In A. Di Vittorio (ed.), *Sale e saline nell'Adriatico (secc. XV-XX)*, 51-67. Napoli, Giannini.

Dondarini, R. 1995. La forzata clandestinità di un'antica risorsa comacchiese alle soglie dell'età moderna. In *Storia di Comacchio nell'età moderna,* Casalecchio di Reno, vol. II, 273-299. Casalecchio di Reno, Grafis Edizioni.

Fazzini, G. 2006. Il sale di Chioggia. *ArcheoVenezia* XVI.
http://www.archeove.com/pubblic/sale/sale.htm

Frega, M. 2003. Dalla Piana di Sibari al Tirreno attraverso i valichi del Pollino e dell'Orsomarso. I greci, i romani ed i normanni commerciavano il sale di Lungo. *Apollinea* 5.
http://www.ungra.it/new/Maria/via_sale/via_sale.htm

Giovannini, A. 1985. Le sel et la fortune de Rome, *Athenaeum* 63, 373-387.

Hocquet, J. C. 1981. Modernità del mercato del sale in Adriatico nel XVI secolo. In A. Di Vittorio (ed.), *Sale e saline nell'Adriatico (secc. XV-XX)*. 3-19. Napoli, Giannini.

Hocquet, J. C. 1982. *Le sel et la fortune de Venise. Production et monopole*, vol. 1. Lille, Publications de l'Université de Lille.

Hocquet, J. C. 1990. *Il sale e il potere dall'anno mille alla Rivoluzione francese*. Genova, ECIG.

Hocquet, J. C. 1991. *Chioggia. Capitale del sale nel Medioevo*. Sottomarina, Libreria Editrice Il Leggio.

Hocquet, J. C. 2001. Prefazione. In S. D'Atri (ed.), *Il sale di Puglia tra marginalità e mercato: monopolio e commercio in età moderna*, 9-16. Salerno, Edizioni del Paguro.

Le Goff, J. 1959. Orientation de recherches sur la production et le commerce du sel en Méditerranée au Moyen Age, *Bulletin Phil. et Historique du Comité des travaux Historiques.* 155-163.

Manca, C. 1966. *Aspetti dell'espansione economica catalano-aragonese nel Mediterraneo occidentale. Il commercio internazionale del sale*. Milano, Giuffrè Editore.

Moinier, B. 1997. *Sel et Société. Une affaire de métier*. Maxéville, Editions Nathan.

Mollat du Jourdin, M. 1993. *L'Europa e il mare*. Roma-Bari, Laterza.

Mondini, G. 1999. *Le saline della provincia di Trapani*. Trapani, Banca del Popolo (anastatic reprint edition, 1881 Trapani, Tipografia Giuseppe Gervasi-Modica).

Pecorari, V. 1784. *Memoria sulla Regia Salina di Barletta*. Napoli, Vincenzo Flauto.

Pira, S. 1997. Il commercio del sale sardo nel Settecento: dal Mediterraneo all'Atlantico (1700-1760). In S. Pira (ed.), *Storia del Commercio del Sale tra Mediterraneo e Atlantico*, 175-206. Cagliari, AM&D Edizioni.

Ruocco, D. 1958. *Le saline della Sicilia con uno sguardo d'insieme sulla produzione di sale in Italia*, vol. XVIII delle Memorie di Geografia Economica. Napoli, C.N.R.

Russo, S. 2001. *Le saline di Barletta tra Settecento e* Ottocento. Foggia, C. Grenzi Editore.

Part V. Linguistic and Philological Approaches

‘Salty’ Geographical Names: A Fresh Look

Alexander Falileyev
University of Wales Aberystwyth, United Kingdom

Abstract

In this article I discuss a range of place-names which have been considered to go back to appellatives meaning ‘salt’ in given languages but which in fact have different etymologies. First, I concentrate on the PIE word for ‘salt’, its semantic shifts and morphological patterns. Then, similar looking IE words are considered, also in conjunction with their reflexes in geographical names. And finally, I pay some attention to the ‘innovative’ words for ‘salt’ and their possible reflexes in the onomastic landscape.

Keywords

Indo-European languages, salt, geographical names, semantics, morphology

Place- and river-names which contain words for ‘salt’ are very well represented on the map of the world. Indeed, there is no doubt that Russian *Sol’Velikaja* (attested 1214), *Sol’Galičskaja* (1335), *Sol’Kamskaja* (now *Solikamsk*, 1553), Romanian *Sărata, Lacul Sărat* belong here, as well as English *Saltfleet* (1086) and *Salthouse* (1086). The majority of ‘salty’ toponyms reflect a common Indo-European word for ‘salt’. It may be the case even for the languages (groups of languages), where the corresponding appellative is obscure, as, e.g., in Hittite or Luvian, where the indigenous word for salt is always rendered by the Sumerogram MUN. Thus, C. Watkins (2005, 35) tentatively notes that ḪURSAG Šaliwana/i may point to a compound ‘Mount Saltrock’ or the like. There are certain discrepancies in the reconstruction of the PIE proto-form, and these are caused by various views on the explanation of the quantity of the vowel. Thus, the traditional reconstruction considers it as a ‘stem-word’ with a short vowel (cf. IEW: 878 **sal-*) which was lengthened under certain conditions in given languages, cf. here the traditional explanation for Lat. *sāl*, Gen. *sălis*. Other scholars prefer to see in the PIE form a long vowel, therefore **seh₂l-* (‘a very archaic word for ‘salt’’, Beekes 1995, 35); see particular F. Kortlandt (1985, 119), who reconstructs a hysterodynamic paradigm (Nom. Sg. **seh₂-l-s*, Gen. Sg. **sh₂-lo-s*, etc.) which accounts for the vocalism. This suggestion has found a few supporters, and several elaborations of this reconstruction have appeared in press. However, as noted by Z. Babik (2005, 144), a belonging here ‘Old Indic *salilá-* (*sarirá-*) ‘Meerflut’ speaks against a **seH₂l-*, which explains neither the vocalism *-a-*, nor the ‘insertion’ of *-i-*’. Therefore Babik thinks that ‘the root can be reconstructed either as **sal-* (monovocalically **sH₂el-*), or as **salHₓ-*, the former possibility being still more plausible’. According to P. Schrijver (1995, 32), a proto-form **sh₂l-* underlines Lat. Gen. Sg. *salis*, **sh₂el-* -- Old Church Slavic *solь*, and **seh₂l-* will account for the Baltic forms. Schrijver also notes that Old Saxon sultia, OHGerman sulza ‘salt water, salted sausage’ need a laryngeal in the underlying form (PIE **sh₂l-d-* > **sHl-d-* > Proto-Germanic **sult*). It is noteworthy that in the most recent fine compendium of Indo-European substantives the authors give a reconstructed form with a short vowel as the head-word for PIE ‘salt’ (Wodtko, Irslinger and Schneider 2008, 586, s.v. **sal-*).

There is a number of derivatives of this stem allegedly existing at a very early period. Already J. Pokorny (IEW, 879) considers a derivation in *-d-* of this stem reflected in Germanic (English *salt*, German *Salz*), Italic (Latin *sallō* ‘to salt, season’, participle *salsus* < **saldt-to-*), Baltic and Slavic (**saldu-* ‘sweet’, Lithuanian *saldùs*, Old Church Slavonic *sladъkъ*; for the adjectival derivation particularly in Baltic and Slavic see Otkupshikov 2001, 346-361 and cf. 283-4, 304-6). For the extension of the stem in *-d-* see also important suggestions in Björvand and Lindeman 2000, 755-6 and cf. Wodtko 2005, 71. Pokorny also notes possible reflexes of the IE word in the toponymy of ancient European onomastic languages, referring to ‘Illyrian’ *Saldae* in Pannonia and Thracian *Salsovia* (< **sal-d-t-ou-*). The former analysis is firmly accepted in modern scholarship (see Anreiter 2001, 117-8), while the latter is doubtful (Detschew 1976, 415). On top of that E. Benveniste (1935) has considered derivations in -n- (Greek ἅλασιν, Old Church Slavic *slanъ* < **solno-*, see a certain amount of skepticism expressed by Chantraine 1968, 65, but cf. Babik 2005, 144, Wodtko, Irslinger and Schneider 2008, 587) and in *-i-* (Old Irish *sail-*, Armenian *i*-stem *ał*; for Armenian a secondary development has been also suggested, see references and further discussion in Wodtko, Irslinger and Schneider 2008, 589). Benveniste also noted a remarkable Italo-Celtic parallel represented by Celtic **salēno-* (Irish *salann*, Old Cornish *haloin*, Welsh *halaen* < **saleino* and Latin *salīnis*), see a discussion with further references in Stifter 2005, 232 and 238. A derivative in *-m-* reflected in Greek ἅλμη ‘eau de mer, saumure’ and several other related words (Chantraine 1968, 65, cf. Wodtko, Irslinger and Schneider 2008, 587) is possibly not restricted to Greek, if Σαλμυδησσὸς αἰγιαλός (Herodotus IV, 93, Xen. Anab. VII, 5, 12; Strabo VII, 6,1) does indeed contain the IE name for salt. More than half a century ago Vladimir Georgiev derived the toponym from **salm-udes-* 'salty water', comparing parts of the compound with Greek ἅλμη and ὕδος 'water'. The problem is that the identification of a language to which the place-name may be assigned to is difficult – the preservation of **-d-* in the form makes its Thracian attribution unlikely, see further Falileyev 2006, 122-3. It is worth mentioning in this respect that the *comparanda* adduced for the discussion of the place-name in Thracia may offer further parallels for a *-m-* derivation from the stem, cf. Σαλμακις (Tomaschek) or Salmuris in Lower Moesia (S. Yanakieva), see references in Falileyev 2006, 123. Further possible examples of **sal-m-* in pre-Greek toponymics in Greece have been considered by S. Yanakieva (2000, 108), e.g., Σαλμώνη, Σαλμών or Σάλμος. However, the very nature of the evidence, which is drawn from onomastics, should be taken into consideration for a discussion of PIE derivation patterns.

These observations give us a certain amount of derivatives of PIE **seh$_2$l-* or **sal-* (see further Wodtko, Irslinger and Schneider 2008, 586-7, and also Björvand and Lindeman 2000, 755-6, Babik 2005), which are also reflected in onomastics. It should be also borne in mind that the PIE word for 'salt' has enjoyed various semantic shifts in given groups of languages or given languages, cf. particularly 'salty (water)' > 'sea' (OInd *sar-irá*, Greek (fem.) ἅλς). T. Gamkrelidze and V. Ivanov (1995, 581) have considered this semantic development attested in Indo-Aryan, Greek, Italic, Celtic and Slavic and postulated PIE **sal-* 'sea' alongside with the familiar **sal-* 'salt'. This does not seem to be necessary at all. First of all, semantic shift 'salt' > 'sea' is trivial (if not universal), cf. explicitly OInd *sal-ilá* in RV 10, 109,1, and thus the reconstruction of an epithet based on the word for 'salt' for the IE period may be misleading although of course probable (cf. Wodtko, Irslinger and Schneider 2008, 587-8). Secondly, the extant forms in various languages which denote 'sea' *vel sim.* show different derivation patterns (cf., e.g., various meanings – including 'zum Meer gehörig' - of the words going back to PIE **sal-imo-* in Wodtko, Irslinger and Schneider 2008, 587), and therefore their coinage can be easily attributed to later (post-IE) epochs, as the semantic shift is totally plausible at any time. There are also some purely semantic difficulties in this presentation: Ivanov and Gamkrelidze (*loc. cit.*) quote here Old Church Slavonic *slanŭ* 'seawater' (sic!), which seems to be more associated with the mineral (rendering Greek ἁλμυρός) than the sea. Old Russian *slan'nik* 'sea fisherman' (cf. G. ἁλιεύς) is considered here as well, notwithstanding the fact that the word, which seems to be a hapax, in its only attestation (Izbornik 1073) is followed by an explanation (*slano bo more*) which directly refers to the salty quality of the sea, rather to the sea itself.

Gamkrelidze and Ivanov (1995, 581 fn. 11) also note that 'in addition to oceans and seas, 'salty' can also be applied to certain lakes, rivers, and springs, evidently those whose waters contain mineral salts. This may be the source of the river name (Gk.) *Halys* (in Asia Minor) (....)'. The hydronym (see also Wodtko, Irslinger and Schneider 2008, 589 with further references) has been long traced to the name of 'salt' following the authority of Strabo (XII, 3,12), who informs us that the river gained its name from salt springs. This is not, however, always the case. For example, Ossetic appellative *xærœn* / *xwærœn* means 'salty springs', and as Vasilij Abaev (1989, 178) notes, it is very tempting to trace it to IE **saleno-* < **sal-* 'salt'. Avaev observes that the (dialect) Ironian form (quoted here first) does indeed allow such interpretation, but Digorian *xwærœn*, however, points to a completely different direction of etymological investigation. The latter is a derivative of a verb *xæryn* / *xwærun* 'to eat', and the formant *-œn* denotes 'place' as in *darœn* 'nomad camp' and *xīzœn* 'pasture'. Therefore, Abaev suggests that the word should be interpreted as 'a place of eating (of salt)', cf. also *syrdty xærœn* 'saline soils', 'places where animals come to eat salt'. Thus, Ossetic place-names and hydronyms *Xærœntœ*, *Xærœnœ sœr, Sau Xærœntœ, Abuzgi xwærœn*, etc (see Tsagaeva 1971, 195; Tsagaeva 1975, 46, 103, 268, 406) have nothing to do with the inherited IE word for 'salt', and some similar cases (consider also Stifter 2005 on Germanic place-name element *hall*) will be noted below.

On top of that it should be noted that some examples normally considered in the analysis of PIE word for 'salt' are difficult and may not belong here at all, as e.g., Welsh *heli* to which we turn our attention later. Moreover, a certain amount of geographical names which on face value are compatible with the reconstructed IE etymon have in fact a different etymology. Ironically, the problem was already well known in Classical antiquity. For example, for Varro (L.L. 5,72) and Festus (436) the divine name *Salacia* was directly related with Lat. *sal* 'salt', while Servius (Aen. 1,72) traced it to *salax*, and the deity was treated therefore as *dea meretricum* (see also, e.g., Ernout and Meillet 1951, 1043). Similar disagreements are found in modern scholarship as well, and contemporary academics sometimes challenge 'salty' etymologies proposed by ancient and mediaeval writers. The best example of this is a Latin name for island *insula* which Latin authors derived from the preposition 'in' and the name for salt. Thus, Martianus Capella (6, 43) notes that *inter fluenta emergentes terras, quae, quod in salo sint, insulae vocitantur*, and this approach is echoed in Isidorus' (Et. XIV, VI) *insulae dictae quod in salo sint, id est in mari.* Despite obvious difficulties, this etymology was acceptable until recently. Modern scholarship, however, does not find this approach convincing, see Jordán Cólera 1997 (bibl.) and note that Wodtko, Irslinger and Schneider 2008, 586-590 aptly do not consider this form.

Interestingly, a similar 'insular' etymology has been suggested recently by K. Stüber (see references in Wodtko, Irslinger and Schneider 2008, 599) for Old Church Slavonic *selo* 'Dorf' on the basis of its semantics. Traditionally the word is considered – alongside alleged Germanic congeners, cf. Old High German *sal* 'Wohnung, Saal, Halle' - under **sel-* 'Wohnraum' (IEW: 898). It may be noted that Lithuanian *salà* quoted by Pokorny is attested in the meaning of 'village' only in some dialects, and also denotes 'island' or 'field surrounded by meadows'. This entry becomes interesting for onomastic observation, as a Thracian place-name Σάλη has been claimed to belong here (see Duridanov 1969, 9, apparently following N. Jokl, see references in Detschew 1976, 413). However, in view of the obvious problems with the semantics of reconstructed PIE form, and also taking into consideration the character of Thracian data, this suggestion does not seem obvious at all. It is notable that W. Tomaschek (see references in Detschew, *loc. cit.*) has compared the town-name in Thracia not only with a toponym Σάλα in Phrygia (which may be coincidental, cf. also Pannonian *Sal(l)a*, for which see Anreiter 2001, 116-7), but also with a Lydian sea-name Σαλόη (Pausanius 7, 24, 13). 'Salty' associations of a lake-name are of course possible, but not necessary. At the same time a toponym located on the Ciconian coast may have something to do with the name for 'salt'. However, this might not be the case, and it may go back to one of several similar-looking PIE stems to be enumerated briefly below.

First, already on PIE level, one should pay attention to a remarkable group represented by the following set of appellatives in various languages. OInd. *sáras* 'sea, pond' has for a long time been compared with Greek ἕλος 'bas-fonds, marais, prairie humide', from IE **selos* (see IEW: 901 and cf. Chantraine 1968, 342). In her most important recent publication V. Kazanskene (2004, 320-2) thoroughly studied the meaning of the Greek word, which incidentally was used as a place-name (dat.-loc. *e-re-i, e-re-e* = Ἕλεσι; note that its interpretations as appellative are also available) in Mycenaean, and this enabled her to compare the Indic and Greek form with a group of words in the Baltic languages. Thus, Old Prussian *salus* 'brook, puddle' has been traced to **sel-* already by K. Būga (cf. Wodtko, Irslinger and Schneider 2008, 588); Lithuanian *salà* 'swampy bank of a river', *ãt-sala* 'creek, backwater', etc., according to Kazanskene (2004, 322), probably belong here as well. P. Schrijver (1995) added to this list more examples from Celtic and Germanic. Thus, Welsh *heledd* 'salt-pit, brine-pit, salt-marsh which is flooded at high tide, salt-boilery', which is also reflected in onomastics, has been long derived from IE word for 'salt'. As Schrijver aptly notes, the vocalism of this derivation is utterly problematic. He argues that 'the semantic element 'salt, saline' of *heledd* is probably secondary', and suggests that the original meaning of the word was 'flooded land', 'tidal marsh', which allows the Welsh word to be connected unproblematically with the set of words discussed here. Schrijver (1995, 33-4) correctly, as I think, rejects a popular comparison of Welsh heli 'brine, pickle' with Lithuanian *sólymas*, as its Breton cognate *hili* 'brine, sauce' prevents this equation. He tentatively notes that to arrive at the Brittonic forms one may postulate an *s*-stem, which as he claims is not attested with the IE word for 'salt', but is well found in the group of words discussed here in conjunction with the PIE **selos*.

According to Schrijver, W. *heli*, Bret. *hili* (<**selesa*) may reflect a semantic development 'marsh water' > 'tidal marsh water' > 'salt water in salt pan'. Generally, it is not unusual for the word for brine to be unconnected historically with a word for 'salt'. A good example comes from Ossetic, where *swar* 'brine' is used in this meaning in some dialects. The word (*swar / sawær*) in fact primarily denotes 'mineral spring', 'mineral water' and is historically connected with Ossetic *swadon / sawædon(æ)* 'spring' (Abaev 1979, 179). The word is well reflected in toponymy, cf. Swar (Tsagaeva 1975, 413), *Swary kom* lit. 'Gorge of a mineral spring' (Tsagaeva 1975, 96), *Suary xussary don* 'Water to the south of the spring' (Tsagaeva 1975, 135), *Suaræg, Suartæ, Suary kom* (Tsagaeva 1975, 163). To explain the meaning 'brine' V. Abaev (1979, 179) refers to an equation 'brine' = 'not fresh (i.e. salted) water'. Ironically, a similar explanation has been used by O. Trubačov in his analysis of a spring and locality in the basin of the South Bug mentioned by Herodotus (IV, 82) – 'Εξαμπαῖος. He notes the information of Herodotus that the water of the spring was 'bitter' (IV, 81), and on its basis offers its semantic treatment as 'not-suitable water'. As Trubačov follows and develops P. Kretchmer's views on the Indo-Aryan presence in the area, he provides a Sanskrit etymology for the place-name (*a-* 'not', *ksamá-* 'fitting, suitable', *páya-* 'water'), which is of course unacceptable; see Otkupshikov 2001a, 306-311, bibl. The semantic parallelism, however, is interesting; note that this semantic shift is attested elsewhere, for example, in Iranian, cf. Afgh. *trix* 'bitter, salty': Saryk. *trāč* 'not tasty' (? < Khovar *trok* 'bitter'), see Steblin-Kamenskij 1999, 359.

P. Schrijver (1995, 34-5) also suggests that Modern Welsh *hêl* 'a meadow along the side of a river, a moor, a marsh' is not borrowed from English *hale* as was commonly maintained, but is an exact cognate of Greek ἕλος. In this context he pays attention to mediaeval Welsh toponymy, particularly to the XII century 'Book of Llandaf', which gives *inter alia* a description of the boundaries of the episcopate. The component hal of a number of place-names found there (*Halunguernen 'Hal*-of-Ash-tree', *Haldu* 'Black *Hal*', etc.) and also in modern Welsh onomastic landscape (e.g., *Llwch-yr-hâl*) is connected by Schrijver (1995, 35) with Cornish *hāl* 'moor' (and possibly Old Irish sal 'dirt', Old Welsh (pl.) *halou* gl. *stercora*). On top of Celtic data quoted above Schrijver (1995, 39) considers here also a Germanic example: he derives Dutch *zeelt* 'tench' from Proto-Germanic **sel-it-* (also considering **-ut-*), and thinks 'if this is correct, zeelt is the only direct reflex of PIE **sel-* in Germanic'.

Now, the entry **selos* 'Sumpf, See' in Pokorny's *Indogermanisches etymologisches Wörterbuch* (IEW:

901), which considered only Indo-Iranian and Greek examples, looks quite different. As modern scholarship shows, the reflexes of the PIE word are found also in Baltic and Celtic, and possibly in Germanic. It has also been claimed (see e.g., Detschew 1976, 415) that a cognate of Old Prussian *salus* is attested in Thracian. However, the form considered here is a place-name Σάλτυς, and, therefore, the relevance of Thracian data may be questioned. The meaning of the PIE word has become clearer (note in this respect the different glosses given for the Old Indian cognate in IEW and in the modern literature, e.g., Mayrhofer 1996, 708) than it was half a century ago. It should be also noted that if P. Schrijver (1995, 35) is correct, the stem must be reconstructed as **selH-*, and the word as **selH-os*. The reflexes of this word are well attested in Indo-Iranian hydronymy (see references in Mayrhofer, loc. cit.) and, as we saw, are also attested in the toponymy. However, identifications of geographical names with this particular stem could also be misleading. Thus, P. Schrijver (1995, 37-39) has suggested that the name of the estuary of Maas and Waal *Helinium* (e.g., Pliny *HN* IV.15: *Helinium*, var. *Hellinium, Elinium, Helinum, Helium*) is Celtic and belongs here as well. This does not seem possible from the point of view of historical phonetics, see Falileyev 2008, 212 (bibl.) for the connection of this Celtic hydronym with PIE **pelh₂-* 'sich nähern'.

Still another PIE root discussed by J. Pokorny is worth considering in the discussion of 'salty' place-names. This is a putative **salo-* (IEW: 879-880) which is glossed as 'etwa 'wogend'' and which is reconstructed by Pokorny on the basis of Lat. *salum, salus* (Ennius) 'hohe See', Ir. *sál* 'see', Old Prussian *salus* 'Regenbach', and a number of hydronyms ('Illyrian' *Salon*, etc.). The stem plays a valuable role in the concept of *Alteuropäisch*, with scores of rivers from Spain to Baltia claimed to be derived from it (e.g. modern *Saale* in Germany and France or *Seile* in Scotland), see Krahe 1964, 49-50. The father of *Alteuropäisch* theory H. Krahe (1964, 49) glosses **sal-* as 'Bach, fließendes Wasser, Strömung' (cf. now Anreiter 2001, 251 with further bibl. on sal-river-names) and also considers here the appellatives listed by Pokorny's in the entry **salo-* (the Irish form is printed as *sal*). As we have seen, the Old Prussian word has a different etymology, and the same is probable for the Latin forms considered here (cf., e.g., Ernout and Meillet, 1951, 1043). What about the Irish word, which is also attested in medieval toponymy, as in *Garbán cain Cinn Sháli* (cf. *Cenn Sáli*)? The noun *sál* means not only 'sea, ocean', but also 'dross upon a stone' and 'seawater, brine', and this set of meanings is easily derived from the PIE **selos*. The comparison with Latin *salus* and *salum*, however, and hence a derivation from the same underlying etymon is impossible due to the quantity of the vowel, see Vendryes 1974, S-17 (for Ir. *sal* 'dirt, filth', *salach* 'dirty' see above but cf. Vendryes 1974, S-16) It is also worth considering that Ir. *sál* is found in compounds (if one is to follow E-DIL), where sometimes it semantically refers to the range of meanings reflected in PIE **selos* (*tar sal-mong mara* 'over the sea's briny mane'), but sometimes is closer to those of the PIE word for 'salt' (with *muir* 'sea' it is rendered 'salt sea', cf. also *salmuir glas* 'ocean'). J. Vendryes (1974, S-17) also considers s. v. *sál* its derivative *sáile* 'sea-water; sea; salt; water; brine'. As noted by Wodtko, Irslinger and Schneider (2008, 588), a connection of the Irish words with the PIE name of 'salt' is also possible. In view of the attempts to reconsider the configuration of the PIE form as far as the quantity of the vowel is concerned, a certain rearrangement of early Irish data presented in Vendryes 1974, S-16-18 seems possible. It may be noted in parenthesis that although there is no doubt in the existence of the Insular Celtic proto-form **saleino-* or the like in the meaning 'salt', the Gaulish **sal-* with the same meaning remains suspicious (see Stifter 2005, 232, 238; Falileyev 2008a, 150, bibl.).

The aim of this article has been to show, that the etymology of geographical names which at some stage were thought to reflect the PIE word for 'salt' could be different, and, *vice versa*, those which have been thought to be nothing to do with this important mineral may in fact be quite salty. This is of course not exclusively connected with the PIE **sal-* (or **sāl-*), but happens in the languages, where the inherited Indo-European word for 'salt' was replaced by innovations. For example, the name of a settlement *Namatgut* (*Nəmətgьt*) located on the right bank of the river Pyandzh in the Pamir mountains is normally explained by the local inhabitants (speakers of eastern Iranian languages – Wakhi and Ishkashmi) as distorted *Namak-kut* 'a lot of salt' or *Namak-kuh* 'salty mountain', cf. Tadjik, Persian namak 'salt'. The appellative 'salt' in several Iranian languages (cf. also Khot. *namva*, Sogd. *nm'δk*) is traditionally derived from the adjective **nam-* 'wet', probably referring to saline solution (cf. Steblin-Kamenskij 1999, 249). It may be noted in parenthesis that a semantic motivation of innovative words for 'salt' as 'wet' may compete with the antonym of the latter, 'dry'. Thus, for example, H. Bailey thought that Ossetic *sūr / sor* 'dry' should be connected with comparable words in several Iranian languages (Persian *šōr* 'salt') on the basis that the salty soil is normally dry. In his discussion of the Ossetic word V. Abaev (1979, 170-1, bibl.) remarks that such an approach cannot explain the meaning 'faint' of the Ossetic word (but note Babik 2005, 144 for the meaning 'to fade'). Instead, Abaev tentatively suggests a possible borrowing from Caucasian language and refers to Kartvelian **šwr-* 'to dry, to get tired', which may account both formally and semantically for the words in question.

The locals base this explanation of *Namatgut* (*Nəmətgьt*) on the story about Ali whose war-cry in the battle against the *infidels* demolished salt-mines which were in the area. This is a folk-etymology, of course. According to I. Steblin-Kamenskij (1976, 183-5), the place-name goes back unproblematically to Iranian **namata-gāt°* 'place of worship / prayer', a model very well attested in Iranian toponymy, cf. Persian place-names with *gāh* (< **gāθ°*, cf. Old Persian *gāθu-* and Avest. *gātav-*) 'place' and their probable east-Iranian correspondences (*-gad, -gat, -god, -gut*) in the onomastic landscape of the Western Pamir. The first component is compared in this analysis with Sogdian *nm-* 'to agree', Choresm. *nmd*, Khot. *nauda* 'worship' (cf. here also Celtic forms in *nemet-*, Gaulish νεμετον or Old Irish *nemed* gl. *sacellum*) to Iranian **nam-* 'to bend' (cf. Avest. *nam-* 'id').

As we know, salt never cries out that it is salty. However, this is not always the case of geographical names, whose 'salty' associations may have been gained sometimes by false analysis or by folk etymology.

References:

Abaev, V. I. 1979. *Istoriko-etimologičeskij slovar' osetinskogo yazyka.* Volume III. Moscow, Nauka.

Abaev, V. I. 1989. *Istoriko-etimologičeskij slovar' osetinskogo yazyka.* Volume IV. Moscow, Nauka.

Anreiter, P. 2001. *Die vorrömischen Namen Pannoniens.* Budapest, Archaeolingua Alapítvány.

Babik, Z. 2005. Old Prussian saltan (EV 376): a Slavist's view. *Rocznik Slawistyczny* 55, 137-147.

Beekes, R. S. P. 1995. *Comparative Indo-European Linguistics.* Amsterdam /Philadelphia, John Benjamins.

Benveniste, E. 1935. *Origines de la formation des noms en indo-européen.* Paris, Adrien-Maisonneuve.

Björvand, H. and Lindeman, F. O. 2000. *Våre arveord. Etymologisk ordbok.* Oslo, Novus.

Chantraine, P. 1968. *Dictionnaire étymologique de la langue grecque. Histoire des mots.* Paris, Klincksieck.

Detschew, D. 1976. *Die thrakischen Sprachreste.* Wien, Verlag der Österreichen Akademie der Wissenschaften.

Duridanov, I. 1969. *Thrakisch-dakische Studien.* Sofia (= *Linguistique balkanique* 13, 2).

Ernout, A. and Meillet, A. 1951. *Dictionnaire étymologique de la langue latine.* Paris, Klincksieck.

Falileyev, A. 2006. *Vostočnye Balkany na karte Ptolemeja. Kritiko-bibliografičeskie izyskanija.* München, Biblion.

Falileyev, A. 2008. Review article, *An Atlas for Celtic Studies.* Ed. J. T. Koch (Oxford, Oxbow: 2007). *Ancient Civilizations from Scythia to Siberia* 14, 208-217.

Falileyev, A. 2008a. Celts on the Margin: Toponymic Notes. In: J. L. García Alonso (ed.), *Celtic and other languages in Ancient Europe.* Salamanca, Universidad, 145-152.

Gamkrelidze. T. and Ivanov. V. 1995. *Indo-European and the Indo-Europeans.* Part 1. Berlin, New York, Mouton de Gruyter.

IEW = J. Pokorny, *Indogermanisches etymologisches Wörterbuch.* Band I. Bern, Francke 1959.

Jordán Cólera, C. 1997. The Etymology of Insula, Aestus and Aestuarium, *Journal of Indo-European Studies* 25, 353-360.

Kazanskene, V. P. 2004. Drevnegrečeskoie ἕλος 'boloto' i yego indo-evropejskije sootvetstvija, *Indoevropejskoie jazykoznanije i klassičeskaja filologija* VIII. St. Petersburg, 320-322.

Kortlandt, F. 1985. Long vowels in Balto-Slavic. *Baltistica* 21/2, 112-124.

Krahe, H. 1964. *Unsere ältesten Flussnamen.* Wiesbaden, Harrassowitz.

Mayrhofer, M. 1996. *Etymologisches Wörterbuch des Altindoarischen.* Band 2. Heidelberg, Winter.

Otkupshikov, Yu. V. 2001. *Opera philologica minora.* St. Petersburg, Nauka.

Otkupshikov, Yu. V. 2001a. *Očerki po etimologiji.* St. Petersburg, University Press.

Schrijver, P. 1995. Welsh *heledd, hêl*, Cornish **heyl*, 'Latin' *helinium*, Dutch *hel-, zeelt. NOWELE* 26, 31-42.

Steblin-Kamenskij, M. I. 1976. Dva vakhanskikh toponyma. In. V. S. Rastorgueva (ed.), *Iranskoie Yazykoznanije,* 182-185. Moscow, Nauka.

Steblin-Kamenskij, M. I. 1999. *Etymologičeskij slovar' vakhanskogo yazyka.* St. Petersburg, Peterburgskoie Vostokovedenije.

Stifter, D. 2005. Hallstatt – In eisenzeitlicher Tradition?. In R. Karl, J. Leskovar (eds.), *Interpretierte Eisenzeiten. Fallstudien, Methoden, Theorie,* 229-240. Linz, Oberösterreichisches Landesmuseum.

Tsagaeva, A. 1971. *Toponimija severnoi Osetii.* Volume 1. Ordžonikidze, Ir.

Tsagaeva, A. 1975. *Toponimija severnoi Osetii.* Volume 2. Ordžonikidze, Ir.

Vendryes, J. 1974. *Lexique étymologique de l'irlandais ancien. Lettres R S.* Paris and Dublin.

Watkins, C. 2005. Luvo-Hittite lapan(a)-. In D. Disterеft, M. Huld and J. Greppin (eds.), *Studies in honor of Jaan Puhvel.* Part One, 29-35, Washington, Institute for the Study of Man.

Wodtko, D. 2005. Nomen und Nominalisierung im indogermanischen Lexicon. *Indogermanische Forschungen* 110, 41-85.

Wodtko, D., Irslinger, B. and Schneider, C. 2008. *Nomina im indogermanischen Lexikon*. Heidelberg, Carl Winter.

Yanakieva, S. 2000. Predgrătskata hidronimija. *Seminarium Thracicum* 4, 85-118.

Etymological and Historical Implications of Romanian Place-Names Referring to Salt

Adrian Poruciuc
Universitatea "Al. I. Cuza" Iaşi, Romania

Abstract
It is common knowledge that salt – obtained by diverse means and used under various forms – has had a role to play in all systems of subsistence, from prehistory up until the present day. Whether the consumers were hunters and fishermen, stockbreeders, or farmers, they all had the exploitation or acquisition of salt among their preoccupations. Under such circumstances, approaches to salt terminology and to onomastic derivations from the latter can reveal significant facts regarding the history of various geographic areas. The present author takes into consideration forms, meanings and implications of certain toponyms that mark places in which salt has been exploited within the territory of today's Romania. Many of those toponyms are based on Romanian terms of Latin origin (such as *sare* 'salt', *sărat* 'salty', *sărătură* 'salt marsh', or *murătoare* 'brine'). But there also are others, of diverse origins, all of them providing precious information about interethnic contacts that occurred, throughout centuries, in the geographic area under discussion.

Keywords
salt, place-names, Romania, Transylvania, Latin, Slavic, Hungarian, German

The international colloquium 'Salt, Practices and Knowledge' (Iaşi – Romania, Oct. 1-5, 2008) was a success, first of all because its spirit was truly interdisciplinary: salt was approached from a number of standpoints, all of intrinsic interest. Inter alia, besides a dominant series of archaeological and ethnoarchaeological presentations, there also were some linguistic ones, including the contribution of the present author. My own approach started from the fact that Romania is a country well-known for its huge salt deposits and numerous sources of salt water. As proved by recent archaeological investigations, salt and natural brine have been exploited in certain regions of today's Romania since prehistory. No wonder that the map of the country under discussion shows abundance of place-names based on terms for 'salt'. Since the great majority of the population of Romania speak Romanian, quite a number of the toponyms under discussion are transparently based on Rm. *sare* 'salt', a term that belongs to the Latin heritage, since it derives – like Italian *sale*, Spanish *sal* and French. *sel* – from Lat. *sāl*.

If we want to go as deep as we can, etymologically, we may observe that Latin *sāl*, in its turn, is considered to derive from a Proto-Indo-European (PIE) term **sal-* 'salt', which may be the same as, or directly related to, **sal-* 'dirty, gray' (cf. AHD-Appendix, s.v. **sal*). The last aspect would imply that PIE **sal-* may originally have referred to the color of rock-salt blocks, extracted from mines. That issue deserves a separate article, just like another issue, namely the apparent lack of terms based on the root **sal-* in Indo-Iranian idioms (Buck 1988, 5.81). Such a lack may raise doubts about the applicability of the label 'Indo-European' to the root under discussion. However, a recent synthetic book on Indo-European matters – Mallory and Adams 2006, 261 – includes comments on the PIE root **seHa-(e)l-* 'salt' (actually the same as the one commonly transcribed as *sal-), as well as on a Sanskrit term that the two authors consider to be based on that root. For the time being, I will remain cautious about the significance of the Sankrit term *salilá-*, which Mallory and Adams add to the list of Indo-European (that is, Armeno-Tocharo-European) terms for 'salt'. As I have already observed in a previous article (Poruciuc 2008), Sanskr. *salilá-* 'sea, flood' may only indirectly refer to salt. The obvious Latin correspondent of the Sanskrit term under discussion would be not *sāl* 'salt', but rather *salum* 'open sea, open roadstead' – cf. Ernout/ Meillet 1985, s.v. *salum* (where a Roman sea-goddess, *Salācia*, is also mentioned). In any case, with or without an Indo-Iranian offshoot, the root **sal-* remains mainly European (or even 'Old European'), since it is represented by abundant lexical families in most languages of Europe (cf. Greek *hals*, Latin *sāl*, Latvian sāls, Old Church Slavonic *solĭ*, Russian *sol'*, Serbian-Croatian *so*, etc.). For the purposes of this article, it is worth mentioning the existence of terms based on very old (possibly PIE) suffixed extensions of the basic root **sal-* (see Gothic *salt*, German *Salz*, Irish Gaelic *salann*, Welsh *halen* – all meaning 'salt' –, as well as numerous other derivatives and compounds meaning 'salty' or 'salted food': Greek *halmurós*, Latin *salsus* (> Spanish *salsa*, French *sauce*), Old Church Slavonic *slanŭ*, etc.

To return to Romania and Romanian, I will continue by observing that about two pages of Iordan's *Toponimia românească* (1963, 125-127) – a book that has been considered as authoritative in the field of Romanian toponymy – contain long lists of place-names that refer to salt. As expected, Iordan first gives the numerous halotoponyms based on Rm. *sărat* ('salty, salted'): *Sărata* (28 village-names and microtoponyms from all over the country), *Săratul* (5), *Sărăţelul* (6), *Sărăţica, Sărăţuica, Lacul Sărat* (literally, 'the salt lake'), *Pîrîul Sărat* ('the salt brook'), *Rîmnicul Sărat* ('the salt pond'), etc. Speaking of Latin prolongations in Romanian toponymy, there also are two halotoponyms, *Murătoarea* and *Murăturile*, recorded as village-names in the Buzău county (as indicated in the dictionary Ghinea 2000). Iordan also mentions *Murătoarea* (1963, 126), as name of three springs and of a brook, all used as sources of natural brine. Romanian inherited *moare* 'brine' from Latin (cf. Lat. *muriēs*, which is translated in the Ernout/ Meillet dictionary by Fr. *saumure* 'brine' – cf. Span. *salmuera* and Rm. *saramură* 'brine' < Vulg.Lat. *salimuria*, or *salemoria*). In Romanian, *moare* has a solid family, including the verb *a mura* 'to pickle' and the derivatives *murat* 'pickled' and *murături* 'pickles'. It is quite clear that Romanian place-names such as *Murătoarea* and *Murăturile* are based on common terms that designated natural brine, which is still used by the local population as preservative for their vegetables and/or meat.

As visible especially in the cases of the Romanian place-names for which I gave literal translations in parentheses, such halotoponyms remained transparent for speakers of Romanian, since the latter are still aware of a connection between those names and certain common words still in use. Another aspect, to which I will return below, is that Iordan's collection of transparent Romanian place-names (1963, 125) includes both 'real' toponyms (or, according to a rule of thumb, designations officially recorded under forms with capitalized initial letters), as well as geographic terms used as such (that is, as common words) by inhabitants of a certain area – cf. *La Sărătură* (literally, 'at the salt marsh') and *La Sărături* ('at the salt marshes'). Even the fact that natives of the areas marked by such designations transmitted them (to the specialists) in forms preceded by prepositions (especially by *la* 'at'), indicate that those natives did not consider those designations to be proper-names proper, but rather 'onymicized entopics', according to a formula used by Diacon (1989). From the latter's book I extracted such 'entopics' as *La Sărătură, La Slatină, La Slatini, La Fîntîna Slatinii* ('at the salt-water well'), which Diacon recorded in northern Moldavia (an area within which, as we can see, Latin-based *Sărătură* occurs side by side with Slavic-based *Slatina*). Other local designations of the same kind as the ones recorded in Diacon 1989 are mentioned in various subchapters of Alexianu, Weller and Brigand 2007, from which I could extract such (capitalized) salt-source designations as *Pârâul Saramurii* (literally, 'the brine brook'), *Fântâna de Slatină* ('the well of brine'), *Salamura de la Dealul Gălean* ('the brine well at the Gălean hill'), *Salamura de la Stroiu* ('the brine well at Stroiu').

Iordan has a second list, which contains what he presents as 'synonyms' of the Latin-based halotoponyms of Romania (1963, 126). Most important among the place-names of that second list are the ones that can be clarified as based on terms of Slavic origin: *Slatina* (13), *Slătinicul* (3), *Slătioara* (8), *Slătineanca, Slătinița, Slătioarele, Slătiorul, Dealul Slatina, Valea Slatinei*; *Slănicul* (4); *Solnoc* (2), *Solona, Soloneț* (6), *Solonț*. In a footnote (p. 126), Iordan discusses a very important northeastern Romanian halotoponym, *Solca*, which is obviously based on a derivative of Slav. *sol'* 'salt'. What may be added here is that the Moldavian village-name *Solca* marks one of the Romanian regions richest not only in salt, but also in archaeological finds that indicate prehistoric exploitation of salt – cf. Monah 2008, 14-17.

In regard to the scientific value of halotoponyms such as the ones included in Iordan's lists, we may, of course, try a dry statistical approach. In that respect (by taking into consideration the present-day Romanian toponyms included in the dictionary Ghinea 2000) I could draw the following conclusion: today's Romania has 47 names of settlements (villages, towns, cities) such as *Sărari, Săreni, Sărata, Sărățel, Sărătura, Murătoarea*, etc., all transparently derived from terms of Latin origin, and it also has 24 names such as *Slatina, Slănic, Slătioara, Solca, Solona, Soloneț*, etc., all suggesting ultimate Slavic origins. We should, however, try to go beyond mere statistics.

Iordan's toponymic 'synonyms' that show Slavic origins can be divided according to at least two kinds of Slavic manifest in their forms (see below). Also, I will point out that I used a rather cautious formula ('suggesting ultimate Slavic origins') in my reference to Iordan's second list of halotoponyms, since we should envisage situations such as the following: (1) in certain cases Slavic immigrants/settlers may have used their own terms as designations of salt-containing elements of landscape (secondary implications being that Slavs could have simply translated local toponyms used by earlier inhabitants, or that non-Slavic inhabitants of the same area, such as the Romanians and the Hungarians, could subsequently adopt place-names produced by local Slavic speakers); (2) in other cases we must take into account that Romanian halotoponyms such as *Slatina* and *Solonț* can actually have been produced by Romanians, after they had come to use Slavic borrowings such as Rm. *slatină* 'salt marsh, brine source' and Rm. *solonț* 'salty ground', respectively. Other correspondences may be visible for etymologists, not for common speakers of Romanian, as in the case of the correspondence between *Slănic* and Rm. *slană, slănină* 'lard, bacon' (the two Romanian terms being obviously based on very early Slavic loans related to Old Church Slavonic *slanŭ* 'salty').

As already suggested above, the Romanian halotoponyms that can be referred to Slavic terms are good indicators for at least two main waves of Slavs that successively immigrated and settled in various regions of today's Romania, approximately between the sixth century and the eleventh. The earliest of the place-names under discussion are obviously the ones that look South Slavic, or, more precisely, Bulgarian. That fact should not be interpreted as implying transportation of Bulgarian terms and/or toponyms from the south of the Danube to the Carpathians, but rather as a reflection of the fact that the earliest Slavs (*Sklavenoi*) that occupied various regions of today's Romania were directly related to the glottal ancestors of the Bulgarians (to whom mediaeval Romanians applied the ethnonyn *schei* < *Sklavenoi* – cf. Albanian *shqerri* 'Slavs'). A part of that early Slavic population subsequently crossed the Danube and managed to preserve their identity in Bulgaria and Macedonia (where they also assimilated most of the local Romance population), whereas another part remained north of the Danube and were gradually assimilated by the early Romanians. The ethnogenetic process I have just referred to is synthetically (and competently) presented by an outstanding specialist in early Slavic matters, Shevelov (1964, 160). In the quotation below, Shevelov's concept of coterritoriality is worth observing, from a theoretical standpoint:

> 'If present day Romania emerged in the Middle Ages as predominantly Rm in its language, it is not because it had been settled by the Romans alone, but because in the course of mutual Sl-Rm influences, Roman speech conquered; and if present-day Bulgaria emerged as a Sl speaking country in the Middle Ages, it is not because the Roman speaking population inhabiting the area before the Sl invasions was expelled or annihilated physically, but because Sl overcame Roman during a prolonged period of competition. Sl was absorbed into Rm in Romania, and Rm was absorbed into Sl in Bulgaria. [...] A large part of population in these areas must have been bilingual for a long period of time. This was a typical case of coterritorial languages, not just contiguous.'

Shevelov's statements provide the only credible explanation for the many and important Old Slavic loans in Romanian, as well as for the many halotoponyms such as *Slatina* and *Slănic* in present-day Romania. In regard to a second Slavic wave of immigration, as indicated by certain place-names of Iordan's second list, that wave is represented by Romanian (mainly northern and northeastern) halotoponyms that contain East Slavic phonological features, such as the change *a* > *o* and the 'pleophonic' vocalism that affected earlier consonantal clusters. In that respect, we may observe that S./W.Slav. *slatina* 'brine, salt marsh, mineral spring' (recorded in Bulgarian, Serbian-Croatian, Slovene, Czech, Slovak) has E.Slav, correspondents such as Russ. *solot', solotina*, and Ukr. *solotva, solotvina.* The phonetic distinctions under discussion are also visible in Romanian halotoponyms based on Slavic terms: see for instance the differences between, on the one hand, Rm. *Slănic* – which shows an earlier kind of Slavic – and, on the other hand, Rm. *Solonț* and *Soloneț* – which show obvious East Slavic features –, all three referring to the Old Church Slavonic *slanŭ* mentioned above.

Last but not least, in his paragraph on Romanian halotoponyms of non-Latin origins, Iordan also mentions (1963, 126) the seven Transylvanian Romanian villages called *Șieu* (plus a diminutive derivative, *Șieuț*), interpretable as based on Hung. *Sajó* (< Hung. *só* 'salt' + Hung. *jo* 'water, river'). That case beautifully illustrates what may happen in situations of coterritorial languages. The fundamental element that should be taken into account (in that case, and others of the same kind) is Transylvanian geology, more precisely, the much coveted salt sources of that province. Transylvanian halotoponyms quite directly illustrate both the ethnic configuration and the successive shifts in political power specific to the history of the area under discussion. Iordan's example, Rm. *Șieu*, firstly reflects a borrowing of a Slavic term for 'salt' (cf. Serbian-Croatian *so* 'salt') into Hungarian, and the subsequent creation of a hybrid compound in the latter language, that compound representing the base of the Hungarian halotoponym *Sajó*. Certainly, we cannot exclude the possibility that, in the case under discussion, the Hungarians should have simply translated a halotoponym created by earlier inhabitants of Transylvania. In that respect, one can observe other cases of semantic continuity. The most 'respectable' of all is the one represented by the name of a salt source recorded as *Salinae* at the time of Dacia as a Roman province. Today the same place is officially known as *Ocna Mureș* (literally 'the salt pit on the Mureș'). There are about ten Romanian *Ocna* halotoponyms (most of them in Transylvania – cf. Ghinea 2000), their base being a Slavic loan, *ocnă* 'salt pit' (< Slav. *okno*). Hungarians also borrowed the term and/or the derived toponym under discussion, as we can see in Hung. *Akana*, whose Romanian name is *Ocna Sibiului*.

Today's Transylvanian halotoponymy displays the following main categories of situations (all reflected in the abundant toponymic material of the Suciu dictionary 1967-1968): (1) certain names were first recorded in mediaeval Latin translations, then in Hungarian, Romanian, and (in certain cases) German versions: for example, a *Villa Salis* recorded in 1236, was later called by local Romanians either *Sărata* or *Somfalău* (the latter representing an adaptation of Hung. *Sófalva* – literally, 'salt-village'), whereas the German colonists of the same region called the village under discussion simply *Salz* ('salt'); (2) other cases did not imply translation, but only adoption (and adaptation, in pronunciation and spelling) of place-names used in coterritorial languages: just as Romanians adopted Hung. *Sajó* and *Sófalva* as *Șieu* and *Somfalău*, respectively, Hungarians adopted Rm. *Sărata* (Sibiu county) as *Száráta* (or *Szarata*) and Rm. *Slatina* (Timiș county) as *Szlatina.* Wherever actual translation of earlier names took place, one major implication is a situation of bilingualism, such as the Romanian-Slavic one discussed by Shevelov (see above). Mere adoptions of toponyms (from coterritorial languages), followed by phonological and/or graphical adaptations, usually reflect shifts in officialdom. Taking such circumstances into account, I consider that toponyms that mark salt sources can be profitably analyzed as reflections of earlier or more recent interethnic contacts in various regions of Romania. The examples given above demonstrate inter alia that Romanian halotoponyms can be used as solid arguments not only in halological studies, but also in studies on coterritoriality, or on multilingualism.

References:

AHD – *The American Heritage Dictionary of the English Language*, 1973, ed. W. Morris. New York, American Heritage Publishing Co.

Alexianu, M., Weller, O. and Brigand, R. 2007. *Izvoarele de apă sărată din Moldova subcarpatică – Cercetări etnoarheologice*. Iaşi, Casa Editorială Demiurg.

Buck, C. D. 1988. *A Dictionary of Selected Synonyms in the Principal Indo-European Languages – A Contribution to the History of Ideas*. Chicago, The University of Chicago Press.

Diacon, V. 1989. *Vechi aşezări pe Suha Bucovineană*. Iaşi, Editura Universităţii "Al. I. Cuza".

Ernout, A. and Meillet, A. 1985 (1959). *Dictionnaire étymologique de la langue latine*. Paris, Klincksieck.

Ghinea, E. and Ghinea, D. 2000. *Localităţile din România–Dicţionar*. Bucureşti, Editura Enciclopedică.

Iordan, I. 1963. *Toponimia romînească*. Bucureşti, Editura Academiei.

Mallory, J. P. and Adams, D. Q. 2006. *The Oxford Introduction to Proto-Indo-European and the Proto-Indo-European World*, Oxford University Press.

Monah, D. 2008. Arheologia preistorică a sării în România. Scurt istoric. In D. Monah, Gh. Dumitroaia, D. Garvăn, (eds.). *Sarea, de la prezent la trecut*, 13-40. Piatra-Neamţ, Constantin Matasă.

Monah, D., Dumitroaia, Gh., Weller, O., Chapman J. (Eds.) 2007. *L'exploitation du sel á travers le temps*. Piatra-Neamţ, Constantin Matasă.

Monah, D., Dumitroaia, Gh., Garvăn, D. (eds.) 2008. *Sarea, de la prezent la trecut*. Piatra-Neamţ, Constantin Matasă.

Poruciuc, A. 2007. Romanian *huscă* ('salt obtained by evaporation of salt water') explained as an Old Germanism. In D. Monah, Gh. Dumitroaia, O. Weller and J. Chapman (eds.), *L'exploitation du sel á travers le temps*, 319-324. Piatra-Neamţ, Constantin Matasă.

Poruciuc, A. 2008. Denumirile sării şi implicaţiile lor în plan (indo)european. In D. Monah, Gh. Dumitroaia, D. Garvăn, (eds.). *Sarea, de la prezent la trecut*, 133-148. Piatra-Neamţ, Constantin Matasă.

Shevelov, G. Y. 1964. *A Prehistory of Slavic – The Historical Phonology of Common Slavic*. Heidelberg, Carl Winter Universitätsverlag.

Suciu, C. 1967-1968. *Dicţionar istoric al localităţilor din Transilvania*, I-II. Bucureşti, Editura Academiei.

Vasmer, M. 1987. *Etimologičeskij slovar' russkogo jazyka*, III. O. N. Trubačev (ed.) Moskva, Progress.

Salt in the Greek and Latin Aphoristic Phrase

Mihaela Paraschiv
Universitatea "Al. I. Cuza" Iaşi, Romania

Motto: *A Proverb is to speech what salt is to food* (S. W. Fallon, *Dictionary of Hindustani Proverbs*, London, 1886)

Abstract
Important and indispensable element to the material life, salt acquires in the aphoristic language of the Greek and Latin Antiquity a noetic and axiological symbolism, which reflects the transfer of the practical experience within the human community in the conceptual and axiological plan. The present study aims to identify a few aphoristic phrases in the Greek and Latin literature and to explain the semantic transfer from the concrete to the abstract, operated by the ancient speakers and writers regarding the noun salt and its derivates.

Keywords
salt, aphoristic phrase, semantic transfer, proverb, metaphoric meaning, noetic symbolism

The popular reflection, the proverb, (gr. *he paroimía*, lat. *proverbium*) and the cult one, the maxim (gr.*he gnome*, lat. *sententia*) were identified as forms of the aphoristic phrase (gr.*ho aphorismós–* 'aphorism', 'maxim') from the very origins of the ancient cultures. Aristotle defines them in *Rhetorica* (II, 1395a) without determining a clear categorical delimitation, as the Peripatetic followers. They say proverbs belong to a popular, common patrimony, and that maxims are written by an author (Tosi 1993, 180; Franga 2006, 241). In our opinion, the origins of such distinction may be traced back to Aristotle who, according to Synesius of Cyrene (*A Eulogy of Baldness*, 22) says that 'the proverbs have been saved [...] as remnants by which to recall the old philosophy that perished in the great destructive revolutions of humanity' (Fitzgerald 1930, 202), while the maxim is 'a statement of the general', 'suitable for one who is advanced in years and in regard to things in which one has experience' (Capps *et al.*, 1926, 283-285). We are talking about both types of reflection, popular and auctorial, about the transfer in the conceptual and axiological plan, through a figurative phrase, of the practical experience of a community or of an individual.

Aristotle and the Peripatetics believed that proverbs and maxims have in common a *stylistic value* – ensuring the concision and economy of the discourse – and an *ethic value*, coming from the fact that they have as subject the human being, the practical actions of life and the objects that people use.

Being such a necessary element to the man kind, naturally salt (gr.; lat. *sal*) was to be found in the ancient aphorisms and to be given a symbolic meaning in the Greek and Latin cultural and spiritual code. We should briefly mention some opinions of the ancient authors regarding the necessity of salt in people's lives.

In Homer's Odyssey (XI, 122) the prophet Tiresias divulges to Odysseus that, after his return to Ithaca, to be forgiven by Posseidon, he would have to travel to the land inhabited by human beings 'who know nothing of the sea/who eat their food unsalted' (Lombardo 2000, 335); Odysseus is thus the symbol of the Greek maritime civilization, opposed to the continental one, who did not know the sea, and thus the salt and its uses (Carusi 2008, 357). It is no wonder that, in the Greek poetic language, ἅλς means both 'sea' and 'salt'. The fact that salt was seen in Homer's time and later as a sea product is also emphasized in the tale' Ἁλῶν δὲ φόρτος ἔνθεν ἦλθεν, ἔνθ' ἔβη', quoted by many authors as the moral of a story – a salt merchant whose ship sunk into the sea as the mariners had fallen asleep (Moulakakis 1994).

A similar explanation to that of Homer regarding the fact that people who lived away from the sea did not know the salt, being below the civilisation level, is that of the Latin historian Sallustius (*De bello Iugurthino*, LXXXIX, 7) referring to the people in Africa *quae procul a mari incultius agebat* ('where the people live rudely and remote from the sea') (Watson 1896, 505); a direct consequence was the fact that they did not need salt or any other taste stimulants (*neque salem, neque alia inritamenta gulae quaerebant*), as the food was a means of satisfying physical needs, hunger and thirst, and not pleasure or voluptuousness (*cibus illis advorsum famem atque sitim, non lubidini neque luxuriae erat*).

A true praise of salt, as one of the most necessary elements of human life, is that of Plutarh *(Moralia* 685 a-b): 'Salt is as useful as either of these [i.e. water, light, the seasons of the year], being a sort of protector to the food as it comes into the body, and making it palatable and agreeable to the appetite' (Goodwin 1874, 337).

The uniqueness that Plutarh gives to salt through the phrase ὧν οὐδενὸς λείπεται χρείᾳ τῶν ἁλῶν is similar to that of Plinius Maior (*Naturalis Historia* XXXI, 88): *Ergo, Hercules vita humanior sine sale non quit degere*) ('We may conclude then, by Hercules that the higher enjoyments of life could not exist without the use of salt' (Bostock and Riley 1856b, 505); the same author quotes an old popular thought:

totis corporibus nihil esse utilius sale et sole dixit (*Naturalis Historia* XXXI, 102: 'there is nothing better for all parts of the body than sun and salt' (Bostock and Riley 1856b, 511).

The first mention of an aphoristic phrase referring to salt is that of Homer's Odyssey (XVII, 455); when addressing Odysseus, dressed as a beggar, Antinoos, who wanted to banish him, thus violating the sacred law of hospitality said: 'You wouldn't give a suppliant even a pinch of salt' (Lombardo 2000, 397); in the phrase οὐδ'ἅλα δοίης, taken over by Homer from the Greek paremiological patrimony, salt represents, by metonymy, the most modest and also the most necessary thing in someone's house, given its circulation and use in the Greek world (Carusi 2008, 361). Not offering somebody even salt meant a flagrant violation of the hospitality law, says Theocritos (*Eidyllia*, XXVII, 60), using the same Homeric phrase. For Archiloch from Paros in the phrase *salt and table* (*Elegeiai*, 96: ἅλας τε καὶ τράπεζα), salt is the symbol of unconditional hospitality towards a friend or a stranger. In relation to salt as a symbol of hospitality and friendship, Greek proverbs also express the idea of a long-lasting relationship between those who consumed salt together for a very long time.

In Ethica Nikomacheia (VIII, 3, 1156b), when talking about friendship, Aristotle said that this relationship needed time and common habits 'for, as the proverb says, it is impossible for people to know one another till they have consumed the requisite quantity of salt together' (Peters 1893, 258). Aristotle makes an allusion to the same proverb in *Ethika Eudemika* (VII, 2, 1238a), stating that the measure he refers to is ὁ μέδιμνος τῶν ἁλῶν ('a *medimnos* of salt', ~48kg n.n.). Plutarh (*Moralia*, 94a) also uses the phrase 'Having consumed with us in the course of time the proverbial bushel of salt' (Babbitt 1962, 51) (τῶν ἁλῶν συγκατεδηδοκέναι μέδιμνον) which meant *being good old friends*. According to this quantification, a friendship was consolidated in over 30 years, taking into account that the annual salt ration necessary to a normal person was no more than 1.50kg (a maximum of 4g a day).

Another traditional aphoristic phrase is the one quoted by Aristotle in Rhetorica (II, 1399a), when talking about the necessity of speaking in public: 'You should speak in public; for if you say what is just, the gods will love you; if you say what is injust, men will'. This is similar to the proverb 'To buy the swamp with the salt' (Capps *et al.,* 1926, 311-313).

The analogy intended by the author, using the phrase τὸ ἕλος πρίασθαι καὶ τοὺς ἅλας (buying the marsh and the salt) makes equal the advantages and the risks of the orator and the merchant. Given the possibility of obtaining salt out of marshes, through evaporation, mentioned by Plinius Maior (*Naturalis historia*, XXXI, 73-74), the phrase that Aristotle uses must have meant 'buying good with evil', which is compatible with the context in which it was mentioned. Aristotle also refers to another analogy, quoting Androcles in Pitheea, Alcibiades' adversary, who compared the apparently paradoxical need to make a law right through another law, with the need of fish for salt, even though it does not seem normal that creatures 'raised in salted water should need salt' ('ἐν ἅλμῃ τρεφομένους δεῖσθαι ἁλός').

A culinary virtue of salt, the conserving one, becomes, for the philosopher Chrysip from Soloi, quoted by Cicero (*De natura deorum*, II, 160), the support of a surprising analogy: he says that the swine was given salt instead of soul, in order to keep its meat fresh (*cui ne putresceret, animam ipsam pro sale datam dicit esse Chrysippus*). This quote comes from a fragment mentioning the many animal uses, according to the divine wish; as the swine has nothing more valuable than its meat, Chrysip believed that this animal had been gifted with a soul in order to keep its meat fresh for the humans.

A similar context of evoking salt for its culinary qualities is one of the epigrams written by Diogenes Laertios, who makes fun of Pythagoras' habit of not eating animal meat as the soul might reincarnate: ('All the food is dead/When boil'd and bak'd and salt-besprinkle-èd// For then it surely is inanimate' (Hicks 1925, 361) (*Bion kai dogmaton synagoghe*, VIII, 1, 23, cf. Diogenes Laertius, VIII, 35: 'of salt he [i.e. Pythagoras] said it should be brought to to table to remind us of what is right; for salt preserves whatever it finds and it arises from the purest sources, sun and sea).

Materially speaking, the culinary quality of salt was metaphorically valorised in the spiritual plan, becoming the source of aphoristic phrases. Thus, the phrase 'spicing with salt' (ἅλατι ἀρτύνειν) means 'talking nicely' in Paul's Letter to Colossians (*Pro Kolassaeis*, 4, 6): ὁ λόγος ὑμῖν πάντοτε ἐν χάριτι, ἅλατι ἠρτυμένος, εἰδέναι, πῶς δεῖ ὑμᾶς ἑνὶ ἑκάστῳ ἀποκρίνεσθαι. ('Let your speech always be gracious, seasoned with salt, so that you may know how you ought to answer every one') (Nestle-Aland 1985, 530). Commenting on this apostolic advice, Augustin explains Paul's culinary metaphor by saying 'Learning has a lot in common with eating: to cater for the dislikes of the majority even the nutrients essential to life must be made appetizing' (Green 1995, 229) (*De doctrina christiana* IV, 11, 26: *inter se habent nonnullam similitudinem vescentes atque discentes*).

Other Latin authors also tried to explain the semantic transfer from the concrete to the abstract operated by speakers when talking about *sal* and its derivatives.

Plinius Maior believes that salt 'Indeed, so highly necessary is this substance to mankind, that the pleasures of the mind, even, can be expressed by no better term than the word 'salt', such name given to all effusions of wit. All the amenities, in fact, of life, supreme hilarity, and relaxation from toil, can find no word in our language to characterize them better than this.' (*Naturalis historia*, XXXI, 88) (Bostock and Riley 1856b, 505-506). Quintilian also explains the metaphoric character of the adjective *salsum* and the plural sales: 'Salt will be a natural seasoning of language, which is perceived by a secret taste, as food is tasted by the palate, and which enlivens discourse and keeps it from becoming wearisome. As salt too, mixed with food rather liberally, but not so as to be in excess, gives it a certain peculiar relish, so salt in language has a certain charm, which creates in us a thirst, as it were, for hearing more' (*Institutio oratoria*, VI, 3, 18-20) (Watson 1903, 434).

From the explanations above, we should also understand the metaphoric meaning of the phrase *cum grano salis*. As it appears at Plinius Maior (*Naturalis historia*, XXIII, 149), *sal* has the meaning of spice, in the recipe of an antidote against poisons discovered by Pompei in king Mithridat's archive: 'Take two dried walnuts, two figs, and twenty leaves of rue; pound them all together, with the addition of a grain of salt' (Bostock and Riley 1856a, 515). The figurative sense of this phrase (with little spirit, with little taste) may be explained through the analogy of the salt grain, able to spice the antidote and make it drinkable and the role of a joke to spice up the speech, making it more appealing to the audience. An equivalent of the syntagm *granum salis* is *mica salis*, which appears at Catul referring to a certain Quintia, of whom the poet thought, despite other opinions, that she was not beautiful, thus arguing this opinion: *Nulla in tam magnost corpore mica salis* (Carmina LXXXVI, 4) ('Never a grain of salt shows in her person so large' (Burton and Smithers 1894, 263). Commenting this phrase, Quintilian says that poet Catul referred to that grain of spirit which spices up a woman's charm (Institutio oratoria, VI, 3, 19). Sal is used with the sense of 'aesthetic sense, good taste, style, elegancy' by Cornelius Nepos regarding Atticus' house, a friend of Cicero's: Ipsum enim tectum antiquitus constitutum plus salis quam sumptus habebat (Liber de excellentibus ducibus, 13: 'For the edifice, constructed after the ancient fashion, showed more regard to convenience than expense' cf. the note: The word 'salis' does not provide a higly satisfactory explanation in this passage. Most interpreters take it for 'gratia', 'venustas', 'ars', 'elegantia' (Watson 1853, 441).

For Horatius, *sal* and its derivates, *salsus* and *salinum*, have symbolical or metaphorical meanings. In one of his satires, he praises the satiric poet Lucilius as he rubbed Rome with a lot of salt (*Sermones*, I, 10, 3-4: *quod sale multo urbem defricuit*); taken over from the culinary and medical language, the phrase *sale defricare* ('rubbing with salt'), means here 'criticizing, moralizing'. In another satire (*Sermones*, II, 2, 17-18), Horatius uses a well-known proverb in order to eulogize the virtue of temperance: *cum sale panis / latrantem stomachum bene leniet* ('Bread and salt will suffice to appease your growling belly' (Fairclough 1942, 137). With the same moralizing intention, Horatius uses the phrase *paternum salinum* ('salt dish inherited from his father') as a symbol of the simplicity of life inherited from the old generation: *Vivitur parvo bene, cui paternum/splendet in mensa tenui salinum* (*Carmina*, II, 16, 13-14: 'A man can please himself with little, a salt dish/ handed down for generations can gleam upon/ his table' (Mc Clatchy 2005, 143). On his friend Aristius Fuscus, who pretended not to observe Horatius' anger when he was accosted by an importuner on Via Sacra, the poet avenges by calling him *male salsus* ('cruel joker') (*Sermones*, I, 9, 65).

At Plaut, in the play *Curculio* (IV, 4, 6), the phrase *salem delingere* ('to lick salt') is used ironically as substitute for *cenare* ('to dine') by Cappadox, who, taking into account the hospitality law, invites to dinner a soldier named Therapontigonus. Nevertheless, his invitation reads as follows: *hic hodie apud me numquam delinges salem* ('This day at my house you won't lick up one grain of salt' (Riley 1880, 556). Plaut uses here a ˜prosdóketon ('unpredictable joke'), by replacing the conventional formula for a dinner invitation with a colloquial phrase suggesting the extreme avarice, the total lack of hospitality; this substitution is one of the ways of ironically treating the topos 'invitation to dinner', recurrent in his comedies. Plaut uses with the same ironic intention the figurative meaning of the antonymic couple *salsus / insulsus*, as we can see in the following context in the comedy *Rudens* (II, 6, 32-33): *Bonamst quod habeas gratiam merito mihi, / Qui te ex insulso salsum feci* ('you have reason deservedly to give me many hearty thanks, who from an insipid morsel by my agency have made you salt' (Riley 1913, 86). This antithesis shows the fact that even since Plaut's time the adjective *salsus* had already been given the metaphorical sense used by Quintilian ('spiritual, 'joking'), a more complex meaning than that of the adjective *ridiculus* ('predisposing to laughter'). These two adjectives were often identified with one another in the colloquial aphoristic phrases (*Institutio oratoria*, VI, 3, 18). Also, at Cicero, in *De oratore*, in the chapter dedicated to the typology of jokes (LXIV-LXIX), we see many occurrences of the adjectives *salsus* and *insulsus* (*salsa verba, salsa reprehensio, non insulsum genus, ab hominibus non insulsis*). The figurative sense of the adjective *salsus* is more obvious as it is identified with the litotes *non insulsus*, emphasizing the meaning given by Quintilian: *salsum igitur erit quod non erit insulsum* (*Institutio oratoria*, VI, 3, 18).

All the examples above regarding the metaphorical uses of the word salt in the Greek and Latin aphoristic phrases are based upon a long and complex experience of obtaining and using salt by the two ancient people. Sometimes however, a Latin or Greek metaphor derives from the experience of other people, and it is only through this experience that we can fully understand its meaning. It is the case for the famous metaphor attributed to Jesus, known in Greek τὸ ἅλας τῆς γῆς and in the Latin translation as *sal terrae* in the *Gospel according to Matthew* (5, 13); the context is that of Jesus' speech on the mountain towards the disciples: 'you are the salt of the earth, but if salt has lost its taste, how shall its saltness be restored? It is no longer good for anything except to be thrown out and trodden under foot by men' (Nestle-Aland 1985, 9).

In *the Gospel according to Mark* (9, 50), when Jesus speaks to the disciples on another occasion, the metaphor doesn't appear anymore, but it may be understood from the context: 'salt is good; but if the salt has lost its saltness, how will you season it? Have salt in yourselves, and be at peace with one another.' (Nestle-Aland 1985, 121). According to the biblical mentions, the Jews extracted the salt from the Dead Sea and from the Salt Hill; the salt extracted from the surface of this hill was not pure, it did not taste like salt, and it was usually thrown away (*Dicţionar biblic*, 1995, 1155); as a consequence of this reality, the phrase 'the salt of the ground' had the original meaning of 'pure salt', obtained from the depths of the ground, the halite, named by Varro *sal fossicius* (*De re rustica*, I, 7, 8), and by Dioscorides ἅλς ὀρυκτος (5, 109) according to the connotative sense used by Jesus, his disciples are the symbol of lasting things, valuable and durable.

One of the issues in the paremiological works is that of the *ethnic paternity* of a proverb; generally, we can explain the presence of a proverb within two people in two ways: 1. each people created the proverb separately; 2. the proverb was created by one people and then it was spread to other peoples (Gheorghe 1986, 26). One such example is that used by Aristotle in *Etika Nikomacheia* and *Etika Eudemika*, previously mentioned, regarding the friendship. Cicero refers to it in *De amicitia* (XIX, 67), attesting its presence in the Roman world: *verumque illud est quod dicitur, multos modios salis simul edendos esse ut amicitiae munus expletum sit* ('We must eat many bushels of salt together with whom we shall perform the parts of friendship.' (Rouse *et al.*, 1906, 42). The fact that this proverb was taken over and used in the Latin paremiology is also proved by its variants, which generally connote the ideas of trust and fidelity (*fides*): *salis absumendus modius priusquam habeas fidem* (Publilius Syrus, *Sententiae*, 869: *Nemini fidas, nisi cum quo prius modium salis absumpseris* ('Trust no man, unless you have eaten a peck of salt with him first.' (*apud* Henderson 1869, 252; Smith 1960, XXX). According to the information given by Cato Maior (*De agricultura*, LXVII), a salt *modius* (around 8,75l) was the annual ration, an impressive quantity if we take into account that modern nutritionists recommend not to exceed 4g a day; but this annual ration for the slave and even for the Roman was not only used for consumption, but also with other culinary therapeutic (conserving food and vegetables) or religious purposes. The same proverb was widely transmitted in the Roman world, in variants more or less resembling to the Latin source (Gheorghe 1986, 329-330):
-French: *Pour connaître un homme, il faut avoir mangé un boisseau de sel avec lui;*
-Italian: *Prima di conoscere uno, bisogna consumare un maggio di sale. / Prima di sceglier l'amico bisogna averci mangiato il sale sett'ani*;
-Spanish: *En tu amigo confiaras, cuando hayas comido con el medio fanega de sal*;
-Portugues: *Não te deves fiar señao daquele com quem j'a comeste un moio de sal*;
-Romanian: *Trebuie să mănânci un car de sare cu cineva ca să-l poţi cunoaşte. / Până nu mănânci cu omul o maje de sare nu-l poţi cunoaşte.*

The ethnical variants of this proverb, separate for many centuries, prove the fact that the axiological transfer from the cultural and spiritual code of the Greek Antiquity towards the conceptual horizon of the modern world, with another set of values and another perspective on the world, is perfectly possible. The coincidence of the same proverb within several peoples, with some inherent differences, is thus explained not only through the loans, but also through the repeatability and constancy of the human lifestyle in its essence. Thus, we cannot attribute to the aphoristic phrase only an artistic value, but also a sapient one, emphasizing its role as philosophemes, given by Aristotle. The current researches regarding the paremiological creation put in evidence the existence of generic affinities between the ancient and modern theories regarding the proverbs and their axiological finality, with continuity from Antiquity to modern times. (Buridant 1976, 163; Greimas 1970, 309-314). We may certainly speak about a complex transfer of thought, mentality and affect structures, of a correspondence network between the ancient and the modern mental.

In conclusion, in the aphoristic phrases quoted in the article, salt acquires a noetic and axiological symbolism, figuratively designing. 1. An element indispensable to the material and spiritual life; 2. A superior form of civilisation; 3. Hospitality, friendship, trust; 4. Good taste; 5. Spirituality; 6. Moral value. Some of these old symbolical meanings can be seen even in today's English: *to eat salt with somebody; above the salt; the salt of the earth.*

At the end of this exegetic study, we can only hope that, talking about salt, we did not risk to be giving the following piece of advice: *Cum sis sine sale, non plura verba loquare*! (Since you don't have the spirit, stop talking!).

References:

Buridant, C. 1976. Nature et function des proverbs. *Revue des sciences humaines*, 156-165.

Carusi, C. 2008. Le sel chez les auteurs grecs et latins. In O. Weller, A. Dufraisse and P. Pétrequin (eds.) *Sel, eau et forêt: hier et aujourd'hui*, 353-364. Cahiers de la MSH Ledoux 12 (coll. Homme et environnement, 1). Besançon, Presses Universitaires de Franche-Comté.

Colas, A. 1985. *Le sel*. Paris, Presses Universitaires de France.

*** 1995. *Dicţionar biblic*. Oradea, Cartea creştină.

Franga, M. 2006. Enunțul paremiologic în cultura literară greacă. In *Antic și modern. In honorem Luciae Wald*, 239-248. București, Humanitas.

Gheorghe, G. 1986. *Proverbele românești și proverbele lumii romanice*. București, Albatros.

Greimas, A. J. 1970. Les proverbes et les dictons. In *Du sens*, 309-314. Paris, Seuil.

Henderson, A. 1869. *Latin Proverbs and Quotations*. London.

Moinier, B. 2008. L'eau et le sel, variations sur un thème. In O. Weller, A. Dufraisse and P. Pétrequin (eds.) *Sel, eau et forêt: hier et aujourd'hui*, 339-351. Cahiers de la MSH Ledoux 12 (coll. Homme et environnement, 1). Besançon, Presses Universitaires de Franche-Comté.

Moulakakis, N. 1994. *Archaies Ellenikes paroimies. Epiloghe apo to Corpus Paroemiographorum Graecorum*, Athena.

Nestle-Aland 1985. *Greek-English New Testament*, in the tradition of Eberhard Nestle and Erwin Nestle, edited by Barbara and Kurt Aland, Johannes Karavidopoulos, Carlo M. Martini, Bruce M. Metzger. English text 2nd Edition of the Revised Standard Version, Deutsche Bibelgesellschaft, Stuttgart.

Smith, W. G. 1960. *The Oxford Dictionary of English Proverbs*, Oxford, Oxford University Press.

Tosi, R. 1993. La lessicografia e la paremiografia in età alessandrina ed il loro sviluppo successivo. *Entretiens sur l'Antiquité classique*, tome XL, 143-197. Vandoeuvres-Génève, Hardt.

Critical editions

Babbitt, F. C. (ed./trans.) 1962. Plutarch's Moralia, vol.II. Cambridge, Massachusetts, Harvard University Press, Loeb Classical Library.

Bostock, J. and Riley, H. T. (eds./transls.) 1856a. The Natural History of Pliny, vol.IV. London, H. G. Bohn, York Street, Covent Garden.

Bostock, J. and Riley, H. T. (eds./transls.) 1856b. The Natural History of Pliny, vol.V. London, H.G. Bohn, York Street, Covent Garden.

Burton, R. F. and Smithers, L. C. (eds./transls.) 1894. The Carmina of Gaius Valerius Catullus. London, R. F. Burton and L. C. Smithers.

Capps, E., Page, T. E., Rouse, W. H. D (eds.) and Freese, J. H. (transl.) 1926. Aristotle: The 'Art' of Rhetoric. London, William Heinemann, New York, G. P. Putnam's Sons.

Fairclough, R. (ed./transl.) 1942. Horace: Satires, Epistles and Ars Poetica. London, William Heinemann Ltd.

Fitzgerald, A. (ed./transl.) 1930. Synesius of Cyrene: A Eulogy of Baldness. Oxford, Oxford University Press.

Goodwin, W. W. (ed.) 1874. Plutarch's Morals Translated from Greek by Several Hands, vol.III. Boston, Little, Brown and Company.

Green, R. P. H. (ed./transl.) 1995. Augustine. De Doctrina Christiana. Oxford Early Christian Texts, Oxford, Clarendon Press.

Hicks, R. D. 1925 (transl.) 1925. Diogenes Laertius: Lives of Eminent Philosophers, vol.II. London, William Heinemann.

Lombardo, S. (ed./trans.) 2000, The Essential Homer. Selections from the Iliad and Odyssey, Indianopolis/Cambridge, Hackett Publishing Company, Inc.

Mc Clatchy, J. D. (ed.) 2005. Horace: The Odes. New Translations from Contemporary Poets. Princeton, Princeton University Press.

Peters, F. H. (transl.) 1893. Nicomachean Ethics: Aristotle. Oxford. (Barnes&Noble 2004).

Riley, H. T. (ed./transl.) 1880. The Comedies of Plautus, vol. I. London, G. Bell and Sons, York Street, Covent Garden.

Riley, H. T. (ed./transl.) 1913. The Comedies of Plautus, vol. II. London, G. Bell and Sons, York Street, Covent Garden.

Rouse. W. H. D. (ed.), Harrington, J. and Newton T. (transls) 1906. Cicero's Books of Friendship, Old Age and Scipio's Dream. London, A. Moring.

Watson, J. S. (ed./transl.) 1853. Justin, Cornelius Nepos and Eutropius. London, H. G. Bohn.

Watson, J. S. (ed./transl.) 1896. Sallust's Conspiracy of Catiline and The Jugurthine War. Philadelphia, David Mc Kay.

Index of Authors

Marius ALEXIANU
(Universitatea "Al. I. Cuza" Iaşi, Romania)
alexianumarius@yahoo.com

Inês AMORIM
(Universidade do Porto, Portugal)
inesamorimflup@gmail.com

Valdo D'ARIENZO
(Università degli Studi di Salerno, Italy)
valdo@unisa.it

Robin BRIGAND
(Université de Franche-Comté, UMR 6249 Laboratoire de Chrono-Environnement, Besançon, France)
robinbrigand@yahoo.fr

Cristina CARUSI
(Università di Parma, Italy)
c.carusi@sns.it

Valeriu CAVRUC
(Muzeul Naţional al Carpaţilor Răsăriteni, Sfântu Gheorghe, Romania)
valer_kavruk@yahoo.com

Jorge Alejandro CEJA ACOSTA
(Universidad Autónoma de México, Mexico), espejo_negro@yahoo.com

Doina CIOBANU
(Muzeul Judeţean Buzău, Romania)
zeubz@net4u.ro

Vasile COTIUGĂ
(Universitatea "Al. I. Cuza" Iaşi, Romania)
vasicot@uaic.ro

Roxana-Gabriela CURCĂ
(Universitatea "Al. I. Cuza" Iaşi, Romania)
roxanigabriela@yahoo.com

Virginie DELRUE
(Université du Littoral Côte d'Opale, Boulogne sur mer, France)
virginie.delrue@univ-littoral.fr

Gheorghe DUMITROAIA
(Complexul Muzeal Judeţean Neamţ, Romania)
muzeupn@yahoo.com

Adela FABREGAS GARCIA
(Universidad de Granada, Spain)
fabregas@ugr.es

Alexander FALILEYEV
(University of Wales Aberystwyth, United Kingdom)
axf@aber.ac.uk

Daniel GARVĂN
(Complexul Muzeal Judeţean Neamţ, Romania)
daniel_garvan@yahoo.com

Anthony F. HARDING
(University of Exeter, United Kingdom)
a.f.harding@exeter.ac.uk

Jesús JIMÉNEZ GUIJARRO
(Universidad Rey Juan Carlos Madrid, Spain)
jesus.jimenez.guijarro@madrid.org

Cornelia-Magda LAZAROVICI
(Institutul de Arheologie, Iaşi, Romania)
magdamantu@yahoo.com

Gheorghe LAZAROVICI
(Universitatea "Lucian Blaga" Sibiu, Romania)
ghlazarovici@yahoo.com

Antonio MALPICA CUELLO
(Universidad de Granada, Spain)
amalpica@ugr.es

María Luisa MARTELL CONTRERAS
(Universidad Veracruzana, Mexico)
yaxche9@hotmail.com

Iulian MOGA
(Universitatea "Al. I. Cuza" Iaşi, Romania)
chairman_me2000@yahoo.co.uk

Bernard MOINIER
(Consultant Salt and Health, Paris, France)
Bernard.Moinier@tele2.fr

Nuria MORÈRE MOLINERO
(Universidad Rey Juan Carlos, Madrid, Spain)
nuria.morere@urjc.es

Dan MONAH
(Institutul de Arheologie, Iaşi, Romania)
danmonah@yahoo.fr

Roxana MUNTEANU
(Complexul Muzeal Judeţean Neamţ, Romania)
e_r_mro@yahoo.com

Dorin NICOLA
(Complexul Muzeal Judeţean Neamţ, Romania)
dorinnico@yahoo.com

Vassil NIKOLOV
(National Institute of Archaeology and Museum, Sofia, Bulgaria)
vassil.nikolov@abv.bg

Laure NUNINGER
(CNRS-UMR 6249 Laboratoire de Chrono-Environnement, Besançon, France)
laure.nuninger@mti.univ-fcomte.fr

Mihaela PARASCHIV
(Universitatea "Al. I. Cuza" Iaşi, Romania)
mihaelaparaschiv_clasice@yahoo.com

Viktoria PETROVA
(National Institute of Archaeology and Museum, Sofia, Bulgaria)
viki5rova@yahoo.com

Adrian PORUCIUC
(Universitatea "Al. I. Cuza" Iaşi, Romania)
apor@uaic.ro

Olivier WELLER
(CNRS-UMR 7041 Protohistoire européenne, Maison de l'Archéologie et de l'Ethnologie, Paris, France)
olivier.weller@mae.cnrs.fr

Nicolae URSULESCU
(Universitatea "Al. I. Cuza" Iaşi, Romania)
n.ursulescu@gmail.com

www.ingramcontent.com/pod-product-compliance
Ingram Content Group UK Ltd.
Pitfield, Milton Keynes, MK11 3LW, UK
UKHW061847041026
15591UKWH00015B/958

9 781407 307541